Electronic Signals and Systems
Analysis, Design and Applications
International Edition

RIVER PUBLISHERS SERIES IN SIGNAL, IMAGE AND SPEECH PROCESSING

Series Editors:

MARCELO SAMPAIO DE ALENCAR
Universidade Federal da Bahia UFBA, Brasil

MONCEF GABBOUJ
Tampere University of Technology, Finland

THANOS STOURAITIS
University of Patras, Greece
and
Khalifa University, UAE

Indexing: all books published in this series are submitted to the Web of Science Book Citation Index (BkCI), to SCOPUS, to CrossRef and to Google Scholar for evaluation and indexing

The "River Publishers Series in Signal, Image and Speech Processing" is a series of comprehensive academic and professional books which focus on all aspects of the theory and practice of signal processing. Books published in the series include research monographs, edited volumes, handbooks and textbooks. The books provide professionals, researchers, educators, and advanced students in the field with an invaluable insight into the latest research and developments.

Topics covered in the series include, but are by no means restricted to the following:

- Signal Processing Systems
- Digital Signal Processing
- Image Processing
- Signal Theory
- Stochastic Processes
- Detection and Estimation
- Pattern Recognition
- Optical Signal Processing
- Multi-dimensional Signal Processing
- Communication Signal Processing
- Biomedical Signal Processing
- Acoustic and Vibration Signal Processing
- Data Processing
- Remote Sensing
- Signal Processing Technology
- Speech Processing
- Radar Signal Processing

For a list of other books in this series, visit www.riverpublishers.com

Electronic Signals and Systems
Analysis, Design and Applications
International Edition

Muhammad Nasir Khan

The University of Lahore, Pakistan

Syed K. Hasnain

Aalborg University, Denmark

Mohsin Jamil

Memorial University of Newfoundland, Canada

Sameeh Ullah

Illinois State University, USA

River Publishers

Routledge
Taylor & Francis Group

LONDON AND NEW YORK

Published 2020 by River Publishers
River Publishers
Alsbjergvej 10, 9260 Gistrup, Denmark
www.riverpublishers.com

Distributed exclusively by Routledge
4 Park Square, Milton Park, Abingdon, Oxon OX14 4RN
605 Third Avenue, New York, NY 10158

First published in paperback 2024

Electronic Signals and Systems Analysis, Design and Applications International Edition / by Muhammad Nasir Khan, Syed K. Hasnain, Mohsin Jamil, Sameeh Ullah.

Routledge is an imprint of the Taylor & Francis Group, an informa business

Publisher's Note
The publisher has gone to great lengths to ensure the quality of this reprint but points out that some imperfections in the original copies may be apparent.

While every effort is made to provide dependable information, the publisher, authors, and editors cannot be held responsible for any errors or omissions.

ISBN: 978-87-7022-170-2 (hbk)
ISBN: 978-87-7004-326-7 (pbk)
ISBN: 978-1-003-33805-5 (ebk)

DOI: 10.1201/9781003338055

To My Loving Family
Atika Nasir,
M. Hamdaan Khan, Ayaan Ahmad Khan and Imaan Khan

To My Parents
Habibullah Khan and Saira Khan

To My Brother
Dr. Muhammad Farooq Khan

Contents

Preface

In recent years, the subject of digital signal processing (DSP) has become so important that it is now taught as a core subject in almost all the undergraduate courses in electronics engineering, electrical engineering, instrument and control engineering, telecommunication engineering and computer science and information technology throughout the world. This technological progress has also led to an enhancement in the availability of information, which is why we have felt the need to enhance certain topics of this book that were presented in the literature.

For many years, we felt the need of an elementary book on DSP. Keeping this requirement of undergraduate course for worldwide and specially for our country, we decided to compile and write a book for the students to build their basic concepts and problem-solving approach in DSP. Our target audience is only the undergraduate level engineering and computer science students who wish to build a basic understanding of this subject, starting from the basics. The worked examples are also specifically targeted for the undergraduate students. Professionals and teaching faculty of the colleges and universities who want to obtain a working knowledge of the DSP will also find this book useful.

The main objective of this textbook is to provide a series of worked examples and problems with solution along with a brief theory of various fields of the DSP. The work in this book is spread over discrete time signals and systems, convolution and correlation, Z-transform, discrete time Fourier transform, discrete Fourier transform including fast Fourier transform and step-by-step design of FIR and IIR digital filters. An attempt has been made to cover a reasonably wide range of subject material in the worked examples. Over 165 worked examples covering theory, 200 problems with solution at the end of chapters, and over 200 review questions with answers and 85 multiple choice questions are included in this book. Further, 85 multiple choice questions and 16 full length test papers at the end of the book can be solved as assignments by the students.

Preface to the International Edition

With the rate at which technology is advancing and the level of research being conducted in various fields, particularly signals and systems (SS), it is vital to keep up with the times.

By the grace of almighty Allah, we are able to complete our book in the form of an international edition, which we feel is a complete course book for undergraduate students. It has been our prime aim to streamline the flow of the book by connecting the numerical problems with the theory in a manner which will be most beneficial to the student. We wish to thank all our students and colleagues in suggesting improvements for this book. The organization of the book is as follows:

Chapter 1 provides comprehensive details regarding continuous time signals and its properties, handling of continuous time signals, transformation from analogue to digital and vice versa, sampling theorem, quantization error, discrete time signals and its properties, manipulation of discrete time signals and energy and power signals in continuous and discrete time.

Chapter 2 provides comprehensive details regarding the solution of differential equation with constant coefficients, transient solution (zero input response) and steady-state solution (zero state response) with different driving functions (forcing functions) such as step, ramp, acceleration, exponential and sinusoidal.

Chapter 3 provides comprehensive details regarding the Laplace transform, inverse Laplace transforms covering three cases: (i) when poles are real and non-repeated, (ii) when the poles are real and repeated and (iii) when the poles are complex, conversion of differential equation into Laplace transform and transfer function.

Chapter 4 provides comprehensive details regarding the systems, continuous time systems and its properties, discrete time system and its properties, block diagram representation of continuous time systems and mathematical modelling of the electrical, mechanical and electromechanical systems.

Chapter 5 provides comprehensive details regarding the response of the first-order and second-order systems, system's step response, system's ramp response, system's dc gain and system's frequency response.

Chapter 6 provides comprehensive details regarding the introduction to stability and Routh–Hurwitz stability criterion.

Chapter 7 provides comprehensive details regarding the introduction to periodic function Fourier synthesis, constructing a waveform, trigonometric form of the Fourier series, use of symmetry, complex form of the Fourier series and Gibbs phenomenon.

Chapter 8 provides comprehensive details regarding the properties of the Fourier transform, some familiar Fourier transform pairs and inverse Fourier transforms.

Chapter 9 provides comprehensive details regarding the homogenous and particular solution and solution of difference equations.

Chapter 10 provides comprehensive details regarding the Z-transforms, properties of Z-transform and inverse Z-transforms.

Chapter 11 begins with general consideration of analogue filters such as low pass, high pass, band pass and band stop, problems associated with passive filters, theory of analogue filter design, The Butterworth approximation, Chebyshev approximation and comparison between Butterworth and Chebyshev filters.

Chapter 12 provides application in the designing of robust higher-order repetitive controller using phase lead compensator.

Audience

This textbook is for a first course on signals and systems. It can be used in both computer science and electrical engineering departments. In terms of programming languages, the book assumes only that the student may have basic experience with MATLAB or C language. Although this book is more precise and analytical than many other introductory signals and systems texts, it uses mathematical concepts that are taught in higher secondary school. We have made a deliberate effort to avoid using most advanced calculus, probability, or stochastic process concepts (although we've included some basic and homework problems for students with this advanced background).

The book is, therefore, appropriate for undergraduate courses. It should also be useful to practitioners in the telecommunications industry.

Unique about This Textbook

The subject of signals and systems is enormously complex, involving many concepts, probabilities and signal processing that are woven together in an intricate manner. To cope with this scope and complexity, many signals and systems texts are often organized around the 'numerical examples' of a communication system. With such an organization, students can see through the complexity of signals and systems they learn about the distinct concepts and protocols in one part of the communication system while seeing the big picture of how all parts fit together. From a pedagogical perspective, our personal experience has been that such approach indeed works well.

Special Features for Students and Instructors

MATLAB includes several signal processing features and is an important tool for illustrating many of the field's applications. The use ofMATLAB has been linked to some aspects of this book to assist students in their understanding and to give them confidence in the subject.

MATLAB is not a prerequisite for this book. Its working is described in sections where it is utilized. For further specifics the help documentation is available online from Mathworks (http://www.mathworks.com), which is easy to use and contains many examples. Our experience has shown that signal processing students completely unfamiliar with MATLAB are able to use MATLAB within a week or two of exposure to tutorial exercises.

Every attempt has been made to ensure the accuracy of all material of the book. Readers are highly welcomed for a positive criticism and comments. Any suggestions or error reporting can be sent to with a CC to nasirhabib1@hotmail.com.

One Final Note: We'd Love to Hear from You

We encourage students and instructors to e-mail us with any comments they might have about our book. It's been wonderful for us to hear from so many instructors and students from around the world about our first international edition. We also encourage instructors to send us new homework problems (and solutions) that would complement the current homework problems. We also encourage instructors and students to create new MATLAB programs that illustrate the concepts in this book. If you have any topic that you think would be appropriate for this text, please submit it to us. So, as the saying goes, 'Keep those cards and letters coming!' Seriously, please *do* continue to send us interesting URLs, point out typos, disagree with any of our claims and tell us what works and what doesn't. Tell us what you think regarding what should or shouldn't be included in the next edition

Acknowledgements

We began writing this book since 2015; we received invaluable help from our colleagues and experts, which proved to be influential in shaping our thoughts on how to best organize the book.We want to say A BIG THANK to everyone who has helped us in drafting this book. Our special thanks to:

- Prof. Dr. Mansoor-u-Zafar Dawood, adjunct professor, King Abdul Aziz University, Kingdom of Saudi Arabia and currently VC, ILMA University, Karachi, Pakistan for continual support and guidance.
- Prof. Dr. Jameel Ahmad, Dean, Faculty of Engineering and Technology, Riphah International University, Islamabad for checking the manuscript for this edition.
- Dr. Ishtiaq Ahmad, associate professor, Department of Electrical Engineering, The University of Lahore, Lahore for being a thorough member and reading the book for final editing.
- Dr. Ghulam Abbas, associate professor, Electrical Engineering Department, The University of Lahore, Lahore for technical support during preparation of this book.
- Dr. Ali Raza, assistant professor, Electrical Engineering Department, The University of Lahore, Lahore for technical support during preparation of this book.
- Dr. Mazhar Hussain Malik, Head, Computer Science Department, Higher Colleges of Technology, Oman for technical support during preparation of this book.
- Dr. Syed Omer Gilani, assistant professor, AI and School of Robotics, National University of Sciences and Technology, Islamabad for continuous support.
- Dr. Junaid Zafar, chairperson, Electrical Engineering Department, GC University of Lahore, Pakistan.

- Dr. Tayyab Mehmood, assistant professor, Electrical Engineering Department, Rachna College of Engineering and Technology, Gujranwala, Pakistan.
- Dr. Qasim Awais, HoD, Electrical Engineering Department, Gift University, Gujranwala, Pakistan.
- Dr. Tosif Zahid, Head, Electrical Engineering Department, Riphah International University, Lahore, Pakistan.
- Dr. Shoaib Usmani, assistant professor, Electrical Engineering Department, National University, Faisalabad, Pakistan.
- Ms. Tarbia Iftikhar, my student of the University of Lahore at graduate level, for helping me in writing and editing my book.
- Dr. Kamran Ezdi, assistant professor, Electrical Engineering Department, University of Central Punjab, Lahore for technical support during preparation of this book.
- Mr. Farhan Abbas Jaffery from OGDCL and Mr. Tajammul Ahsan Rasheed from SNGPL for designing the cover of this edition of the book.
- Mr. Qaiser Mahmood for typing this book for first and second editions with great zeal and enthusiasm.
- Mr Hasnain Kashif, my PhD student of the University of Lahore at graduate level, for helping me in writing and editing my book.
- Dr. Hamid Khan, DVM, University of Veterinary and Animal Sciences, Lahore for his moral support and prayers.

We also want to thank the entire publisher team, who has done an absolutely outstanding job on this international edition. Finally, most special thanks go to the editor. This book would not be what it is (and may well not have been at all) without their graceful management, constant encouragement, nearly infinite patience, good humour and perseverance.

Dr. Muhammad Nasir Khan
Dr. S K Hasnain
Dr. Mohsin Jamil
and
Dr. Sameeh Ullah

List of Figures

List of Tables

List of Abbreviations

CT	continuous-time
AD	analogue-to-digital
DA	digital-to-analogue
DT	discrete-time
DSP	digital signal processing
ADC	analogue-to-digital converters
ECG	electrocardiogram
DTS	discrete-time system
BIBO	bounded-input bounded-output
LTI	linear time-invariant
LHP	Left half-plane
RHP	Right half-plane
FT	Fourier transform
IFT	Inverse Fourier transform
IIR	Infinite impulse response
RoC	region of convergence
LPF	Low-pass filters
HPF	High-pass filters
BPF	Band-pass filters
BRF	Band-reject filters
KVL	Kirchhoff's voltage law
ROI	region of interest
CNN	convolutional neural networks
ABC	artificial bee colony
DTM	Delaunay triangulation method
ISL	image-wise supervised learning
MSCA	multiscale super pixel based cellular automata
FCN	fully convolutional network
CDNN	convolution deconvolutional neural network
ISIC	International Skin Imaging Collaboration
ROC	receiver operative characteristics

IPSs	Indoor positioning systems
VLC	visible light communication
LED	light emitting diode
OWC	optical wireless communication
GPS	global positioning system
RFID	radio frequency identification
FoV	field of view
FDM	frequency division multiplexing
TDOA	time-difference-of-arrival
LOS	line-of-sight
TDM	time division multiplexing
KF	Kalman filter
PF	Particle filter

1

Signals

This chapter provides comprehensive details regarding continuous-time (CT) signals and its properties, handling of CT signals, transformation from analogue-to-digital (AD) and then back from digital-to-analogue (DA), sampling theorem, quantization error, discrete-time (DT) signals and its properties, manipulation of DT signals and energy and power signals in CT and DT. At the end of the chapter, relevant problems and solutions are also given for more understanding.

1.1 Introduction

In the modern era of microelectronics, signals and systems have shown tremendous involvement and play vital roles. To talk about the signals, it would be better to say that signals might be a vast range of physical phenomena. More generally signals can be denoted by different methods but for every case signals contained some form of information with a given pattern. More precisely, it might be better to say that a function of one or more independent variables, which contains some information, is called a signal. Signals can be either continuous-time (CT) or discrete-time (DT). Signals, which normally occur in nature (e.g., speech) are continuous in time as well as in amplitude and are called analogue signals. DT signals have values given at only discrete instants of time. These time instants need not be equidistant, but in practice, they are usually taken at equally spaced intervals for computational convenience and mathematical tractability. If the amplitude of DT signals is also made discrete through process of quantization or rounding off, then the signal becomes a digital signal.

1.2 CT Signals

An analogue signal (CT signal) has infinite range of values with the time goes on and changes continuously (i.e., smoothly) over time and is denoted as x(t). Such a signal is often known as analogue, but a suitable name is continuous signal. The following are some examples of CT signals for positive values of time (t ≥ 0):

(a): $x_1(t) = 1$,   (b): $x_2(t) = t$,   (c): $x_3(t) = t^2$,   (d): $x_4(t) = e^{-t}$,
(e): $x_5(t) = \sin(\omega t)$

1.2.1 Frequency-CT Sinusoid Signals

A mathematical representation of a basic harmonic oscillation is given by

$$x_a(t) = A \cos(\Omega t + \theta) \quad \text{for} \quad -\infty < t < \infty \tag{1.1}$$

Equation (1.1) is a CT sinusoidal signal, where Ω (which is used to represent an analogue form) has been used just to differentiate ω (which is used for digital frequency in upcoming chapters of the book), and subscript 'a' is used to denote an analogue signal. The signal in (1.1) is measured by three parameters: A → amplitude, Ω → frequency in radian per second and θ → phase in radians. Expression (1.1) can be written in terms of frequency F (hertz) instead of Ω (i.e., $\Omega = 2\pi F$), where F denotes the frequency in cycle per second as

$$x_a(t) = A \cos(2\pi F t + \theta); -\infty < t < \infty \tag{1.2}$$

1.2.2 Periodic and Aperiodic Signals

The concept of periodic and aperiodic signals is important to know in signals and systems course for using it at later stages. For CT signal it is easier to determine the periodicity or non-periodicity of the signal. A periodic signal has a definite pattern and repeats over and over with a repetition period of T seconds. For more simplicity, a CT signal is called periodic if it exhibits periodicity as

$$x(t + T) = x(t), -\infty < t < \infty, \tag{1.3}$$

where T denotes the period of the signal. Whereas a signal, which does not fulfil the requirement of periodicity given in (1.3) or repeats its pattern over a period of time is known as aperiodic signal or non-periodic signal. The smallest value of period T, which satisfies (1.3), is called fundamental period T_0. The CT sinusoids are characterized with the help of the following properties:

(a) Periodic functions are supposed to exist for each value of time. In x(t + T) = x(t), t is not limited.
(b) A periodic function with period T is periodic as well with period nT, where n denotes an integer. Hence, we may write for a periodic function as x(t) = x (t + T) = x(t + nT) for any value of integer n.
(c) The fundamental period T_0 can be defined as the minimum value of the period T > 0 that satisfies x(t) = x(t + T) and the fundamental frequency is given by $\Omega = 2\pi F = 2\pi/T$, where T is in seconds.

The following examples have been added to gain the knowledge of periodic signals in CT. In which T has been calculated to determine the periodicity of the signals using $\Omega = 2\pi F = 2\pi/T$.

Example 1.1
Calculate the fundamental period and periodicity of the following given sinusoids.

$$\text{(a) } x(t) = \sin 15\pi t$$
$$\text{(b) } x(t) = \sin \sqrt{2}\pi t$$

Solution 1.1

(a) The fundamental period T is given as $T = \frac{2\pi}{\Omega} = \frac{2\pi}{15\pi} = 0.1333 \ x(t) = \sin 15\pi t$ (periodic signal).

$$x(t + T) = \sin 15\pi(t + T) = \sin 15\pi(t + 0.1333) = \sin 15\pi t + \sin 2\pi$$
$$= \sin 15\pi t$$

(b) The fundamental period T is given as

$$T = \frac{2\pi}{\Omega} = \frac{2\pi}{\pi\sqrt{2}} = \frac{2}{\sqrt{2}} = \sqrt{2} = 1.414$$

$x(t) = \sin \pi\sqrt{2}t$ is a periodic signal

$$x(t + T) = \sin \sqrt{2}\pi t = \sin \sqrt{2}\pi(t + \sqrt{2}) = \sin \sqrt{2}\pi t + \sin 2\pi$$
$$= \sin \sqrt{2}\pi t$$

Example 1.2

Suppose that $x_1(t)$ and $x_2(t)$ are periodic signals with periods T_1 and T_2, respectively. Under what conditions the sum $x(t) = x_1(t) + x_2(t)$ can be periodic? What would be the period of $x(t)$ if it will be periodic?

Solution 1.2

Given that $x_1(t)$ and $x_2(t)$ are periodic signals with periods T_1 and T_2, respectively.

Thus, $x_1(t)$ and $x_2(t)$ may be written as

$$x_1(t) = x_1(t + T_1) = x_1(t + mT_1) \text{ where m is an integer}$$

$$x_2(t) = x_2(t + T_2) = x_2(t + nT_2) \text{ where n is an integer}$$

Now, if T_1 and T_2 are such that $mT_1 = nT_2 = T$. Then

$$x(t + T) = x_1(t + T) + x_2(t + T)$$

$$x(t + T) = x_1(t) + x_2(t) \text{ i.e., } x(t) \text{ is periodic in this case.}$$

Therefore condition of $x(t)$ to be periodic is $\frac{T_1}{T_2} = \frac{n}{m}$ is a rational number.

Example 1.3

The sinusoidal signal $x(t) = 3\cos(200t + \pi/6)$ is passed through a square-law device defined by the input–output relation $y(t) = x^2(t)$.

Show that the output $y(t)$ consists of a dc component and a sinusoidal component.

(a) Specify the dc component.
(b) Specify the amplitude and fundamental frequency of the sinusoidal component in the output $y(t)$.

Solution 1.3

Using the trigonometric identity $\cos^2 \theta = \frac{1}{2}(\cos 2\theta + 1)$

$$y(t) = x^2(t) = \left[3\cos\left(200t + \frac{\pi}{6}\right)\right]^2 \quad y(t) = \frac{9}{2}\left[1 + \cos\left(400t + \frac{\pi}{3}\right)\right]$$

(a) The d.c. component is $\frac{9}{2}$
(b) The amplitude $= \frac{9}{2}$, fundamental frequency $= \frac{\pi}{200}$

1.3 Manipulation of CT Signals

1.3.1 Reflection/Folding/Flipping

Reflection is the first property of the signal manipulation both in CT and DT. For the modification signal with respect to time base, is to replace the independent variable t by −t. The resultant operation is known as folding or a reflection of the signal about the origin along y-axis.

Example 1.4
Consider again the signals x(t) as shown in the figure provided. Find x(2 − t/3).

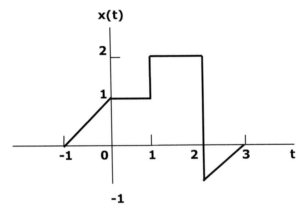

Solution 1.4
In this example x(t) is given, and we have to find out x(2 − t/3). It means first a reflection is required in this case because −t is there in the expression. In this example only reflection has been explained.

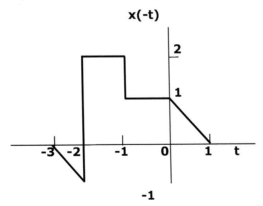

1.3.2 Shifting (Advanced and Delayed)

Any signal x(t) can be shifted in time by substituting the independent variable
't' by 't −τ', where τ denotes an integer. If τ comes out to be a positive, then
by shifting a value of time results a delay in the signal by τ value of time.
It shows that the new shifted version of the original signal shifts towards the
right side by τ amount positively. If τ comes out to be a negative, then by
shifting, it results in an advancement of the signal by τ units and the new
shifted version shifts to the left by τ amount.

Example 1.5
Let y(t) be given as in the figure provided. Carefully sketch the signals
y(−t − 1).

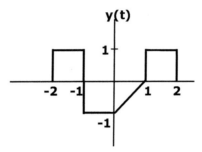

Solution 1.5
In this example y(−t−1) has to be calculated. In this case first reflection is
carried out, and y(−t) has to be calculated first. Then equating −t−1 = 0
means t = −1 in that t has to be shifted one unit to the left.

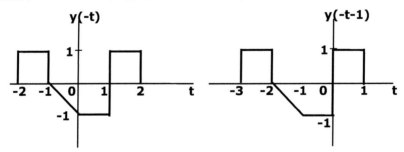

1.3.3 Scaling (Time and Magnitude)

This property is represented by two types of scaling, time scaling and magnitude scaling. In *time scaling* also known as *down sampling*, t is replaced by *ct*

where c is an integer. In *magnitude scaling*, it is to be multiplied by numbers with every value of signals and is represented by

$$y(t) = Ax(t); -\infty < n < \infty$$

1.3.4 Rule for Reflection, Shifting and Time Scaling

Precedence rule for reflection, shifting and time scaling is to be followed strictly to get the correct solution. Let y(t) denotes a CT signal, which is derived from another CT signal x(t) by the application of time shifting and time scaling; that is

$$y(t) = x(at - b) \tag{1.4}$$

The relationship between y(t) and x(t) satisfies the conditions:

$$y(0) = x(-b) \tag{1.5}$$

and

$$y(b/a) = x(0), \tag{1.6}$$

which provides for useful check on y(t) in terms of corresponding value of x(t).

Example 1.6 Precedence Rule for CT Signal
Consider the signal x(t) as shown in the figure provided. Find x(3 − t/2).

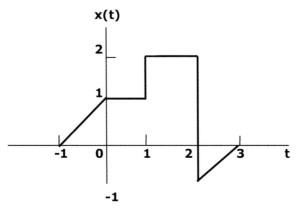

Solution 1.6
In this example x(3 − t/2) has to be found out. The expression contains −t , it means a reflection is required first.

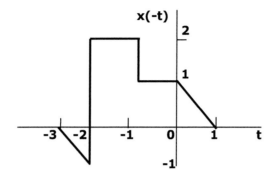

The precedence rule describes that a reflection should be taken first, then shifting in this example $3 - t = 0$ means $t = 3$, three unit to the right (positive).

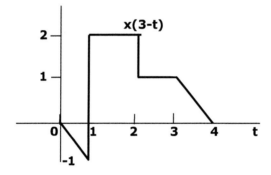

Once the shifting is done, the rule describes scaling operation should be carried out. In this example, $t/2$ is the time scaling applied. It means the signal on the time scale is expanded by a factor of two.

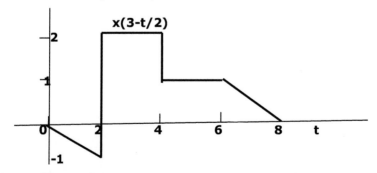

Example 1.7

A triangular pulse signal x(t) is depicted in the figure provided. Sketch each of the following signals derived from x(t):

    (a) $x(3t)$
    (b) $x(3t + 2)$
    (c) $x(-2t - 1)$
    (d) $x(2(t + 2))$
    (e) $x(2(t - 2))$

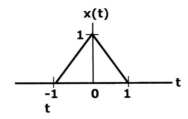

Solution 1.7

  (a) In this example a $= 3$ and b $= 0$. Hence, no shifting of the given pulse x(t). Next follow the precedence rule in that scaling the variable t by 3 units gets the final solution.

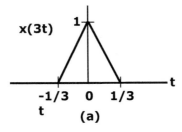

(a)

  (b) In this example a $= 3$ and b $= -2$. Hence, shifting the given pulse x(t) to the left by 2 units. Next follow the precedence rule in that scaling the variable t by 3 units gets the final solution.

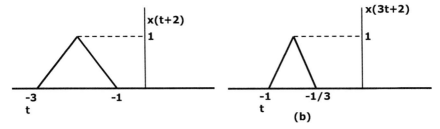

(b)

(c) In this example a $= -2$ and b $= -1$. $-t - 1 = 0$ means t $= -1$, Hence, reflection first then shifting the given pulse x(t) to the left by 1 unit to the left (negative). Next follow the precedence rule in that scaling the variable t by 2 units compresses the signal.

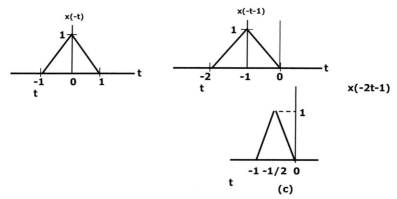

(c)

(d) In this example a $= 2$ and b $= 4$. Hence, shifting the given pulse x(t) by 4 units to the left. Next follow the precedence rule in that scaling the variable t by 2 units gets the final solution.

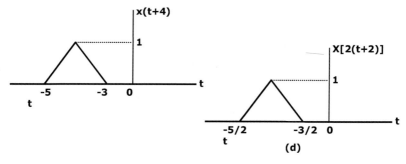

(d)

(e) In this example a $= 2$ and b $= -4$. Hence, shifting the given pulse x(t) by 4 units to the right. Next follow the precedence rule in that scaling the variable t by 2 units gets the final solution.

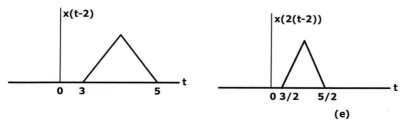

(e)

Example 1.8

Let x(t) and y(t) signals be given in the form of following mathematical expression.

$$x(t) = \begin{cases} -1 - t, & -1 \leq t \leq 0 \\ t, & 0 \leq t \leq 1 \\ 1, & 1 \leq t \leq 2 \\ 3 - t, & 2 \leq t \leq 3 \\ 0, & otherwise \end{cases} \qquad y(t) = \begin{cases} 1, & -2 \leq t \leq -1 \\ -1, & -1 \leq t \leq 0 \\ t - 1, & 0 \leq t \leq 1 \\ 1, & 1 \leq t \leq 2 \end{cases}$$

Carefully sketch the signals x(t), y(t), y(t − 1) and g(t) = x(t) y(t − 1).

Solution 1.8

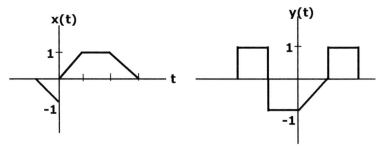

$$x(t) = \begin{cases} -1 - t, & -1 \leq t \leq 0 \\ t, & 0 \leq t \leq 1 \\ 1, & 1 \leq t \leq 2 \\ 3 - t, & 2 \leq t \leq 3 \\ 0, & otherwise \end{cases} \qquad y(t-1) = \begin{cases} 1, & -1 \leq t \leq 0 \\ -1, & 0 \leq t \leq 1 \\ t - 2, & 1 \leq t \leq 2 \\ 1, & 2 \leq t \leq 3 \\ 0, & otherwise \end{cases}$$

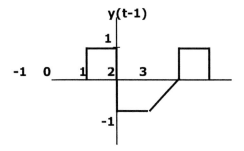

Now, $g(t) = x(t)y(t-1) = \begin{cases} -1-t, & -1 \le t \le 0 \\ -t, & 0 \le t \le 1 \\ t-2, & 1 \le t \le 2 \\ 3-t, & 2 \le t \le 3 \\ 0, & otherwise \end{cases}$

The sketch of g(t) is shown as follows:

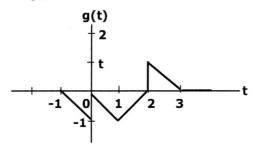

Example 1.9

Let x(t) and y(t) signals be given in the form of the following mathematical expression:

$$x(t) = \begin{cases} -1-t, & -1 \le t \le 0 \\ t, & 0 \le t \le 1 \\ 1, & 1 \le t \le 2 \\ 3-t, & 2 \le t \le 3 \\ 0, & otherwise \end{cases} \qquad y(t) = \begin{cases} 1, & -2 \le t \le -1 \\ -1, & -1 \le t \le 0 \\ t-1, & 0 \le t \le 1 \\ 1, & 1 \le t \le 2 \end{cases}$$

Carefully sketch the signals h(t) = x(t) y(t).

Solution 1.9

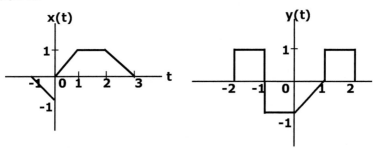

It is advisable for students in this example or any other example in which signal addition, multiplication or subtraction is required, it can be done in the following ways, generating a tabular form, in which there is less chances of errors. If there is final expression after multiplication, the value h(t) has an expression of t^2, different values have to be calculated to draw the final result, which will generate a curve.

The sketch of h(t) is shown as follows:

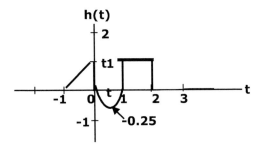

Time axis	X(t)	y(t)	$h(t) = x(t) \cdot y(t)$
$-1 \leq t \leq 0$	$x(t) = -1-t$ $x(-1) = 0$ $x(0) = -1$	$y(-1) = -1$ $y(-1) = -1$ $y(0) = -1$	$h(t) = 1+t$ $h(-1) = (0) \cdot (-1) = 0$ $h(-1) = (-1) \cdot (-1) = 1$
$0 \leq t \leq 1$	$x(t) = t$ $x(0) = 0$ $x(1) = 1$ $x(0.5) = 0.5$ $x(0.25) = 0.25$ $x(0.75) = -0.75$	$y(t) = -1 +t$ $y(0) = -1$ $y(1) = 0$ $y(0.5) = -0.5$ $y(0.25) = -0.75$ $y(0.75) = -0.75$	$h(t) = t(t-1) = -t + t^2$ $h(0) = (0) \cdot (-1) = 0$ $h(1) = (1) \cdot (0) = 0$ $h(0.5) = -0.50 + 0.25$ $= -0.25$ $h(0.25) = -0.1875$ $h(0.75) = -0.1875$
$1 \leq t \leq 2$	$x(t) = 1$ $x(1) = 1$ $x(2) = 1$	$y(t) = 1$ $y(1) = 1$ $y(2) = 1$	$h(t) = 1$ $h(1) = (1) \cdot (1) = 1$ $h(2) = (1) \cdot (1) = 1$
$2 \leq t \leq 3$	$x(t) = 3-t$ $x(2) = 1$ $x(3) = 0$	$y(t) = 0$ $y(2) = 0$ $y(3) = 0$	$h(t) = 0$ $h(2) = (1) \cdot (0) = 0$ $h(3) = (0) \cdot (0) = 0$

1.3.5 Use of Step and Ramp Function in Signal Processing

Step and ramp function plays a vital role in signal formation; it is, therefore, suggested that it is better to learn these techniques in order to draw the signal, or to calculate the expression in terms of step and ramp functions.

Example 1.10

Sketch the waveforms of the following signals, where u(t) is a unit step signal and r(t) is unit ramp signal:

(a) $x(t) = u(t) - u(t-2)$

(b) $x(t) = u(t+1) - 2u(t) + u(t-1)$

(c) $x(t) = -u(t+3) + 2u(t+1) - 2u(t-1) + u(t-3)$

(d) $y(t) = r(t+1) - r(t) + r(t-2)$

(e) $y(t) = r(t+2) - r(t+1) - r(t-1) + r(t-2)$

Solution 1.10

(a) $x(t) = u(t) - u(t-2)$

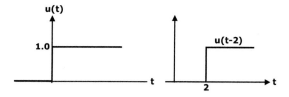 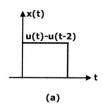

(b) $x(t) = u(t+1) - 2u(t) + u(t-1)$
   or, $x(t) = u(t+1) - u(t) - u(t) + u(t-1)$

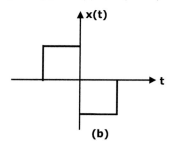

(c) $x(t) = -u(t+3) + 2u(t+1) - 2u(t-1) + u(t-3)$

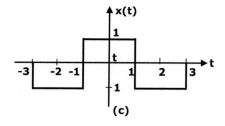

(d) $y(t) = r(t+1) - r(t) + r(t-2)$

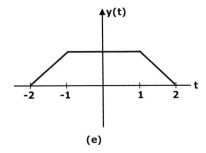

(d)

(e) $y(t) = r(t+2) - r(t+1) - r(t-1) + r(t-2)$

(e)

Example 1.11

Illustrates with labelled sketches the following signals:

$$x_1(t) = u(t) - 2u(t-1) + u(t-2)$$

$$x_2(t) = u(t-1) + u(t-2) - 3u(t-3) + u(t-4)$$

$$x_3(t) = -3u(t) + u(t-1) + u(t-2) + u(t-3)$$

Express the signals shown in the figure provided in terms of scaled delayed versions of the unit step signal.

Solution 1.11

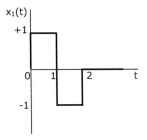

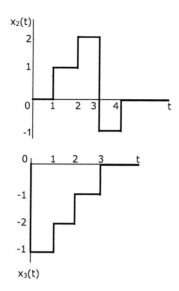

1.3.6 Even and Odd Signals

An even signal exhibits symmetry in time domain and are identical about the origin. Mathematically any signal $x(t)$ is known to be an even (i.e., symmetric) if it satisfies $x(t) = x(-t)$ and an odd (i.e., anti-symmetric) if it satisfies $x(t) = -x(-t)$.

If a signal is neither an even signal and nor an odd signal, it can be decomposed into its corresponding odd $x_o(t)$ and even $x_e(t)$ components. The following important relationships given by (1.7) and (1.8) are used to find the odd and even components. Using the $x(t) = x_e(t) + x_o(t)$, relationship the same original signal can be recovered.

$$x_e(t) = \frac{1}{2}[x(t) + x(-t)] \tag{1.7}$$

$$x_o(t) = \frac{1}{2}[x(t) - x(-t)] \tag{1.8}$$

Example 1.12
Consider signal h(t) as shown in the figure provided. Find the odd and even function of h(t).

Solution 1.12

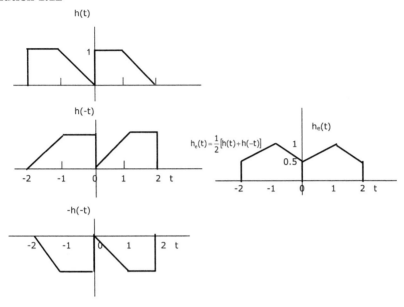

Once $h_e(t)$ has been drawn after adding $h(t)$ and $h(-t)$, divide its additive magnitude by 2. It can be verified using $h_e(t) = h_e(-t)$, which could give a student confidence about correctness of this solution.

$$h_o(t) = \frac{1}{2}[h(t) - h(-t)]$$

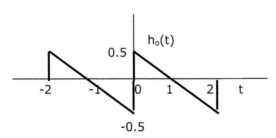

In the same way, once $h_o(t)$ has been drawn after adding $h(t)$ and $-h(-t)$ divide its additive magnitude by 2. It can be verified using $h_o(t) = h_o(-t)$, which could give a student confidence about correctness of this solution.

1.4 DT Signals

To form up a basis of digital signal processing (DSP), DT signals manipulation has been introduced. To accentuate its DT nature, a DT signal is written as x[n] and is shown in Figure 1.1, instead of x(t) representation for CT signals. A DT signal x[n] is a function of an independent variable n, where n is an integer. A DT signal is not defined at the instants between any two successive samples. DT signals are defined only at the discrete values of time. Following are the few examples of DT signals. Substituting t by nT alters a signal from CT domain to DT domain. In (a) and (b) given hereunder, T is not used with n and is suppressed for easiness while on the right-hand side t is replaced by nT.

(a) $X[n] = e^{-nT}$ for $n = 0, 1, 2, 3. \ldots \ldots$

(b) $x[n] = \begin{cases} 0.8^n & for \quad n \geq 0 \\ 0 & for \quad n < 0 \end{cases}$

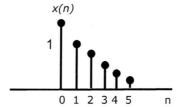

Figure 1.1   Discrete time signal

1.4.1 Continuous Versus Discrete Signals

The values of a CT or DT signal may be continuous or discrete. If a signal takes on all likely values on a finite or infinite range, it is known to be a continuous valued signal. On the other hand, if the signal takes on values from finite set of possible values, it is known to be a discrete-valued signal. A DT signal having a set of discrete values is called a digital signal.

1.4.2 Concept of Frequency – DT Signals

A DT sinusoidal signal can be given by

$$X[n] = A\cos(\omega n + \theta), \quad -\infty < n < \infty \quad (1.9)$$

where n denotes an integer variable or (sample number), A is the amplitude of sinusoid, ω is the frequency in radians per sample and θ represents the phase in radians. Rather than ω, we use the frequency variable f defined by $\omega = 2\pi f$, where f is digital frequency. The relation becomes

$$X[n] = A\cos(2\pi fn + \theta), -\infty < n < \infty \tag{1.10}$$

To find the periodicity of a DT signal, one has to be familiar with the following properties of DT sinusoids:

(a) A DT sinusoid is periodic only if its frequency f is a rational number. By definition, DT signal x[n] is periodic with period N (N > 0) if and only if x [n + N] = x[n] for all n. The smallest value of N is called the fundamental period or $\cos[2\pi f(n + N) + \theta] = \cos[2\pi fn + \theta]$.

The relationship is true if and only if there exists an integer k such that $2\pi fN = 2k\pi$ or $f = k/N$. Thus a DT signal is periodic only if its frequency f can be expressed as the ratio of two integers (i.e. f is rational). This property is the main property for recognizing DT signals' periodicity and dominates over the two other properties explained hereunder.

(b) DT sinusoids whose frequencies are separated by an integer multiple of 2π are alike.

$$\cos[(\omega + 2\pi)n + \theta] = \cos[\omega n + 2\pi n + \theta] = \cos[\omega n + \theta]$$

As a consequence, all the sequences are

$$x_k(n) = A\cos[\omega_k n + \theta]; k = 0, 1, 2. \ldots$$

where $\omega_k = \omega + 2k\pi; -\pi < \omega < \pi$ are same.

(c) The peak rate of oscillation in a DT sinusoid is achieved when

$$\omega = \pi(\text{or } \omega = -\pi) \text{ or equivalently } f = \frac{1}{2} \text{ or}$$

$$f = \frac{-1}{2} \text{ or } x[n] = \cos \omega n$$

where the frequency varies from 0 to π. To make a simpler argument, we take the value of $\omega = 0, \pi/8, \pi/4, \pi/2, \pi$ corresponding to f = 0, 1/16, 1/8, 1/4, 1/2, which consequences in periodic sequence having period N = ∞, 16, 8, 4, 2. It is to be noted that period of sinusoidal declines as the frequency growing. The rate of oscillation increases as the frequency increases. Frequency range of DT sinusoids is finite with duration 2π. Usually the range of $0 \le \omega \le 2\pi$ or $-\pi \le \omega \le \pi$ or $-1/2 \le f \le 1/2$ is chosen, which is also called the fundamental range.

The following examples cover periodic signals in DT, in which, N has been calculated to determine the periodicity of the signals using f = k/N, where N(or N_P) is period of the signal.

Example 1.13

Compute the fundamental period (N_P) and determine the periodicity of the signal.

(a) $\cos 0.01\pi n$ (b) $\cos\left(\pi\frac{30n}{105}\right)$

Solution 1.13

(a) $f = \frac{0.01\pi}{2\pi} = \frac{1}{200} \Rightarrow periodic\ with\ N_p = 200$

   $\cos .01\,\pi(n+200) = \cos .01\pi n + \cos 2\pi = \cos 2\pi$. It is a periodic signal.

(b) $f = \frac{30\,\pi}{105}\left(\frac{1}{2\pi}\right) = \frac{1}{7} \Rightarrow periodic\ with\ N_p = 7$

$$\cos\left(\pi\frac{30n}{105}\right) = \cos\left(\pi\frac{30}{105}(n+7)\right) = \cos\left(\pi\frac{30n}{105}\right) + \cos 2\pi$$

$$= \cos\left(\pi\frac{30n}{105}\right)$$

1.4.3 Time Domain and Frequency Domain

The signals introduced so far are all time dependent; we calculate the signal's amplitude at different time instants, which shows signal monitoring at different instants of time. If we made a sketch of all calculated results, this will be known as time-domain representation of the signal. The time-domain representation is very beneficial for many applications such as calculating the average value of a signal or defining when the amplitude of the signal surpasses certain limits.

There are certain applications for which alternative representation such as frequency-domain representation is more suitable. Sine waves occur frequently in nature and are building block of many other waveforms. The frequency-domain representation shows us the amplitudes and frequencies of the constituent sinusoidal waves present in the signals being measured. This representation is also known to be the spectrum of the signal.

1.5 AD and DA Conversion

Many signals of practical interest, such as speech signals, biological signals, seismic signals, radar signals, sonar signals and various communication signals such as audio and video signals are analogue in nature. To process analogue signals by digital means, it is first essential to alter them into digital form, i.e., to translate them to a sequence of number having finite samples. This procedure is called analogue-to-digital (AD) conversion, and the corresponding devices are called analogue-to-digital converters (ADCs). Once the digital signal is processed in the processor the output is then organized to drive a certain unit, now a digital-to-analogue (DA) conversion is required. The following steps are used to translate a CT signal into a DT signal.

1.5.1 Processing Steps for AD Conversion

1.5.1.1 Sample and hold

The conversion of a CT signal into a DT signal is obtained by taking samples of the CT signal at DT instants. Thus if $x_a(t)$ is the input to the sampler, the output is $x_a(nT) = x[n]$, where T is called the sampling interval. For the sake of simplicity, T is dropped. In sample and hold operation the previous value is held, using a zero-order hold till the next value is produced.

1.5.1.2 Quantization

This is the conversion of DT continuous-valued signal into a DT discrete-valued (digital) signal. The value of each signal sample is denoted by a value selected from a finite set of possible values. The difference between the unquantized sample $x[n]$ and the quantized output $x_q[n]$ is called the quantization error.

1.5.1.3 Coding

In the coding process, each discrete value $x_q[n]$ is offered by a b-bit binary sequence. An AD converter is demonstrated as a sampler followed by a quantizer and coder as shown in Figure 1.2, in practice the AD conversion is performed by a single device that takes $x_a(t)$ and creates a binary coded number. The operation of sampling and quantization can be performed in either order in practice; however, sampling is always performed before quantization.

In many cases, it is appropriate to have a signal in an analogue form such as in speech signal processing because sequence of samples cannot be understood. So the conversion is mandatory from digital to analogue form. All DA converters convert the discrete values into an analogue signal by performing some kind of interpolation such as linear interpolation and quadratic interpolation.

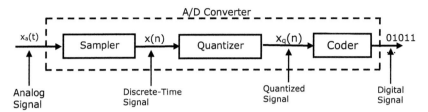

Figure 1.2   Basic parts of an AD converter

1.5.2 Sampling of Analogue Signals

There are many ways to sample an analogue signal. Our discussion is limited here to periodic or uniform sampling, which is the type of sampling used most often in practice. This is designated by the following relation

$$X[n] = x_a(nT), -\infty < n < \infty \qquad (1.11)$$

where x[n] is the DT signal obtained by taking samples of the analogue signal $x_a(t)$ at every T seconds. This procedure is explained with the help of Figure 1.3. The time interval T between successive samples is called the sampling period or sample interval and its reciprocal $1/T = F_s$ is called the sampling rate (samples per second) or the sampling frequency (hertz).

Periodic sampling forms a relationship between the time variables t and n of CT and DT signals, respectively. Indeed, these variables are linearly related via the sampling period T or, equivalently, the sampling rate $F_s = 1/T$, as

$$t = nT = \frac{n}{F_s} \qquad (1.12)$$

Figure 1.3(a)   Sampler

There is a relationship between the frequency variable F (or Ω) for CT signals and the frequency variable f (or ω) for DT signals. To form the relationship, consider a CT sinusoidal signal of the form

$$x_a(t) = A \cos(2\pi Ft + \theta) \qquad (1.13)$$

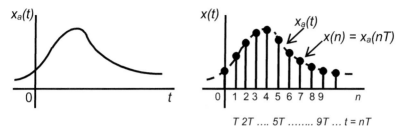

Figure 1.3(b) Periodic sampling of an analog signal

when sampled periodically at a rate $F_s = 1/T$ samples per second, yields

$$x_a(nT) = A \cos{(2\pi nFT + \theta)} \qquad (1.14)$$

$$x_a(n) = A \cos\left(\frac{2\pi nF}{F_s} + \theta\right)$$

Comparing (1.4) with (1.3), it is to be noted that the frequency variables F and f are linearly related as

$$f = \frac{F}{F_s} = \frac{T}{T_p} \text{ or } \omega = \Omega T \qquad (1.15)$$

where *f is* named as the normalized frequency; it can be used to determine the frequency F in hertz only if the sampling frequency F_s is known.

Example 1.14
The implications of frequency conversion can be considered by the analogue sinusoidal signal x(t) = cos (4000π)t, which is sampled at a rate $F_s = 4000$ Hz and $F_s = 2000$ Hz.

Solution 1.14
The corresponding DT signal at $F_s = 4000$ Hz is

$$x_{(n)} = \cos 2\pi \left(\frac{2000}{4000}\right) n = \cos 2\pi (1/2) n$$

$$x_{(n)} = \cos 2\pi (1/2) n$$

Using (1.15), it follows that digital frequency f = 1/2.

Since only the frequency component at 4000 Hz is present in the sampled signal, the analogue signal we can recover is $y_a(t) = \cos 4000\pi t$.

The corresponding DT signal at $F_s = 2000$ Hz is

$$x_{(n)} = \cos2\pi \left(\frac{2000}{2000}\right) n = \cos2\pi(1)n$$

$$x_{(n)} = \cos2\pi(1)n$$

From (1.15) it follows that the digital frequency $f = 1$ cycle/sample. Since only the frequency component at 2000 Hz is present in the sampled signal, the analogue signal we can recover is $y_a(t) = x(F_s t) = \cos 2000\pi t$. This example shows that by taking a wrong selection of the sampling frequency, the original signal cannot be recovered.

1.6 The Sampling Theorem

The sampling theorem states that a continuous signal can be sampled very well, if and only if it does not hold any frequency components exceeding one-half of the sampling rate. Figures 1.4(a–d) specify several sinusoids before and after digitization. The continuous line denotes the analogue signal incoming to the ADC, while the square markers are the digital signal exiting the ADC.

The signal is a constant DC value as shown in Figure 1.4(a), a zero frequency. Since the analogue signal is a series of straight lines between each of the samples, all of the information needed to reconstruct the analogue signal is contained in the digital data. According to classification, this is well sampled.

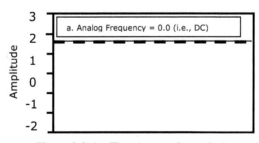

Figure 1.4(a)   Time (or sample number)

The sine wave shown in Figure 1.4(b), a 90° cycle per second sine wave being sampled at 1000 samples/s, has a frequency of 0.09 of the sampling rate. Stating in another way, the result is only 11.1 samples per sine wave cycle.

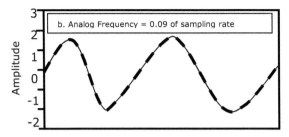

Figure 1.4(b)   Time (or sample number)

The state in Figure 1.4(b) is more complex than the previous case in Figure 1.4(a), because the analogue signal cannot be reconstructed by simply drawing straight lines between the data points. Do these samples correctly represent the analogue signal? The answer is yes, because no other sinusoid, or combination of sinusoids, will give this pattern of samples. These samples agree to only one analogue signal, and therefore the analogue signal can be exactly reconstructed. Again, it is an example of correct sampling.

In Figure 1.4(c), a 310 cycle/s sine wave being sampled at 1000 samples/s, the condition is made tougher by increasing the sine wave's frequency to 0.31 of the sampling rate, and this result in only 3.2 samples per sine wave cycle.

Here the samples are so sparse that they do not even appear to follow the general trend of the analogue signal. Do these samples correctly represent the analogue waveform? Again, the answer is yes, and for exactly the same reason. The samples are a unique representation of the analogue signal.

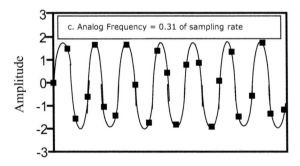

Figure 1.4(c)   Time or sample number

All the information desired to reconstruct the continuous waveform is contained in the digital data. Perceptibly, it must be more refined than just drawing straight lines between the data points. As strange as it seems, this is accurate sampling according to the sampling theorem.

In Figure 1.4(d), a 950 cycle/s sine wave being sampled at 1000 samples/s, the analogue frequency is pressed even higher to 0.95 of the sampling rate, with a simple 1.05 samples per sine wave cycle. Do these samples accurately represent the data? No, they do not. The samples denote a different sine wave from the one contained in the analogue signal. In particular, the original sine wave of 0.95 frequencies falsifies itself as a sine wave of 0.05 frequencies in the digital signal.

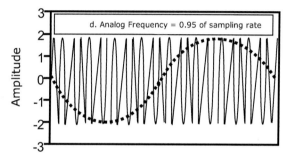

Figure 1.4(d)   frequencies in the digital signal.

The occurrence of sinusoids changing frequency during sampling is called *aliasing*. The sinusoid adopts another frequency that is not the original ones. Since the digital data is no longer exclusively linked to a particular analogue signal, an unambiguous reconstruction is impossible. There is nothing in the sampled data to suggest that the original analogue signal had a frequency of 0.95 rather than 0.05.

The sine wave has veiled its original identity completely; the perfect crime has been committed! According to the definition, this is an example of incorrect sampling. This reasoning leads to a milestone in DSP, the *sampling theorem*. This is called the Shannon sampling theorem or the Nyquist sampling theorem. For instance, a sampling rate of 2,000 samples/s requires the analogue signal to be composed of frequencies below 1000 cycles/s. If frequencies above this limit are present in the signal, they will be *aliased* to frequencies between 0 and 1000 cycles/s, combining with whatever information that was legitimately there.

Two terms are widely used when discussing the sampling theorem: Nyquist frequency and Nyquist rate. Unfortunately, their meaning is not standardized. To understand this, consider an analogue signal composed of frequencies between DC and 3 kHz. To accurately digitize this signal, it must be sampled at 6,000 samples/s (6 kHz) or higher. Suppose we choose to sample at 8,000 samples/s (8 kHz), allowing frequencies between DC and 4 kHz to be correctly represented. In this situation, there are four important frequencies:

(a) The highest frequency in the signal, 3 kHz;
(b) Twice of this frequency, 6 kHz;
(c) The sampling rate should be 8 kHz and
(d) One-half the sampling rate, 4 kHz.

Example 1.15
Consider the analogue signal, $x_a(t) = 3\cos 100\pi t$.

(a) Determine the minimum required sampling rate to avoid aliasing.
(b) Suppose that the signal is sampled at the rate $F_s = 200$ Hz. What is the DT signal obtained after sampling?
(c) Suppose that the signal is sampled at the rate $F_s = 75$ Hz. What is the DT signal obtained after sampling?
(d) What is the frequency $F < F_s/2$ of a sinusoid that yields samples identical to those obtained in part (c)?

Solution 1.15

(a) The frequency of the analogue signal is $F = 50$ Hz. Hence the minimum sampling rate required to avoid aliasing is $F_s = 100$ Hz.
(b) The signal is sampled at $F_s = 200$ Hz, the DT signal is

$$x(n) = 3\cos\frac{100\pi}{200}n = 3\cos\frac{\pi}{2}n$$

(c) If the signal is sampled $F_s = 75$ Hz, the DT signals is

$$x(n) = 3\cos\frac{100\pi}{75}n = 3\cos\frac{4\pi}{3}n$$

$$x(n) = 3\cos\left(2\pi - \frac{2\pi}{3}\right)n = 3\cos\frac{2\pi}{3}n$$

(d) For the sampling rate of $F_s = 75$ Hz, we have

$$F = fF_s = 75f$$

The frequency of the sinusoid in part (c) is $f = \frac{1}{3}$. Hence, F = 25 Hz.

Obviously, the sinusoidal signal $y_a(t) = x(F_s t) = 3\cos 2\pi F t = 3\cos 50\pi t$ sampled at $F_s = 75$ samples/s gives identical samples. Hence F = 50 Hz is an alias of F = 25 Hz over the sampling rate $F_s = 75$ Hz.

1.7 Quantization Error

It is well explained previously that the process of converting a DT continuous-amplitude signal into a digital signal by articulating each sample value as a finite number of digits is called **quantization**. Precision of the signal representation is directly proportional to how many discrete levels are permitted to denote the magnitude of the signal. The error presented in representing the continuous-valued signal by a finite set of discrete value levels is called **quantization error** or **quantization noise.** Quantization error is the difference between the original and the quantized value.

Normally, number of bits controls the resolution and the degree of quantization error or noise. It is common practice in signal processing to define the ratio of the largest undistorted signal to the smallest undistorted signal that a system can process. The ratio is called the **dynamic range**.

The noise voltage and the maximum signal-to-noise ratio are both significant measures of quantization error. They are linked with the following expression:

$$V_{noise(rms)} = \frac{V_{full\,scale}(0.289)}{2^n} \tag{1.16}$$

$$Maximum\ signal\ to\ noise\ ratio\ (dB) = 6.02n + 1.76 \tag{1.17}$$

where n is the number of bits. Quantizer operation is completed on the samples x[n] as Q[x(n)] and let $x_q(n)$ represent the sequence of quantized samples at the output of the quantizer.

$$x_q(n) = Q[x(n)] \tag{1.18}$$

Then the quantization error is a sequence $e_q(n)$ defined as the difference between the quantized value and the actual sample value.

$$e_q(n) = x_q(n) - x(n) \tag{1.19}$$

The quantization process is illustrated with an example. Let us consider the DT signal

$$x(n) = \begin{cases} 0.9^n, & n \geq 0 \\ 0, & n < 0 \end{cases} \tag{1.20}$$

which was attained by sampling the analogue exponential signal $x_a(t) = 0.9^t$, $t \geq 0$ with a sampling frequency $F_s = 1$ Hz (Figure 1.5(a)), the quantization is completed by rounding, although it is easy to treat truncation, the rounding process is graphically explained in Figure 1.5(b).

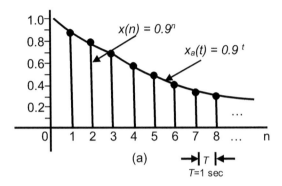

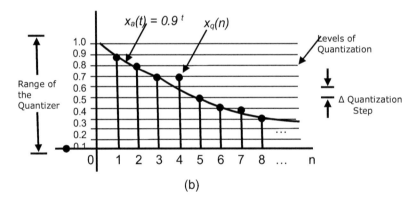

Figure 1.5   Illustration of quantization.

It is to be noted that if x_{min} and x_{max} denote the minimum and maximum value of $x(n)$ and L is the number of quantization level, then

$$\Delta = \frac{x_{max} - x_{min}}{L - 1} = \frac{range}{L - 1} \tag{1.21}$$

where $x_{min} - x_{max}$ is called the range of the signal and L is the quantization level.

Example 1.16
Determine the root mean square (rms) noise quantization noise voltage for 8 and 12 bit systems when the signal-to-noise ratio is from 0 to 5 V.

Solution 1.16
Applying (1.16)

$$V_{noise(rms)} = \frac{(5V)(0.289)}{2^8} = 5.64\text{mv}$$

$$V_{noise(rms)} = \frac{(5V)(0.289)}{2^{12}} = 353\mu\text{v}$$

Example 1.17
Find the maximum signal-to-noise ratio for a 12 bit DSP system.

Solution 1.17
Applying (1.17) $Maximum\ signal\ to\ noise\ ratio = 6.02.(12) + 1.76 = 74\ dB$.

1.8 Representing DT Signal

A DT signal x[n] is a function of an independent variable, i.e., an integer. A DT signal is not defined at instants between two successive samples. The following approaches are used to demonstrate the following digital signals

$$X[n] = \{\ldots, 2, 1, -2, -2, 3, 2, 2, -2, 1, \ldots\}$$

1.8.1 Graphical Representation

The DT signal is graphically represented in Figure 1.6.

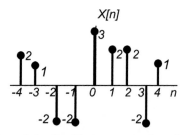

Figure 1.6 Graphical representation of x(n)

1.8.2 Functional Representation

$$x(n) = \begin{cases} 3 & \text{for} & n = 0 \\ 1 & \text{for} & n = -3,\ 4 \\ 2 & \text{for} & n = 1,\ 2,\ -4 \\ -2 & \text{for} & n = -1,\ -2,\ 3 \\ 0 & \text{elsewhere} \end{cases}$$

1.8.3 Sequence Representation

$$x(n) = \{\ldots, 2, 1, -2, -2, 3, 2, 2, -2, , 1, \ldots\}$$

1.8.4 Tabular Representation

n	−4	−3	−2	−1	0	1	2	3	4....
x(n)	2	1	−2	−2	3	2	2	−2	1

1.9 Elementary DT Signals

1.9.1 Unit Impulse

Unit impulse shown in Figure 1.6 (a), also known as an unit sample sequence, and in the DT system is defined by the following mathematical relationship.

$$\delta(n) = \begin{cases} 1 & \text{for} & n = 0 \\ 0 & \text{for} & n \neq 0 \end{cases}$$

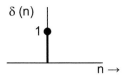

Figure 1.6 (a) Unit impulse function

1.9.2 Unit Step Signal

Unit step sequence shown in Figure 1.6 (b), in the DT system is defined by the following mathematical relationship:

$$u(n) = \begin{cases} 1 & \text{for} & n \geq 0 \\ 0 & \text{for} & n < 0 \end{cases}$$

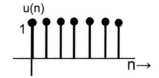

Figure 1.6   (b) Unit step function

Unit step signal is truly the sum of unit impulses:

$$u(n) = \delta(n) + \delta(n-1) + \delta(n-2) + \delta(n-3) \qquad (1.22)$$

Expression (1.22) can be defined in the closed form as

$$u(n) = \sum_{k=0}^{n} \delta(n-k)$$

1.9.3 Unit Ramp Signal

Unit ramp sequence shown in Figure 1.6 (c), in the DT system is defined by the following mathematical relationship.

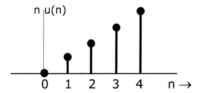

Figure 1.6   (c) Unit ramp function

$$u_r(n) = \begin{cases} n & \text{for} \quad n \ge 0 \\ 0 & \text{for} \quad n < 0 \end{cases}$$

1.9.4 Exponential Signal

Exponential sequence shown in Figure 1.6 (d), in the DT system is defined by the following mathematical relationship. The value of 'a' mentioned hereunder can be real and/or imaginary:

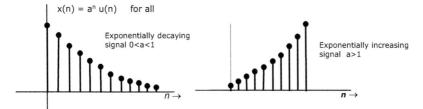

Figure 1.6 (d) Exponentially decaying and increasing signals

1.9.5 Sinusoidal Signal

A sinusoidal sequence shown by Figure 1.6 (e), in the DT system is defined by the following mathematical relationship $x(n) = A \sin \omega n$.

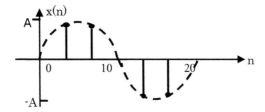

Figure 1.6 (e) Sinusoidal signals.

1.10 Simple Manipulations of DT Signal

There are few procedures, which are mandatory to be handled at different stages in DSP. Any signal y(n) on which an operation has to be performed can be easily understood and calculated by substituting the different values of *n* in the original signal and getting the new values of y(n). The following examples are given to elaborate the point.

1.10.1 Reflection/Folding/Flipping

The modification of time base is to replace the independent variable n by $-n$. The result of this procedure is a folding or a reflection of the signal about the origin as shown in Figure 1.7. If for $x(n) = \{-1, 1, 2, 2, 1, 1\}$, to find x(−n), replacing y(n) = x(−n), and substituting the value of n results as $\{-2, -1, 0, 1, 2, 3\}$, because these are the values of n for which the original signal x(n) exists.

We get y(n) = x(−n), n = −2, y(−2) = x (2) , y (−2) indicates the location of new signal and its value would be x (2) and for n = 3, y(n) = x(−(3)) = x(−3) = 0 and so on.

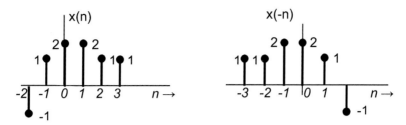

Figure 1.7   (a) Signal x(n) and (b) Reflected signal x(−n).

1.10.2 Shifting (Advanced and Delayed)

A signal x(n) can be shifted in time by substituting independent variable n by n − k as shown in Figure 1.8, where k is an integer. If k is a positive integer, then the time shift results in a delay of the signal by k units of time. It means the new signal shifts towards right side by k amount. If k is a negative integer, then the time shift results in an advance of the signal by |k| units in time. It means the new signal shifts towards the left side by k amount.

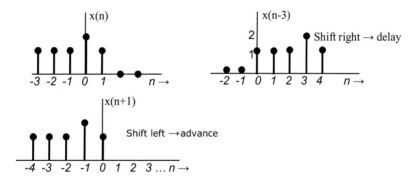

Figure 1.8   Shifted version of a signal

1.10.3 Scaling (Time and Magnitude)

Scaling can be done using the two properties: *time scaling or down sampling,* in which n is replaced by *cn,* where *c* is an integer and *magnitude scaling,* which is done by multiplying numbers with the every value of signals.

$$y(n) = Ax(n); -\infty < n < \infty$$

Example 1.18

Consider the signal, $x(n) = \{\ldots 0, 2, 1, 2, -2, -2, 1, 0\ldots..\}$.

   Find (a) $x(2n)$

           (b) $4x(n)$

Solution 1.18

  (a) Let $y(n) = x(2n)$

$$y(0) = x(0) = 0; y(1) = x(2) = 1; y(2) = x(4) = -2;$$
$$y(3) = x(6) = 1; y(-1) = x(-2) = 0; y(-2) = x(-4) = 0$$
$$y(n) = x(2n) = \{\ldots 0, 1, -2, 1\ldots..\}$$

    The value of y gives the location and the value of $x(2n)$ gives its magnitude.

  (b) $y(n) = 4x(n) = \{\ldots 0, 8, 4, 8, -8, -8, 4, 0\ldots..\}$

1.10.4 Addition and Multiplication

The sum of two signals $x_1(n)$ and $x_2(n)$ is a signal $y(n)$, whose value at any instant is equal to sum of values of these signals at that instant:

$$y(n) = x_1(n) + x_2(n); -\infty < n < \infty$$

    The product of two signals is similarly defined on a sample-to-sample basis as

$$y(n) = x_1(n)x_2(n); -\infty < n < \infty$$

Example 1.19

A DT signal $x[n]$ is shown in Figure.

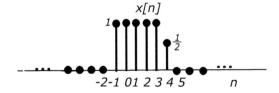

Sketch and label carefully each of the following signals.

    (i) $x[n-1]\delta[n-3]$

    (ii) $\frac{1}{2}x[n] + \frac{1}{2}(-1)^n x[n]$

    (iii) $x[n^2]$

Solution 1.19

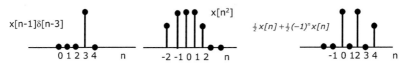

1.10.5 Even and Odd Signals

Similar to the CT, it is also same for DT that an even signal is that type of signal, which displays symmetry in time domain. This type of signal is identical about the origin. Mathematically a signal x(n) is known as an even (symmetric) signal as shown in Figure 1.9 (b), if $x(n) = x(-n)$ and odd (anti-symmetric) as shown in Figure 1.9 (a) if $x(n) = -x(-n)$.

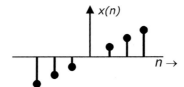

Figure 1.9 (a) Odd signal

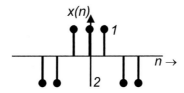

Figure 1.9 (b) Even signal

If a signal is neither an even nor an odd, it can be disintegrated into its odd $x_o(n)$ and even $x_e(n)$ components. The following imperative relationship is used to find the odd and even components. Using this $x(n) = x_e(n) + x_o(n)$ relationship the same original signal can be recovered:

$$x_e(n) = \frac{1}{2}[x(n) + x(-n)] \tag{1.23}$$

$$x_o(n) = \frac{1}{2}[x(n) - x(-n)] \tag{1.24}$$

Example 1.20

A DT signal x(n) is shown in the figure provided. Sketch and label carefully each of the following signals.

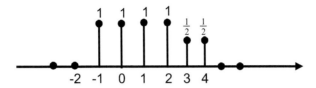

(a) even part of x(n)
(b) odd part of x(n)

Solution 1.20

$$x(n) = \{\ldots 0, 1, \underset{\uparrow}{1}, 1, 1, \tfrac{1}{2}, \tfrac{1}{2}, 0 \ldots\}$$

(a) $x_e(n) = \frac{x(n)+x(-n)}{2}$, $x(-n) = \{\ldots 0, \tfrac{1}{2}, \tfrac{1}{2}, 1, 1, \underset{\uparrow}{1}, 1, 0, 0, 0, \ldots\}$

$$= \{\ldots 0, \tfrac{1}{4}, \tfrac{1}{4}, \tfrac{1}{2}, 1, \underset{\uparrow}{1}, 1, \tfrac{1}{2}, \tfrac{1}{4}, \tfrac{1}{4}, 0, \ldots\}$$

(b) $x_0(n) = \frac{x(n)-x(-n)}{2}$,

$$= \{\ldots 0, -\tfrac{1}{4}, -\tfrac{1}{4}, -\tfrac{1}{2}, 0, \underset{\uparrow}{0}, 0, \tfrac{1}{2}, \tfrac{1}{4}, \tfrac{1}{4}, 0, \ldots\}$$

1.11 Energy and Power Signals for CT and DT Signals

Signals can be categorized as energy and power signals. Since we often called a signal as a function of varying amplitude in time, it appears to say that a good measurement of the strength of a signal would be area under the curve. However, this area may have negative part. The negative part does not have less strength than a positive signal of the same size. This proposes squaring the signal or taking its absolute value, and then finding the area under the curve. We can call the energy of a signal is the area under curve (square signal). The energy signal is one, which has finite energy and zero average power. However, there are some signals, which can neither be classified as energy signals or power signals.

In CT, x(t) is an energy signal, if $0 < E < \infty$ and $P = 0$, where E is the energy and P is the power of the signal x(t). The power signal is one, which has finite average power and infinite energy.

$$E = \int_{-\infty}^{\infty} |x(t)|^2 \, dt \qquad (1.25)$$

Hence x(t) is an power signal, if $0 < P < \infty$ and $E = \infty$. However, if the signal does not satisfy any of the above conditions, then it is neither energy nor power signal.

For DT signals the area under the squared signal makes no sense, so we have to provide another energy definition. The energy can be defined as the sum of the squared magnitude of the samples.

$$E = \sum_{n=-\infty}^{\infty} |x(n)|^2 \qquad (1.26)$$

In DT, x(n) is an energy signal, if $0 < E < \infty$ and $P = 0$, where E is the energy and P is the power of the signal x(n). Squared values of x(n) can be applied to both complex-values signal and real signals. Energy of signal may be finite or infinite. If E is finite, then x(n) is an energy signal.

For analogue signals, it is stated that the power as energy per time interval. The power of a signal x(t) is given as

$$P = \lim_{T \to \infty} \frac{1}{T} \int_{-T/2}^{T/2} |x(t)|^2 \, dt \qquad (1.27)$$

For discrete signals, it is stated that the power as energy per sample. The power in a signal x(n) is given as

$$P = \lim_{N \to \infty} \frac{1}{N} \sum_{-N/2}^{N/2} |x(n)|^2 \qquad (1.28)$$

The signal power P is equal to the mean square value of x(n). Table 1.1 shows a comparison of energy and power signal.

Example 1.21
Consider the sinusoidal signal

$$x(t) = A\cos(\omega t + \phi)$$

Determine the average power of x(t).

Table 1.1   A comparison of energy and power signal

	Energy Signal	Power Signal
1	Total normalized energy is finite and non-zero	Total normalized average power is finite and non-zero
2	The energy is obtained by $$E = \int_{-\infty}^{\infty} |x(t)|^2 \, dt$$ $$E = \sum_{n=-\infty}^{\infty} |x(n)|^2$$	The average power is obtained by $$P = \lim_{T \to \infty} \frac{1}{T} \int_{-T/2}^{T/2} |x(t)|^2 \, dt$$ $$P = \lim_{N \to \infty} \frac{1}{N} \sum_{n=-N/2}^{N/2} |x(n)|^2$$
3	Non-periodic signals are energy signals.	Practically, periodic signals are power signals.
4	These signals are time limited.	These signals can exist over infinite time.
5	Power of energy signal is zero.	Energy of power signal is infinite.

Solution 1.21

$$p = \frac{1}{T} \int_{-T/2}^{T/2} x^2(t) dt$$

where T is the fundamental period equal to

$$T = \frac{2\pi}{\omega}$$

Now, $p = \dfrac{\omega}{2\pi} \displaystyle\int_{-\pi/\omega}^{\pi/\omega} A^2 \cos^2(\omega t + \phi) dt = \dfrac{\omega}{2\pi} \displaystyle\int_{-\pi/\omega}^{\pi/\omega} [1 + \cos 2(\omega t + \phi)] dt$

$$= \frac{\omega A^2}{4\pi} \left[t + \frac{\sin 2(\omega t + \phi)}{2\omega} \right] \Big|_{-\frac{\pi}{\omega}}^{\frac{\pi}{\omega}} = \frac{\omega A^2}{4\pi} \left(\frac{2\pi}{\omega} \right) = \frac{A^2}{2}$$

Example 1.22
Determine the signal $x(t) = A[u(t + a) - u(t - a)]$. Determine whether the signal is power signal or energy signal.

Solution 1.22
It is clear that x(t) is a finite duration signal and therefore x(t) is an energy signal. Energy of a signal is expressed as

$$E = \int_{-\infty}^{\infty} |x(t)|^2 \, dt = \int_{-a}^{a} |A|^2 \, dt \quad x(t) = A \quad for \quad -a < t < a$$

$$E = 2 \int_{0}^{a} |A|^2 \, dt \quad = a = A^2 |t|_0^a = 2aA^2$$

Example 1.23
Compute the energy of the length-N sequence

$$x(n) = \cos\left(\frac{2\pi kn}{N}\right) \quad 0 \le n \le N-1$$

Solution 1.23

$$x(n) = \cos\left(\frac{2\pi kn}{N}\right) \quad 0 \le n \le N-1$$

$$E = \sum_{n=0}^{N-1} |x(n)|^2 = \sum_{n=0}^{N-1} \cos^2(2\,\pi kn/N) = \frac{1}{2}\sum_{n=0}^{N-1}(1+\cos(4\,\pi kn/N))$$

$$= \frac{N}{2} + \frac{1}{2}\sum_{n=0}^{N-1}\cos(4\,\pi kn/N)$$

Let $C = \sum_{n=0}^{N-1}\cos(4\pi kn/N)$ and $S = \sum_{n=0}^{N-1}\sin(4\,\pi k n/N)$

Therefore $C + jS = \sum_{n=0}^{N-1} e^{j4\pi kn/N} = \frac{e^{j4\pi k}-1}{e^{j4\pi k/N}-1} = 0$ Thus,

$C = \text{Re}\{C+jS\} = 0$.

As $C = \text{Re}\{C+jS\} = 0$, it follows that $E = \frac{N}{2}$.

Example 1.24
Categorize each of the following signals as an energy or power signal, and find the energy or power of the signal.

(a) $x(t) = \begin{cases} t, & 0 \le t \le 1 \\ 2-t, & 1 \le t \le 2 \\ 0, & otherwise \end{cases}$

(b) $x(t) = 5\cos(\pi t) + \sin(5\pi t), -\infty < t < \infty$

(c) $x(t) = \begin{cases} 5\cos(\pi t) & -1 \le t \le 1 \\ 0, & otherwise \end{cases}$

(d) (d) $x[n] = \begin{cases} \sin(\pi t) & -4 \le n \le 4 \\ 0, & otherwise \end{cases}$

Solution 1.24

(a) $E = \underset{T\to\infty}{Lt} \int_{-T/2}^{T/2} x^2(t)dt = \int_{-\infty}^{\infty} x^2(t)dt$

$= \int_0^1 t^2 dt + \int_1^2 (2-t)^2 dt = \frac{1}{3} + \frac{1}{3} = \frac{2}{3}$

Since $0 < E < \infty$, x(t) is an energy signal.

(b) $x(t) = 5\cos(\pi t) + \sin(5\pi t)$

$$E = \int_{-\infty}^{\infty} X^2(t)dt$$

$$= \int_{-\infty}^{\infty} [25\cos^2(\pi t) + \sin^2(5\pi t) + 10\sin(5\pi t)\cos \pi t]dt = \infty$$

$x(t)$ can't be an energy signal

$$p = \underset{T\to\infty}{Lt} \int_{-T/2}^{T/2} x^2(t)dt = \frac{1}{T}\int_{-T/2}^{T/2}[5\cos(\pi t) + \sin(5\pi t)]^2 dt$$

where T is the fundamental period equal to 2

$$= \frac{1}{2}\int_{-1}^{1} [25\cos^2 \pi t + \sin^2 5\pi t + 5\sin(6\pi t) + \sin(4\pi t)]dt$$

$$= \frac{1}{2}\int_{-1}^{1} \left[\frac{25}{2}(1 + \cos 2\pi t) + \frac{(1 - \cos 2\pi t)}{2} + 5\sin(6\pi t) + \sin(9\pi t)\right] dt$$

$$= \frac{13}{2}$$

Since $0 < P < \infty$, p is a power signal.

(c) $x(t) = \begin{cases} 5\cos(\pi t), & -1 \le t \le 1 \\ 0, & otherwise \end{cases}$

$$E = \int_{-\infty}^{\infty} x^2(t)dt = \int_{-\infty}^{\infty} 25\cos^2(\pi t)dt = \frac{25}{2}\int_{-1}^{1}(1 + \cos 2\pi t)dt$$

$$= \frac{25}{2}\left\{t + \frac{\sin 2\pi t}{2\pi} \quad \Big|_{-1}^{1}\right\} = 25$$

Since $0 < E < \infty$, x(t) is an energy signal.

$$p = \frac{1}{T}\int_{-T/2}^{T/2} x^2(t)dt \quad here, \quad T = 2$$

$$= \frac{1}{2}\int_{-1}^{1} 25\cos^2(\pi t)dt = \frac{25}{2}\int_{-1}^{1}\{1 + \cos(2\pi t)\}\, dt = \frac{25}{2}$$

Since $0 < P < \infty$, x(t) is a power signal also.

(d) $E = \sum\limits_{n=-4}^{4} \sin^2(\pi n) = \sum\limits_{n=-4}^{4} \left(\frac{1-\cos 2\pi n}{2}\right) = 4.$

Since $0 < E < \infty$, x[n] is an energy signal.

$$P = \frac{1}{N} \sum_{n=0}^{N-1} x^2[n]$$

where N is the fundamental period, here N = 2

$$p = \frac{1}{2} \sum_{n=0}^{1} \sin^2(\pi n) = \frac{\sin^2 \pi}{2} = 0$$

x(n) cannot be a power signal.

1.12 Problems and Solutions

Problem 1.1
Determine which of the following sinusoids are periodic and compute their fundamental period.

$$\text{(a) } \cos 3\pi n \quad \text{(b) } \sin 3n \quad \text{(c) } \sin\left(\pi \frac{62n}{10}\right)$$

Solution 1.1
(a) $f = \frac{3\pi}{2\pi} = \frac{3}{2} \Rightarrow periodic\ with\ N_p = 2$

  $\cos 3\pi(n + N) = \cos 3\pi(n + 2) = \cos 3\pi n + \cos 2\pi = \cos 3\pi n$

(b) $f = \frac{3}{2\pi} \Rightarrow non - periodic$

(c) $f = \frac{62\pi}{10}\left(\frac{1}{2\pi}\right) = \frac{31}{10} \Rightarrow periodic\ with\ N_p = 10$
  $\sin\left(\pi \frac{62\,n}{10}\right) = \sin\left(\pi \frac{62}{10}(n + 10)\right) = \sin\left(\pi \frac{62\,n}{10}\right) \Rightarrow periodic\ with$
  $N_p = 10$

Problem 1.2
Determine whether or not each of the following signals is non-periodic. In case a signal is periodic, specify its fundamental period.

(a) $x_a(t) = 3\cos(5t + \pi/6)$
(b) $x(n) = 3\cos(5n + \pi/6)$
(c) $x(n) = 2\exp[j(n/6 - \pi)]$
(d) $x(n) = \cos(n/8)\cos(\pi n/8)$
(e) $x(n) = \cos(\pi n/2) - \sin(\pi n/8) + 3\cos(\pi n/4 + \pi/3)$

Solution 1.2

(a) Analogue signal: Periodic with period $T_p = 2\pi/\Omega = 2\pi/5$
(b) Discrete signal: $f = k/N = 5/2\pi \Rightarrow$ non-periodic
(c) Discrete signal: $f = k/N = 1/12\pi \Rightarrow$ non-periodic
(d) Discrete signal: $\cos n/8$ is non-periodic; $\cos(\pi n/8)$ is periodic. Their product is non-periodic.
(e) Discrete signal: $\cos(\pi n/2)$ is periodic with period $N_p = 4$; $f = k/N = 1/4$
$\sin(\pi n/8)$ is periodic with period $N_p = 16$; $f = k/N = 1/16$
$\cos(\frac{\pi n}{4} + \frac{\pi}{3})$ is periodic with period $N_P = 8$, Because it is a composite signal having multiple frequencies, therefore, $x(n)$ is periodic with period $N_p = 16$ [$16 =$ least common multiple of 4, 8, 16].

Problem 1.3

Consider the following analogue sinusoidal signal $x_a(t) = 3\sin(100\pi t)$.

(a) Sketch the signal $x_a(t)$ for $0 \le$ to $\le$ 30 ms.
(b) The signal $x_a(t)$ is sampled with a sampling rate $F_s = 300$ samples. Determine the frequency of the DT signal $x(n) = x_a(nT)$, $T = 1/F_s$, and show that it is periodic.
(c) Compute the sample values in one period of $x(n)$. Sketch $x(n)$ on the same diagram with $x_a(t)$. What is the period of the DT signal in milliseconds?
(d) Can you find a sampling rate F_s such that the signal $x(n)$ reaches its peak value of 3? What is the minimum F_s suitable for this task?

Solution 1.3

(a)

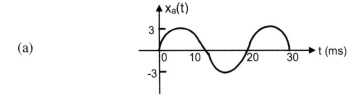

(b) $x(n) = x_a(nT) = x_a(n/Fs) = 3\sin\left(\frac{\pi}{3}n\right) \Rightarrow f = \frac{1}{2\pi}\left(\frac{\pi}{3}\right) = \frac{1}{6}, N_p = 6$

(c)

$$x(n) = \left\{0, \frac{3\sqrt{3}}{2}, \frac{3\sqrt{3}}{2}, 0, -\frac{3\sqrt{3}}{2}, -\frac{3\sqrt{3}}{2}\right\}, \qquad N_p = 6$$

(d)   *Yes.* $x(1) = 3 = 3\sin\left(\frac{100\pi}{F_s}\right) \Rightarrow F_s = 200$ *samples/sec.*

Problem 1.4

An analogue signal contains frequency up to 10 kHz.

(a) What range of sampling frequencies will allow exact reconstruction of this signal from its sample?

(b) Suppose that we sample this signal with a sampling frequency $F_s = 8$ kHz. Examine what will happen to the frequency $F_1 = 5$ kHz.

(c) Repeat part (b) for a frequency $F_2 = 9$ kHz.

Solution 1.4

(a) $F_{max} = 10$ kHz $\Rightarrow$ Fs $\geq 2F_{max} = 20$ kHz

(b) For $F_s = 8$ kHz, $F_{fold} = F_s/2 = 4$ kHz $\Rightarrow 5$ kHz will alias to 3 kHz.

(c) $F = 9$ kHz will alias to 1 kHz.

Problem 1.5

An analogue electrocardiogram (ECG) signal contains useful frequencies up to 100 Hz.

(a) What is the Nyquist rate for this signal?

(b) Suppose that we sample this signal at a rate of 250 samples/s. What is the highest frequency that can be represented uniquely at this sampling rate?

Solution 1.5

(a) $F_{max} = 100$ Hz, Fs $\geq 2\, F_{max} = 200$ Hz
(b) $F_{fold} = \frac{F_s}{2} = 125$ Hz

Problem 1.6

An analogue signal $x_a(t) = \sin(480\pi t) + 3\sin(720\pi t)$ is sampled 600 times per second.

(a) Determine the Nyquist sampling rate for $x_a(t)$.
(b) Determine the folding frequency.
(c) What are the frequencies, in radians, in the resulting DT signal x(n)?
(d) If x(n) is passed through an ideal D/A converter, what is the reconstructed signal $y_a(t)$?

Solution 1.6

(a) $F_{max} = 360$ Hz, $F_N = 2F_{max} = 720$ Hz
(b) $F_{fold} = \frac{F_s}{2} = 300$ Hz
(c) $x(n) = x_a(nT) = x_a(n/F_s) = \sin(480\pi n/600) + 3\sin(720\pi n/600)$
$x(n) = \sin(4\pi n/5) - 3\sin(4\pi n/5) = -2\sin(4\pi n/5)$
Therefore, $\omega = 4\pi/5$.
(d) $y_a(t) = x(F_s t) = -2\sin(480\pi t)$

Problem 1.7

A digital communication link carries binary-coded word representing samples of an input signal $x_a(t) = 3\cos 600\pi t + 2\cos 1800\pi t$.

The link is operated at 10,000 bits/s and each input sample is quantized into 1024 different voltage levels.

(a) What is the sampling frequency and the folding frequency?
(b) What is the Nyquist rate for the signal $x_a(t)$?
(c) What are the frequencies in the resulting DT signal x(n)?
(d) What is the resolution Δ?

Solution 1.7

(a) Number of bits/sample $= \log_2 1024 = 10$

$$Fs = [10,000 \text{ bits/s}]/[10 \text{ bits/sample}]$$
$$= 1000 \text{ sample/s.}; F_{fold} = 500 \text{ Hz}$$

(b) $F_{max} = 1800\pi/2\pi = 900$ Hz; $F_N = 2F_{max} = 1800$ Hz

(c) $f_1 = \frac{600\pi}{2\pi} \left(\frac{1}{F_s}\right) = 0.3; f_2 = \frac{1800\pi}{2\pi} \left(\frac{1}{F_s}\right) = 0.9$

But $f_2 = 0.9 > 0.5 \Rightarrow f_2 = 0.1$.

Hence, $x(n) = 3 \cos[(2\pi)(0.3)n] + 2 \cos[(2\pi)(0.1)n]$.

(d) $\Delta = \frac{x_{max} - x_{min}}{L-1} = \frac{5-(-5)}{1023} = \frac{10}{1023}$, where L is quantization level.

Problem 1.8

Consider the simple signal processing shown in the figure given hereunder. The sampling periods of the A/D and D/A converters are T = 5 ms and T = 1 ms, respectively. Determine the output $y_a(t)$ of the system, if the input is

$$x_a(t) = 3 \cos 100\pi t + 2 \sin 250\pi t \text{ (t in seconds)}$$

The post-filter removes any frequency component above $F_s/2$.

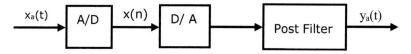

Solution 1.8

$$x(n) = x_a(nT) = 3 \cos\left(\frac{100\pi n}{200}\right) + 2 \sin\left(\frac{250\pi n}{200}\right)$$

$$= \cos(\pi n/2) - 2 \sin(3\pi n/4)$$

$$T^{-1} = \frac{1}{1000} \quad y_a(t) = x\left(\frac{1}{T}\right) = 3 \cos\left(\frac{\pi}{2}1000t\right) - 2 \sin\left(\frac{3\pi}{4}1000t\right)$$

$$y_a(t) = 3 \cos(500\pi t) - 2 \sin(750\pi t)$$

Problem 1.9

The DT signal $x(n) = 6.35 \cos(\pi/10)n$ is quantized with a resolution

(a) $\Delta = 0.1$

(b) $\Delta = 0.02$

How many bits are required in the A/D converter in each case?

Solution 1.9

(a) Range $= x_{max} - x_{min} = 12.7$

$L = 1 + \frac{range}{\Delta} = 127 + 1 = 128 \Rightarrow \log_2 128 = 7$ bits

(b) $L = 1 + \frac{12.7}{0.02} = 636 \Rightarrow \log_2 636 \Rightarrow 10$ bit A/D

Problem 1.10

How many bits are required for the storage of a seismic signal if the sampling rate is $F_s = 20$ samples/s and we use an 8-bit A/D converter? What is the maximum frequency that can be present in the resulting digital seismic signal?

Solution 1.10

$R = (20 \text{ sample/s}) \times (8 \text{ bits/sample}) = 160 \text{ bits/s}.$
$\quad F_{\text{fold}} = \frac{F_s}{2} = 10 \text{ Hz}. F_s = 20$ samples per second (given)

Problem 1.11

A DT signal x(n) is defined as

$$x(n) = \begin{cases} 1 + \frac{n}{3} & -3 \le n \le -1 \\ 1, & 0 \le n \le 3 \\ 0, & elsewhere \end{cases}$$

(a) Determine its values and sketch the signal x(n).
(b) Sketch the signals that result if we:
   (1) First fold x(n) and then delay the resulting signal by four samples.
   (2) First delay x(n) by four samples and then fold the resulting signal.
(c) Sketch the signal x(−n + 4).
(d) Compare the results in parts (b) and (c) and derive a rule for obtaining the signal x(−n + k) from x(n).
(e) Can you express the signal x(n) in terms of signals δ(n) and u(n)?

Solution 1.11

(a) $x(n) = \{\ldots 0, \frac{1}{3}, \frac{2}{3}, \underset{\uparrow}{1}, 1, 1, 1, 0 \ldots\}$

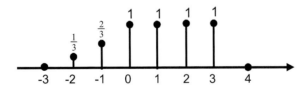

(b) After folding x(n), we have

$$x(-n) = \{\ldots 0, 1, 1, 1, \underset{\uparrow}{1}, \frac{2}{3}, \frac{1}{3}, 0 \ldots\}$$

After delaying the folded signal by four samples, we have $x(-n+4) = \{\ldots 0, 1, 1, 1, 1, \frac{2}{3}, \frac{1}{3}, 0 \ldots\}$ on the other hand, if we delay x(n) by 4 samples we have x(n − 4) = {...0, 0, 0, $\frac{1}{3}$, $\frac{2}{3}$, 1, 1, 1, 1, ...} now, if we

fold x(n−4) we have x(−n-4) = {..., 0, 1, 1, 1, 1, $\frac{2}{3}$, $\frac{1}{3}$, 0, 0, ...}.
(c) x(−n + 4) = {...0, 1, 1, 1, 1, $\frac{2}{3}$, $\frac{1}{3}$, 0...}

(d) To obtain x(−n + k), first we fold x(n), this yields x(−n); then, we shift x(−n) by k samples to the right if k > 0, or k samples to the left if k < 0.

Problem 1.12
A DT signal x(n) is shown in the figure hereunder. Sketch and label carefully each of the following signals.

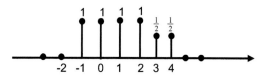

(a) x(n−2)   (b) x(4−n)   (c) x(n+2)   (d) x(n)u(2−n)   (e) x(n-1)δ(n−3)
(f) x(n²)   (g) even part of x(n)   (h) odd part of x(n)

Solution 1.12
$x(n) = \{\ldots 0, 1, \underset{\uparrow}{1}, 1, 1, \frac{1}{2}, \frac{1}{2}, 0\ldots\}$

(a) $x(n - 2) = \{\ldots 0, \underset{\uparrow}{\frac{1}{2}}, 1, 1, 1, 1, \frac{1}{2}, \frac{1}{2}, 0\ldots\}$

(b) $x(4 - n) = \{\ldots 0, \underset{\uparrow}{\frac{1}{2}}, \frac{1}{2}, 1, 1, 1, 1, 0, \ldots\}$

(c) $x(n + 2) = \{\ldots 0, 1, 1, 1, \underset{\uparrow}{1}, \frac{1}{2}, \frac{1}{2}, 0\ldots\}$

(d) $x(n)u(2 - n) = \{\ldots 0, 1, \underset{\uparrow}{1}, 1, 1, 0, 0, \ldots\}$

(e) $x(n - 1)\delta(n - 3) = \{\ldots 0, \underset{\uparrow}{0}, 0, 0, 1, 0, \ldots\}$

(f) $x(n^2) = \{\ldots 0, x(4), x(1), x(0), x(1), x(4), 0, \ldots\}$
$= \{\ldots 0, \frac{1}{2}, 1, \underset{\uparrow}{1}, 1, \frac{1}{2}, 0\ldots\}$

(g) $x_e(n) = \dfrac{x(n) + x(-n)}{2}, x(-n) = \{\ldots 0, \frac{1}{2}, \frac{1}{2}, 1, 1, \underset{\uparrow}{1}, 1, 0, 0, 0, \ldots\}$
$= \{\ldots 0, \frac{1}{4}, \frac{1}{4}, \frac{1}{2}, 1, \underset{\uparrow}{1}, 1, \frac{1}{2}, \frac{1}{4}, \frac{1}{4}, 0, \ldots\}$

(h) $x_0(n) = \dfrac{x(n) - x(-n)}{2},$
$= \{\ldots 0, -\frac{1}{4}, -\frac{1}{4}, -\frac{1}{2}, 0, \underset{\uparrow}{0}, 0, \frac{1}{2}, \frac{1}{4}, \frac{1}{4}, 0, \ldots\}$

Problem 1.12

Show that any signal can be decomposed into an even and an odd component. Is the decomposition unique? Illustrate your arguments using the signal

$$x(n) = \{2, 3, \underset{\uparrow}{4}, 5, 6\}$$

Solution 1.12

Let $x_e(n) = \frac{1}{2}[x(n) + x(-n)]$, $x_0(n) = \frac{1}{2}[x(n) - x(-n)]$. Since $x_e(-n) = x_e(n)$ and $x_0(-n) = -x_0(n)$, it follows that $x(n) = x_e(n) + x_0(n)$.
  The decomposition is unique.

For $x(n) = \{2, 3, \underset{\uparrow}{4}, 5, 6\}$, we have

$$x_e(n) = \{4, 4, \underset{\uparrow}{4}, 4, 4\}, \text{ and } x_0(n) = \{-2, -1, \underset{\uparrow}{0}, 1, 2\}.$$

Problem 1.13

Determine the energy of the following sequence:

$$x(n) = \left(\frac{1}{2}\right)^n \quad for \quad n \geq 0$$
$$x(n) = 0 \quad for \quad n < 0$$

Solution 1.13

We know that for a DT signal, the energy is expressed as

$$E = \sum_{n=-\infty}^{\infty} |x(n)|^2$$

$E = \sum_{n=0}^{\infty} (\frac{1}{2})^2$ Therefore, summing the infinite series $E = 2$

Problem 1.14

Show that the energy (power) of a real-valued energy (power) signal is equal to the sum of the energies (powers) of its even and odd components.

Solution 1.14

$$First, \, we \, prove \, that \quad \sum_{n=-\infty}^{\infty} x_e(n) \, x_0(n) = 0$$

$$\sum_{n=-\infty}^{\infty} x_e(n)x_0(n) = \sum_{m=-\infty}^{\infty} x_e(-m)x_0(-m) = \sum_{m=-\infty}^{\infty} x_e(m)x_0(m)$$

$$= -\sum_{n=-\infty}^{\infty} x_e(n)x_0(n) \implies \sum_{n=-\infty}^{\infty} x_e(n)x_0(n) = 0$$

Then

$$\sum_{n=-\infty}^{\infty} x^2(n) = \sum_{n=-\infty}^{\infty} [x_e(n) + x_0(n)]^2$$

$$= \sum_{n=-\infty}^{\infty} x_e^2(n) + \sum_{n=-\infty}^{\infty} x_0^2(n) + 2\sum_{n=-\infty}^{\infty} x_e(n)x_0(n)$$

$$= E_e + E_0$$

Problem 1.15

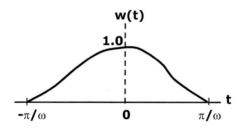

w(t)

1.0

$-\pi/\omega$ **0** π/ω **t**

Figure P 1.15

The raised-cosine pulse x(t) is shown in Figure P 1.15 is defined as

$$x(t) = \begin{cases} \frac{1}{2}[\cos(\omega t) + 1], & -\pi/\omega \le t \le \pi/\omega \\ 0, & otherwise \end{cases}$$

Determine the total energy of x(t).

Solution 1.15

$$E = \int_{-\infty}^{\infty} x^2(t)dt = \int_{-\pi/\omega}^{\pi/\omega} \frac{1}{2}[\cos \omega t + 1]^2 dt$$

$$= \frac{1}{2}\int_{-\pi/\omega}^{\pi/\omega} [1 + \cos^2 \omega t + 2\cos \omega t]dt$$

$$= \frac{1}{2} \int_{-\pi/\omega}^{\pi/\omega} \left[1 + \frac{1 + \cos 2\omega t}{2} + 2\cos \omega t \right] dt$$

$$= \frac{1}{2} \int_{-\pi/\omega}^{\pi/\omega} \left[t + \frac{\sin 2\omega t}{4\omega} + \frac{t}{2} + 2\sin \omega t \right] dt = \frac{1}{2} \frac{3\pi}{\omega} = \frac{3\pi}{2\omega} Ans.$$

Problem 1.16

The trapezoidal pulse x(t) shown in Figure P 1.16 is defined as

$$x(t) = \begin{cases} 5 - t, & 4 \le t \le 5 \\ 1, & -4 \le t \le 4 \\ t + 5, & -4 \le t \le -5 \\ 0, & \text{otherwise} \end{cases}$$

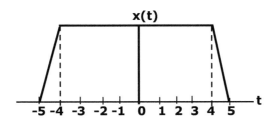

x(t)

Figure P 1.16

Determine the total energy of x(t).

Solution 1.16

$$E = \int_{-\infty}^{\infty} x^2(t)dt = + \int_{-5}^{-4} (t+5)^2 dt + \int_{-4}^{4} dt + \int_{4}^{5} (5-t)^2 dt$$

$$= \frac{(t+5)^3}{3} \Big|_{-5}^{-4} + t \Big|_{-4}^{4} - \frac{(5-t)^3}{3} \Big|_{4}^{5} = \frac{1}{3} + 8 + \frac{1}{3} = \frac{26}{3} \ units$$

Problem 1.17

The trapezoidal pulse x(t) of Figure P1.16 is applied to a differentiator, defined by

$$y(t) = \frac{d}{dt} x(t)$$

(a) Determine the resulting output y(t) of the differentiator.
(b) Determine the total energy of y(t).

Solution 1.17

(a) $y(t) = \begin{cases} -1, & 4 \le t \le 5 \\ 0, & -4 \le t \le 4 \\ 1, & -5 \le t \le -4 \\ 0, & otherwise \end{cases}$

(b) $E = \int_{-\infty}^{\infty} x^2(t)dt = +\int_{-5}^{-4} 1 \cdot dt + \int_{4}^{5} -1 \cdot dt = 1 - 1 = 0$

Problem 1.18

A rectangular pulse x(t) is defined by

$$x(t) = \begin{cases} A, & 0 \le t \le T \\ 0, & otherwise \end{cases}$$

The pulse x(t) is applied to an integrator defined

$$y(t) = \int_{0}^{t} x(\tau)d\tau$$

Find the total energy of the output y(t).

Solution 1.18

$$x(t) = \begin{cases} A, & 0 \le t \le T \\ 0, & otherwise \end{cases}$$

$$y(t) = \int_{0}^{t} x(\tau)d\tau = \int_{0}^{t} A\, d\tau = At, \ 0 < t \le T$$

and, the total energy, $E = \int_{-\infty}^{\infty} y^2(t)dt = \int_{0}^{T} A^2 t^2 dt = \frac{A^2 t^3}{3} \Big|_{0}^{T} = \frac{A^2 t^3}{3}$

Problem 1.19

Find the even and odd components of each of the following signals:

(a) $x(t) = \cos(t) + \sin(t)\cos(t)$
(b) $x(t) = 1 + t + 3t^2 + 5t^2 + 9t^4$
(c) $x(t) = 1 + t\cos(t) + t^2\sin(t) + t^3\sin(t)\cos(t)$
(d) $x(t) = (1 + t^3)\cos^3(10t)$

Solution 1.19

For even part

$$x_e(-t) = x_e(t) \quad \text{for all} \quad t$$

and for odd part
$$x_0(t) = x_0(t) \quad \text{for all} \quad t$$

(a) $x_e(t) = \cos t$

$x_0(t) = \sin(t)\cos(t)$

(b) $x_e(t) = 1 + 3\,t^2 + 9t^4$

$x_0(t) = t + 5\,t^3$

(c) $x_e(t) = 1 + t^3\,\sin t \cos t$

$x_0(t) = t \cos t + t^2 \sin\ t$

(d) $x_e(t) = \cos^3(10t)$

$x_0(t) = t^3 \cos^3(10t)$

Problem 1.20

Consider the analogue signal $x_a(t) = 3\cos 2000\pi t + 5\sin 6000\pi t + 10\cos 12,000\pi t$.

(a) Assume now that the sampling rate $F_s = 12000$ samples/s. What is the DT signal obtained after sampling?

(b) What is the analogue signal $y_a(t)$ we can reconstruct from the samples if we use ideal interpolation?

Solution 1.20

(a) The frequencies existing in the analogue signal are

$$F_1 = 1\ \text{kHz}, F_2 = 3\ \text{kHz} \quad \text{and} \quad F_3 = 6\ \text{kHz}.$$

Thus $F_{max} = 6$ kHz, and according to the sampling theorem

$$F_s \geq 2\,F_{max} = 12\ \text{kHz (sampled correctly)}$$

The Nyquist rate is $F_N = 12$ kHz.
Since we have chosen $F_s = 12$ kHz, the folding frequency is

$$\frac{F_s}{2} = 6\ \text{kHz}$$

and this is the maximum frequency that can be represented uniquely by the sampled signal. The three frequencies F_1, F_2 and F_3 are below or equal to the folding frequency and they will not be changed by the aliasing effect.

From (1.10) it follows that the three digital frequencies $f_1 = \frac{1}{12}$, $f_2 = \frac{3}{12}$ and $f_3 = \frac{6}{12}$.

Again using (1.10) we obtain

$$x(n) = x_a(nT) = x_a \left(\frac{n}{F_s} \right)$$

$$x(n) = 3\cos 2\pi(\tfrac{1}{12})n + 5\sin 2\pi(\tfrac{3}{12})n + 10\cos 2\pi(\tfrac{6}{12})n$$

which are in agreement with the result obtained by (1.17) earlier. The frequencies F_1, F_2 and $F_3 \leq F_s/2$ and thus it is not affected by aliasing.

(b) Since all the frequency components at 1 kHz, 3 kHz and 6 kHz are present in the sampled signal, the analogue signal we can recover is

$$x_a(t) = x(F_s t) = 3\cos 2000\pi t + 5\sin 6000\pi t + 10\cos 12,000\pi t$$

which is obviously different from the original signal $x_a(t)$. This distortion of the original analogue signal was caused by the aliasing effect, due to the low sampling rate used.

Problem 1.21
The DT sequence

$$x(n) = \cos\left(\frac{\pi}{4}\right)n \quad -\infty < n < \infty$$

was obtained by sampling analogue signal

$$x_a(t) = \cos(\Omega t) \quad -\infty < t < \infty$$

at a sampling rate of 1000 samples/s. What are the two possible values of Ω that could have resulted in the sequence $x(n)$?

Solution 1.21
The DT sequence

$$x(n) = \cos\left(\frac{\pi}{4}\right)n$$

results by sampling the CT signal $x_a(t) = \cos(\Omega t)$. Using (1.10)

$$\omega = \frac{\Omega}{F_s} \quad \text{or} \quad \Omega = \omega F_s \quad \text{or} \quad \Omega = \left(\frac{\pi}{4}\right)1000 = 250\pi$$

or possibly $\Omega = (2\pi + \frac{\pi}{4})\,1000 = 2250\pi$

Problem 1.22

Consider the analogue signal $x_a(t) = 3\cos 2000\pi t + 5\sin 6000\pi t + 10\cos 12,000\pi t$

(a) What is the Nyquist rate for this signal?
(b) Assume the sampling rate $F_s = 5000$ samples/s. What is the DT signal obtained after sampling?
(c) What is the analogue signal $y_a(t)$ we can reconstruct from the samples if we use ideal interpolation?

Solution 1.22

(a) The frequencies existing in the analogue signal are

$$F_1 = 1 \text{ kHz}, F_2 = 3 \text{ kHz}, F_3 = 6 \text{ kHz}$$

Thus $F_{max} = 6$ kHz, and according to the sampling theorem

$$F_s \geq 2F_{max} = 12 \text{ kHz}$$

The Nyquist rate is

$$F_N = 12 \text{ kHz}$$

(b) Since we have chosen $F_s = 5$ kHz, the folding frequency is

$$\frac{F_s}{2} = 2.5 \text{ kHz}$$

and this is the maximum frequency that can be represented uniquely by the sampled signal.

From $F_k = F_0 + kF_s$, we have $F_0 = F_k - kF_s \cdot F_1 = 1$ kHz, the other two frequencies F_2 and F_3 are above the folding frequency and they will be changed by the aliasing effect.

Indeed, $F'_2 = F_2 - F_s = -2$ kHz and $F'_3 = F_3 - F_s = 1$ kHz.
From (1.10) it follows that the three digital frequencies $f_1 = \frac{1}{5}, f_2 = -\frac{2}{5}$ and $f_3 = \frac{1}{5}$.
Again using (1.10) we obtain

$$x(n) = x_a(nT) = x_a\left(\frac{n}{F_s}\right)$$

$$= 3\cos 2\pi\left(\tfrac{1}{5}\right)n + 5\sin 2\pi\left(\tfrac{3}{5}\right)n + 10\cos 2\pi\left(\tfrac{6}{5}\right)n$$

$$= 3\cos 2\pi\left(\tfrac{1}{5}\right)n + 5\sin 2\pi\left(1 - \tfrac{2}{5}\right)n + 10\cos 2\pi\left(1 + \tfrac{1}{5}\right)n$$

$$= 3\cos 2\pi\left(\tfrac{1}{5}\right)n + 5\sin 2\pi\left(-\tfrac{2}{5}\right)n + 10\cos 2\pi\left(\tfrac{1}{5}\right)n$$

Finally, we obtain

$$x(n) = 13\cos 2\pi(\tfrac{1}{5})n - 5\sin 2\pi(\tfrac{2}{5})n$$

which are in agreement with the result obtained by (1.17).

Thus F_0 can be obtained by subtracting from F_k an integer multiple of F_s such that $-F_s/2 \le F_0 \le F_s/2$.

The frequency F_1 is less than $F_s/2$ and thus it is not affected by aliasing.

(c) Since only the frequency components at 1 kHz and 2 kHz are present in the sampled signal, the analogue signal we can recover is

$$y_a(t) = y(F_s t) = 13\cos 2000\pi t - 5\sin 4000\pi t$$

which is obviously different from the original signal $x_a(t)$. This distortion of the original analogue signal was caused by the aliasing effect, due to the low sampling rate used.

Problem 1.23

Let x[n] and y[n] be as given in the following figures, respectively.

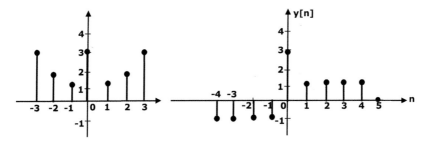

Carefully sketch the following signals.

(a) $y[2 - 2n]$
(b) $x[n - 2] + y[n + 2]$
(c) $x[2n] + y[n - 4]$
(d) $x[n + 2]y[n - 2]$

Solution 1.23

(a) $y[2 - 2n]$

$$y[2 - 2n] = \begin{cases} 1, & n = 0, -1 \\ -1, & n = 2, 3 \\ 3, & n = 1 \end{cases}$$

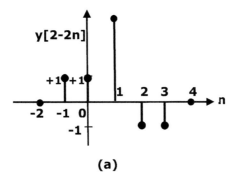

(a)

(b) $x[n-2] + y[n+2]$

$$x[n-2] = \begin{cases} 1, & n = 1,3 \\ 2, & n = 0,4 \\ 3, & n = -1,2,5 \\ 0, & n = \text{rest} \end{cases} \qquad y[n+2] = \begin{cases} 1, & n = -1,0,1,2 \\ -1, & n = -3,-4,-5,-6 \\ 3, & n = -2 \end{cases}$$

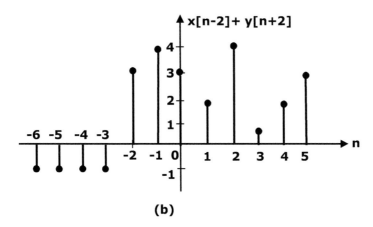

(b)

(c) $x[2n] + y[n-4]$

$$x[2n] = \begin{cases} 2, & n = \pm 1 \\ 0, & n = 3 \end{cases}$$

$$y[n-4] = \begin{cases} 1, & n = 5,6,7,8 \\ -1, & n = 0,1,2,3 \\ 3, & n = 4 \end{cases}$$

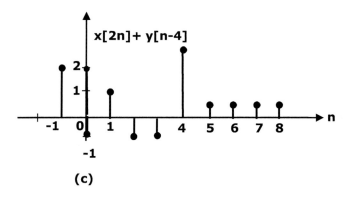

(c)

(d) $x[n+2]y[n-2]$

$$x[n+2] = \begin{cases} 1, & n = -3, -1 \\ 2, & n = -4, 0 \\ 3, & n = -5, -2, 1 \end{cases} \qquad y[n-2] = \begin{cases} 1, & n = 3, 4, 5, 6 \\ -1, & n = 1, 0, -1, -2 \\ 3, & n = 2 \end{cases}$$

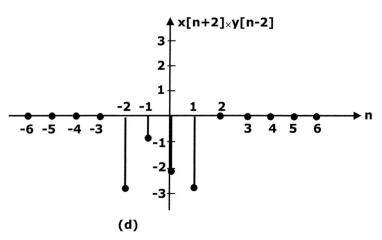

(d)

Problem 1.24

Consider the length -7 sequences defined for $-3 \le n \le 3$:

$x(n) = \{3\ -2\ 0\ 1\ 4\ 5\ 2\}$; $y(n) = \{0\ 7\ 1\ -3\ 4\ 9\ -2\}$ and $w(n) = \{-5\ 4\ 3\ 6\ -5\ 0\ 1\}$.

Generate the following sequences:

(a) $u(n) = x(n) + y(n)$
(b) $v(n) = x(n) \cdot w(n)$
(c) $s(n) = y(n) - w(n)$
(d) $r(n) = 4.5\ y(n)$

Solution 1.24

(a) $u(n) = x(n) + y(n) = \{\,3\ 5\ 1\ -2\ 8\ 14\ 0\,\}$
(b) $v(n) = x(n) \cdot w(n) = \{\,-15\ -8\ 0\ 6\ -20\ 0\ 2\,\}$
(c) $s(n) = y(n) - w(n) = \{5\ 3\ -2\ -9\ 9\ 9\ -3\}$
(d) $r(n) = 4.5\,y(n) = \{\,0\ 31.5\ 4.5\ -13.5\ 19\ 40.5\ -9\}$

Problem 1.25

Categorize the following signals as an energy or power signal, and find the energy or power of the signal.

(a) $x[n] = \begin{cases} n, & 0 \le n \le 5 \\ 10 - n, & 6 \le n \le 10 \\ 0, & otherwise \end{cases}$

(b) $\quad E = \displaystyle\sum_{n=-4}^{4} \cos(\pi n)$

Solution 1.25

(a) $E = \displaystyle\sum_{n=-\infty}^{\infty} x^2[n] = \sum_{n=0}^{5} n^2 + \sum_{n=6}^{10} (10 - n)^2 = 85$

Since $0 < E < \infty$, x[n] is an energy signal.

(b) $E = \displaystyle\sum_{n=-4}^{4} \cos^2(\pi n) = \sum_{n=-4}^{4} \left(\tfrac{1+\cos 2\pi n}{2}\right) = 4$

Since $0 < E < \infty$, x[n] is an energy signal:

$$P = \frac{1}{N} \sum_{n=0}^{N-1} x^2[n]$$

where N is the fundamental period, here $N = 2$

$$p = \frac{1}{2} \sum_{n=0}^{1} \sin^2(\pi n) = \frac{\sin^2 \pi}{2} = 0$$

x(n) cannot be a power signal

Problem 1.26

The angular frequency ω of the sinusoidal signal $x[n] = A\cos(\omega n + \phi)$ satisfies the condition for x[n] to be periodic. Determine the average power of x[n].

Solution 1.26

Average power,

$$P = \frac{1}{N} \sum_{n-0}^{N-1} x^2[n]$$

Here,

$$N = \frac{2\pi}{\omega} (fundamental\ peroid)$$

Then,

$$P = \frac{\omega}{2\pi} \sum_{n=0}^{\left(\frac{2\pi}{\omega} - 1\right)} \frac{A^2}{2} \{\cos(\omega n + \phi) + 1\}$$

$$= \frac{A^2\omega}{4\pi} \{1 + \cos 2\phi + 1 + \cos[2(2\pi - \omega) + 2\phi]\}$$

$$= \frac{A^2\omega}{2\pi} \{1 + \cos 2\phi + 1 + \cos(2\phi - 2\omega)\}$$

Problem 1.27

Let x(t) and y(t) signals be given in the form of the following mathematical expression:

$$x(t) = \begin{cases} -1 - t, & -1 \leq t \leq 0 \\ t, & 0 \leq t \leq 1 \\ 1, & 1 \leq t \leq 2 \\ 3 - t, & 2 \leq t \leq 3 \\ 0, & otherwise \end{cases} \qquad y(t) = \begin{cases} 1, & -2 \leq t \leq -1 \\ -1, & -1 \leq t \leq 0 \\ t - 1, & 0 \leq t \leq 1 \\ 1, & 1 \leq t \leq 2 \end{cases}$$

Carefully sketch the signals h(t) = x(t) y(t).

Solution 1.27

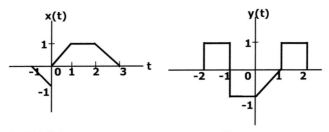

It is advisable for the students that this example or any other example in which signal addition, multiplication or subtraction is required can be done in the following way, generating a tabular form, in which less chances of error can occur. If there is a final expression after multiplication the value h(t) has an expression of t^2, different values have to be calculated to draw the final result, which will generate a curve.

Time axis	x(t)	y(t)	h(t) = x(t).y(t)
$-1 \leq t \leq 0$	$x(t) = -1 - t$ $x(-1) = 0$ $x(0) = -1$	$y(t) = -1$ $y(-1) = -1$ $y(0) = -1$	$h(t) = 1 + t$ $h(-1) = (0) \cdot (-1) = 0$ $h(-1) = (-1) \cdot (-1) = 1$
$0 \leq t \leq 1$	$x(t) = t$ $x(0) = 0$ $x(1) = 1$ $x(0.5) = 0.5$	$y(t) = -1 + t$ $y(0) = -1$ $y(1) = 0$ $y(0.5) = -0.5$	$h(t) = t(t-1) = -t + t^2$ $h(0) = (0) \cdot (-1) = 0$ $h(1) = (1) \cdot (0) = 0$ $h(0.5) = -0.50 + 0.25$ $= -0.25$
	$x(0.25) = 0.25$ $x(0.75) = -0.75$	$y(0.25) = -0.75$ $y(0.75) = -0.75$	$h(0.25) = -0.1875$ $h(0.75) = -0.1875$
$1 \leq t \leq 2$	$x(t) = 1$ $x(1) = 1$ $x(2) = 1$	$y(t) = 1$ $y(1) = 1$ $y(2) = 1$	$h(t) = 1$ $h(1) = (1) \cdot (1) = 1$ $h(2) = (1) \cdot (1) = 1$
$2 \leq t \leq 3$	$x(t) = 3 - t$ $x(2) = 1$ $x(3) = 0$	$y(t) = 0$ $y(2) = 0$ $y(3) = 0$	$h(t) = 0$ $h(2) = (1) \cdot (0) = 0$ $h(3) = (0) \cdot (0) = 0$

The sketch of h(t) is shown hereunder:

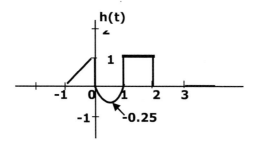

Practice Problem 1.28

The figure provided shows part of a control system where the error signal e(t) is formed by the difference between the input signal x(t) and the feedback signal y(t):

$$e(t) = x(t) - y(t)$$

If

$$x(t) = 3\sin(\omega t + 20°)$$

$$y(t) = 5\cos(\omega t - 20°)$$

Determine e(t) and express it in the form A sin($\omega t + \phi$).

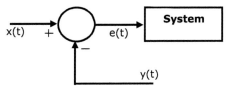

Practice Problem 1.29

(a) A CT signal x(t) is shown in Figure P1.29(a).
Sketch and label carefully each of the following signals:

(i) x(2t + 2)
(ii) x(t − t/3)
(iii) [x(t) + x(2 − t)]u(1 − t)
(iv) x(t)[δ(t + 3/2) − δ(t − 3/2)]

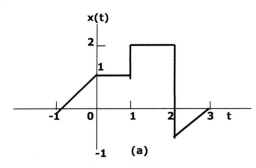

Figure P1.29(a)

(b) For the signal h(t) depicted in Figure P 1.4(b), sketch and label carefully
each of the following signals.

(i) $h(t/2 - 2)$
(ii) $h(1 - 2t)$
(iii) $4h(t/4)$
(iv) $h(t)u(t) + h(-t)]u(t)$
(v) $h(t/2)\delta(t + 1)$
(vi) $h(t)[u(t + 1) - u(t - 1)]$

(c) Consider again the signals x(t) and h(t) shown in Figures P1.29(a) and
(b), respectively.

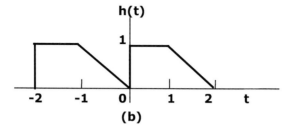

(b)

Figure P1.29(b)

(i) x(t)h(t + 1)
(ii) x(t)h(−t)
(iii) x(t − 1)h(1 − t)
(iv) x(1 − t)h(t − 1)
(v) x(2 − t/2)h(t + 4)

Practice Problem 1.30

(a) A DT signal x[n] is shown in Figure P1.30. Sketch and label carefully each of the following signals.

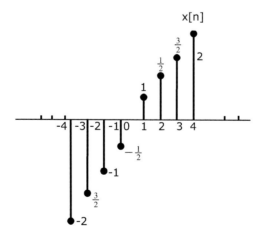

Figure P1.30

(i) x[n − 2]
(ii) x[4 − n]
(iii) x[2n]
(iv) x(1−n/3)
(v) x[x]u[2 − n]
(vi) $x_{\frac{1}{2}}[n − 1]\delta_{\frac{1}{2}}[n − 3]$
(vii) $x[n] + (−1)^n x[n]$
(viii) $x[n^2]$

Practice Problem 1.31

The Figure P1.31 provided shows a CT signal x(t). Make labelled sketches of the following time signals:

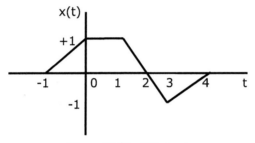

Figure P1.31   CT signal.

(a) $2x(t)$
(b) $0.5x(−t)$
(c) $x(t − 2)$
(d) $x(2t)$
(e) $x(2t + 1)$
(f) $x(1 − t)$

Practice Problem 1.32

The Figure P1.32 provided shows a DT signal x(n). Make labelled sketches of each of the following time signals:

(a) $2x(n)$
(b) $3x(−n)$
(c) $x(n − 2)$
(d) $x(2 − n)$

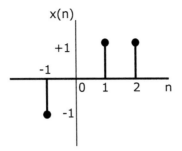

Figure P1.32 DT signal.

Practice Problem 1.33

The Figure P1.33 provided shows two continuous signals $x_1(t)$. Make labelled sketches of each of the following signals:

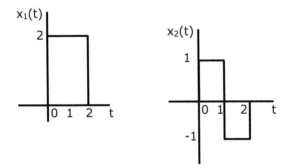

Figure P1.33 2 continuous signals.

(a) $x_1(t) + x_2(t)$
(b) $x_1(t) - 2x_2(t)$
(c) $0.5x_1(2t) - x_2(t)$
(d) $x_1(t - 2) + x_2(4 - t)$

Practice Problem 1.34

The Figure P1.34 provided shows two discrete signals $x_1(n)$. Make labelled sketches of each of the following signals:

(a) $x_1(n) + x_2(n)$
(b) $x_1(-n) + x_2(n)$
(c) $x_1(n - 1) + x_2(n)$
(d) $x_1(2 - n) - x_2(n)$

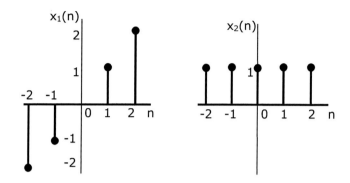

Figure P1.34   2 discrete signals.

Practice Problem 1.35

Determine and make a labelled sketch of the odd and even components of the signals shown in the Figure P1.35 provided. Verify that addition of the components produces the original signal.

  (i) $h[n]x[-n]$
 (ii) $x[n+2]\,h\,[1-2n]$
(iii) $x[1-n]\,h\,[n+4]$
(iv) $x[n-1]\,h\,[n-3]$

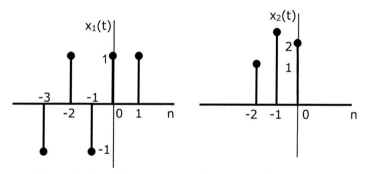

Figure P1.35   2 discrete signals with odd and even components.

Practice Problem 1.36

Determine the amplitude, frequency (Hz) and phase with respect to $\sin \omega t$ of the following signals:

(a) $20 \sin(30t + \pi/4)$
(b) $50 \cos(100t - \pi/4)$
(c) $100 \cos 20t + 20 \sin 20t$
(d) $\text{Re}\{e^{j(3t+/4)}\}$
(e) $\text{Im},\{e^{j(3t+/4)}\}$
(f) $Ae^{j2t} + A^*e^{-j2t}$, where $A = 1 + 2j$

2

Differential Equations

This chapter provides comprehensive details regarding the solution of differential equation with constant coefficients, transient solution (zero input response) and steady-state solution (zero state response) with different driving functions (forcing functions) such as step, ramp, acceleration, exponential and sinusoidal. Practice problems are also included.

2.1 Introduction

The dynamic behaviour of continuous time physical systems may be phrased in terms of linear differential equations containing constant coefficients. Consider for example an extremely simple problem involving an elastic, inertia-less shaft connected to an inertia-less viscous damping paddle at the far end. The near end is being driven so that it moves with sinusoidal, angular oscillations of constant angular frequency ω and constant peak angular amplitude $\hat{\theta}_i$, then the motion of the near end θ_i is described by

$$\theta_i = \hat{\theta}_i \sin \omega t \tag{2.1}$$

The arrangement is represented diagrammatically in Figure 2.1.

The object is to set up an equation relating to the motion of the paddle (θ_0) with respect to the input motion (θ_i).

Let θ = angle of twist in the shaft, $\theta = \theta_i - \theta_0$.

Torque transmitted through the shaft to paddle is assumed to be $K\theta$, which must be balanced by the retarding torque due to the damping paddle $Bd\theta_0/\mathrm{dt}$ and is given by

$$B\frac{d\theta_0}{dt} = K\theta$$

$$B\frac{d\theta_0}{dt} = K(\theta_i - \theta_0)$$

69

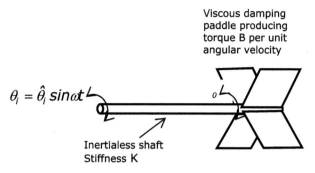

Figure 2.1   Simple mechanical system.

hence

$$B\frac{d\theta_0}{dt} + K\theta_0 = K\theta_i$$

But

$$\theta_i = \hat{\theta}_i \sin \omega t$$

$$\therefore \qquad B\frac{d\theta_0}{dt} + K\theta_0 = K\hat{\theta}_i \sin \omega t \qquad (2.2)$$

The differential Equation (2.2) describes the output motion of the system. The left-hand side of (2.2) is a description of the physical response of the system and the right-hand side represents the driving function, which is responsible for motion.

The term driving function (also known as forcing function) is used to take the mean of the right-hand side of (2.2).

$$\theta_0 = \theta_{0t} + \theta_{0ss} \qquad (2.3)$$

The differential equation may be solved by using certain rules and its solution (which is called the response of the system) consists of two distinct parts:

(i) **The transient response** θ_{0t} (which mathematicians call the complementary function); this part of the response occurs near to $t = 0$ and subsequently decays.

(ii) **The steady-state response** θ_{0ss} (which mathematicians call the particular integral); this part of the response is continuously present from $t = 0$ to $t = \infty$, but at $t = \infty$, it is in fact the complete solution. The driving function is entirely responsible for the existence and nature of θ_{0ss}.

The complete response θ_0 is the sum of the transient and steady-state solution, which is given by (2.3), where θ_{0t} can be deduced as

$$\theta_{0t} = Ae^{-Kt/F} \qquad (2.4)$$

where A is an arbitrary constant,

$$\theta_{oss} = \frac{K\hat{\theta}_i}{\sqrt{\{(B\omega)^2 + K^2\}}}\sin(\omega t - \tan^{-1} B\omega/K) \qquad (2.5)$$

The complete general solution is thus

$$\theta_o = Ae^{-Kt/F} + \frac{K\hat{\theta}_i}{\sqrt{\{(B\omega)^2 + K^2\}}}\sin(\omega t - \tan^{-1} B\omega/K) \qquad (2.6)$$

The arbitrary constant may only be found if the value of θ_0 at the constant $t = 0$ is specified. This value is termed an 'initial condition'. In the present problem let us specify that at $t = 0$, $\theta_0 = 0$, then (2.6) becomes

$$\theta_o = \frac{B\omega K\hat{\theta}_i}{\sqrt{\{(B\omega)^2 + K^2\}}}e^{-Kt/F} + \frac{K\hat{\theta}_i}{\sqrt{\{(B\omega)^2 + K^2\}}}\sin(\omega t - \tan^{-1} B\omega/K)$$

$$(2.7)$$

The transient, steady-state and complete particular responses of the system are represented graphically in Figure 2.2. It is noted in Figure 2.2(a) that how the transient response decays leaving only the steady-state response.

The general differential equation to be discussed in this chapter is denoted by

$$a_n\frac{d^n\theta}{dt^n} + a_{n-1}\frac{d^{n-1}\theta}{dt^{n-1}} + \cdots + a_1\frac{d\theta}{dt} + a_0\theta = f(t) \qquad (2.8)$$

Figure 2.2(a) Transient solution $\theta_{0t} = \left\{\frac{B\omega K\hat{\theta}_i}{(B\omega)^2 + K^2}\right\}e^{-Kt/B}$

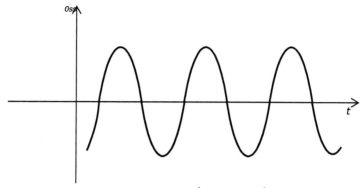

Figure 2.2(b)   Steady-state solution $\theta_{oss} = \left\{\dfrac{K\theta_i}{\sqrt{(B\omega)^2+K^2}}\right\} \sin(\omega t - \tan^{-1} B\omega/K)$

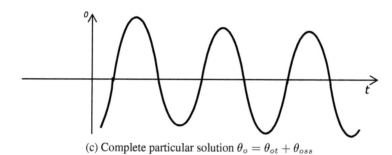

(c) Complete particular solution $\theta_o = \theta_{ot} + \theta_{oss}$

Figure 2.2   Response of simple system.

where $f(t)$ is a function of time t, the driving function or forcing function, θ is a physical quantity, the a's are constants and n is a positive integer. The solution to such a differential equation represents the response of a physical system, and is composed of a transient component θ_t and a steady-state component, θ_{ss}.

There are only a few types of driving function that occur in system design and analysis and it will be possible to give in this chapter suitable procedures for solving equations which features them.

2.2 Determination of the Transient Response, θ_t

To find θ_t from the differential equation given by (2.8), the following procedure is employed:

i. The auxiliary equation in λ is constructed. This is the algebraic equation formed by replacing $d^n\theta/dt^n$ by λ^n, $d^{n-1}\theta/d^{n-1}t$ by λ^{n-1} and so on down to $d\theta/dt$ by λ and θ by 1 and taking the right-hand side as zero, i.e. $a_n\lambda^n + \ldots a_1\lambda + a_0 = 0$. This equation is solved, and in cases where $n \leq 2$ this is easy.

ii. For $n > 2$, a simple solution is only possible when the equation factorizes. However, the solution can always be obtained to some specified degree of accuracy and the values of λ (roots of the equation) will fall into one of the following, out of three, categories:

(a) Different real roots,

$$4\lambda^2 + 9\lambda + 5 = 0$$

has roots $\lambda = -\frac{5}{4}$ or -1

(b) Repeated roots,

$$4\lambda^2 + 9\lambda + \tfrac{81}{16} = 0$$

has the root $\lambda = -1\frac{1}{8}$ repeated twice; this is because the equation factorized into $(2\lambda + \frac{9}{4})^2 = 0$ and each bracket gives $\lambda = -1\frac{1}{8}$.

(c) Complex conjugate roots,

$$4\lambda^2 + 12\lambda + 13 = 0$$

has roots $\lambda = -\frac{-3\pm2j}{2}$, notice that an equation of degree higher than 2 may have roots in all these categories, for instance the equation formed by multiplying the three examples together, i.e. $(4\lambda^2 + 9\lambda + 5)(4\lambda^2 + 9\lambda + \frac{81}{16})(4\lambda^2 + 9\lambda + 13) = 0$, has six roots, two in each category.

iii. Form θ_t as the sum of a number of terms as follows:

(a) For each different real root $\lambda = \alpha$, a term of the form $Ae^{\alpha t}$.

(b) For any number (say p) of repeated roots $m = \alpha$, a term of the form

$$(A_1 + A_2 t + \ldots + A_p t^{p-1})e^{\alpha t}$$

(c) for each pair of conjugate complex roots, $\lambda = \alpha \pm j\omega$, a term of the form

$$e^{\alpha t}(A\cos\omega t + B\sin\omega t)$$

Here $A, B, A_1, A_2\ldots$ are arbitrary constants of integration. The actual value of these constants may be found by using the initial conditions. These are the values of $d^{n-1}\theta/dt^{n-1}$, $d^{n-2}\theta/dt^{n-2}, \ldots\theta$ when $t = 0$. If the arbitrary constants are to be determined the initial conditions must be known. In general, if the equation is nth order, there must be n arbitrary constant.

Example 2.1

Find the transient solution of the differential equation

$$5\frac{d\theta}{dt} + 2\theta = 3\sin\omega t$$

Solution 2.1

Auxiliary equation is

$$5\lambda + 2 = 0$$

$$\therefore \quad \lambda = -\frac{2}{5}$$

$$\therefore \quad \theta_t = Ae^{-2t/5}$$

Example 2.2

Find the transient solution of the equation

$$\frac{d^2\theta}{dt^2} + 3\frac{d\theta}{dt} + 2\theta = 5t$$

Solution 2.2

Auxiliary equation is

$$(\lambda + 2)(\lambda + 1) = 0$$

Solutions are

$$\lambda = -2 \text{ and } \lambda = -1$$

$$\therefore$$

$$\theta_t = Ae^{-2t/5} + Be^{-t}$$

Example 2.3

Find the transient solution of the equation

$$5\frac{d^2\theta}{dt^2} + 3\frac{d\theta}{dt} + \theta = te^{-3t}$$

Solution 2.3

Auxiliary equation is

$$5\lambda^2 + 3\lambda + 1 = 0$$

Solutions are

$$\lambda = \frac{-3 \pm \sqrt{(9-20)}}{10}$$

$$= -0.3 \pm j0.332$$

$$\theta_t = e^{-0.3t}(A \cos 0.332t + B \sin 0.332t)$$

Example 2.4
Find the transient solution of the equation

$$\frac{d^2\theta}{dt^2} + 4\frac{d\theta}{dt} + 4\theta = 5e^{-3t} \sin 30t$$

Solution 2.4
Auxiliary equation is

$$\lambda^2 + 4\lambda + 4 = 0 \quad \text{or} \quad (\lambda+2)^2 = 0$$

$$\therefore \quad \lambda = -2$$

Thus, $\qquad\qquad \theta_t = (A + Bt)e^{-2t}$

Example 2.5
Find the transient solution of the equation

$$4\frac{d^2\theta}{dt^2} + \theta = 3$$

Solution 2.5
Auxiliary equation is

$$4\lambda^2 + 1 = 0$$

$$\therefore \quad \lambda = 0 \pm j0.5$$

Thus, $\qquad\begin{aligned} \theta_t &= e^{0t}(A \cos 0.5t + B \sin 0.5t)\\ &= A \cos 0.5t + B \sin 0.5t\end{aligned}$

Example 2.6
Find the transient solution of the equation

$$\frac{d^2\theta}{dt^2} + 3\frac{d\theta}{dt} = 0$$

Solution 2.6

Auxiliary equation is

$$\lambda^2 + 3\lambda = 0,$$

$$\therefore \quad \lambda(\lambda + 3) = 0$$

$$\therefore \quad \lambda = 0 \text{ or } \lambda = -3$$

Thus,

$$\theta_t = Ae^{0t} + Be^{-3t} = A + Be^{-3t}$$

2.3 Determination of the Steady-State Response, θ_{ss}

The method of determining the steady-state solution, θ_{ss} depends largely on the type of driving function or forcing function, $f(t)$. In control systems engineering, some important types of driving functions are well known as

(i) Zero-driving function ($f(t) = 0$ for $t > 0$).
(ii) A step-driving function ($f(t) = k$ for $t > 0$).
(iii) A ramp-driving function ($f(t) = kt$ for $t > 0$).
(iv) A constant acceleration driving function ($f(t) = kt^2$ for $t > 0$).
(v) An exponential-driving function ($f(t) = ke^t$ for $t > 0$).
(vi) A sinusoidal-driving function ($f(t) = k \sin \omega t$ for $t > 0$).

where k is a constant. In the determination of the steady-state solution we shall use the operator D which stands for $\frac{d}{dt}$, D^2 will be used for $\frac{d^2}{dt^2}$, D^3 for $\frac{d^3}{dt^3}$, and so on.

The expression $2\frac{d^2\theta}{dt^2} + 5\frac{d\theta}{dt} + 2\theta$ may be written as $2D^2\theta + 5D\theta + 2\theta$.

Taking out a common factor θ this becomes $(2D^2 + 5D + 2)\theta$. This may be factorized and written as $(2D + 1)(D + 2)\theta$. It is thus intended to manipulate D as if it were an algebraic quantity. This approach is justified in the literature. The operation 1/D will be defined as the reverse of the process of differentiation, which is integration. It is, however, only integration in form and the arbitrary constant may always be omitted.

2.3.1 Zero- or Constant-Driving Function

The differential equation

$$a_n \frac{d^n\theta}{dt^n} + a_{n-1}\frac{d^{n-1}\theta}{dt^{n-1}} + \cdots + a_1\frac{d\theta}{dt} + a_0\theta = k$$

becomes, using the D-operator,

$$(a_n D^n + a_{n-1} D^{n-1} + \cdots + a_1 D + a_0)\theta = k$$

To find θ_{ss}, D is equated to zero, the value of θ remaining is then θ_{ss}, i.e.

$$(a_n \times 0 + a_{n-1} \times 0 + \cdots a_1 \times 0 + a_0)\theta_{ss} = k$$

or
$$\theta_{ss} = k/a_0$$

This method is still applicable when $k = 0$, giving $\theta_{ss} = 0$.

The Failing Case: There is a failing case which gives an indeterminate form; this occurs when there is no term $a_0\theta$ and the equation takes the form

$$(a_n D^n + a_{n-1} D^{n-1} + \cdots a_1 D)\theta = k$$

Here we may take out a factor D,

$$(a_n D^{n-1} + a_{n-1} D^{n-2} + \cdots + a_1)D\theta = k$$

or
$$(a_n D^{n-1} + a_{n-1} D^{n-2} + \cdots + a_1)\theta = \tfrac{1}{D}k$$

Put $D = 0$ on left-hand side only. In putting $\frac{1}{D}$ as the process of integration we have

$$a_1\theta_{ss} = kt + A$$

where A is an arbitrary constant of integration. The value of A can only be determined if the complete solution is found and the initial conditions known (see Section 5.4).

$$\theta_{ss} = \frac{kt}{a_1} + \frac{A}{a_1}$$

Example 2.7
Find the steady-state solution of the equation

$$\left(\frac{d^2\theta}{dt^2} + 40\right) = 3$$

Solution 2.7
In D-form, then $D = 0$

$$(D^2 + 4)\theta = 3$$

$$\theta_{ss} = 3/4$$

Example 2.8
Determine the steady-state solution of the equation

$$\frac{d^2\theta}{dt^2} - 4\frac{d\theta}{dt} + 3\theta = 5$$

Solution 2.8
In D-form the equation becomes

$$(D^2 - 4D + 3)\theta = 5$$

thus
$$\theta_{ss} = \tfrac{5}{3}$$

2.3.2 Ramp- or Acceleration-Driving Function

A function having constant acceleration may be represented by the equation

$$f(t) = kt^2$$

Both these functions are special cases of the general equation

$$f(t) = kt^p$$

where p is an integer.
    A differential equation of form

$$(a_n D^n + a_{n-1} D^{n-1} + \cdots + a_1 D + a_0)\theta = kt^p$$

may be written more briefly as

$$\{F(D)\}\theta = kt^p$$

and the steady-state solution is given (formally) by dividing by F(D), i.e.

$$\theta_{ss} = \frac{1}{F(D)} kt^p$$

The form of θ_{ss} is now obtained by expanding $\frac{1}{F(D)}$ in a series of ascending powers of D; the method is illustrated in the following examples.

Example 2.9
Find the steady-state solution of the equation

$$\left(\frac{d^2\theta}{dt^2} + 4\theta\right) = 3t$$

Solution 2.9
In D-form

$$(D^2 + 4)\theta = 3t$$

$$\theta_{ss} = \frac{1}{D^2 + 4} 3t$$

$$= \frac{1}{4\left(1 + \frac{D^2}{4}\right)} 3t \quad = \frac{1}{4}\left(1 + \frac{D^2}{4}\right)^{-1} 3t$$

Then, using binomial expansion

$$\theta_{ss} = \frac{1}{4}\left(1 - \frac{D^2}{4} + \frac{D^4}{16} \cdots\right) 3t$$

now $D.3t = \frac{d(3t)}{dt} = 3$

and $D^2.3t = \frac{d^2(3t)}{dt^2} = 0$

also higher powers of D acting on 3t all disappear.

$$\therefore \quad \theta_{ss} = \frac{3t}{4}$$

Example 2.10
Find the steady-state solution of the equation

$$\frac{d^2\theta}{dt^2} + 4\theta = 4t^2$$

Solution 2.10
In D-form

$$(D^2 + 4)\theta = 4t^2$$

$$\theta_{ss} = \frac{1}{D^2 + 4} 4t^2 \quad = \frac{1}{4\left(1 + \frac{D^2}{4}\right)} 4t^2 \quad = \frac{1}{4}\left(1 + \frac{D^2}{4}\right)^{-1} 4t^2$$

Then, using binomial expansion

$$\theta_{ss} = \frac{1}{4}\left(1 - \frac{D^2}{4} + \frac{D^4}{16} \cdots\right) 4t^2$$

now
$$D.4t^2 = \frac{d(4t^2)}{dt} = 8t$$
and
$$D^2.4t^2 = \frac{d^2(4t^2)}{dt^2} = 8$$

also higher powers of D acting on $4t^2$ all disappear.

$$\therefore \quad \theta_{ss} = \frac{8t}{4} + \frac{8}{16}$$

Example 2.11

Determine the steady-state solution of the equation

$$\frac{d^2\theta}{dt^2} - 4\frac{d\theta}{dt} + 3\theta = 5t^2$$

Solution 2.11

In D-form, the equation becomes

$$(D^2 - 4D + 3)\theta = 5t^2$$

thus $\quad \theta_{ss} = \frac{1}{D^2-4D+3}5t^2$

$$= \frac{1}{3\{1 + \frac{1}{3}(-4D + D^2)\}}5t^2$$
$$= \frac{1}{3}\{1 + \frac{1}{3}(-4D + D^2)\}^{-1}\{5t^2\}$$
$$= \frac{1}{3}\{1 - \frac{1}{3}(-4D + D^2) + \frac{1}{9}(-4D + D^2)^2 + \cdots\}\{5t^2\}$$
$$= \frac{1}{3}\{5t^2 - \frac{1}{3}(-40t + 10) + \frac{1}{9}(160)\}$$
$$= \frac{5t^2}{3} + \frac{40t}{9} + \frac{130}{27}$$

2.3.3 Exponential-Driving Function

A differential equation

$$a_n\frac{d^n\theta}{dt^n} + a_{n-1}\frac{d^{n-1}\theta}{dt^{n-1}} + \cdots + a_1\frac{d\theta}{dt} + a_0\theta = ke^{\alpha t}$$

may be written as

$$\{F(D)\}\theta = ke^{\alpha t}$$

The steady-state solution is obtained by dividing by F(D), i.e.

$$\theta_{ss} = \frac{1}{F(D)}.\{ke^{\alpha t}\}$$

The form of θ_{ss} is now obtained by substituting α (the substitution rule), the coefficient of the exponent, for D, i.e.

$$\theta_{ss} = \frac{1}{F(\alpha)} \times ke^{\alpha t}$$

If, however, $F(\alpha)$ is zero this result is without meaning and the shift rule must be used.

The shift rule is used when $F(\alpha)$ (as defined earlier) is zero.
We have

$$\theta_{ss} = \frac{1}{F(D)}\{ke^{\alpha t}\}$$

$$= e^{\alpha t} \times \frac{1}{F(D+\alpha)}\{k\}$$

It is then necessary to convert $1/F(D + \alpha)$ into a series in ascending powers of D.

The shift rule as quoted here is not the most general case. If $V(t)$ is any function of time t, then the most general statement of the shift rule is

$$\{F(D)\}\{e^{\alpha t}V(t)\} = e^{\alpha t}\{F(D+\alpha)\}\{V(t)\}$$

Example 2.12
Find the steady-state solution of the equation

$$\frac{d^2\theta}{dt^2} + 6\frac{d\theta}{dt} + 9\theta = 50e^{2t}$$

Solution 2.12
In D-form, the equation becomes

$$(D^2 + 6D + 9)\theta = 50e^{2t}$$
$$\theta_{ss} = \frac{1}{D^2+6D+9}\{50e^{2t}\}$$

We can use the substitution rule, hence we have

$$\theta_{ss} = e^{2t}\frac{1}{4+12+9}\{50\}$$

$$\theta_{ss} = e^{2t}\frac{1}{25}\{50\} = e^{2t}(2)$$

Example 2.13
Find the steady-state solution of the equation

$$\frac{d\theta}{dt} + 2\theta = 2e^{-2t}$$

Solution 2.13
In D-form, the equation becomes

$$(D + 2)\theta = 2e^{-2t}$$

$$\theta_{ss} = \frac{1}{D+2}\{2e^{-2t}\}$$

We cannot use the substitution rule because *F(D)* would disappear. Hence using the shift rule, we have

$$\theta_{ss} = e^{-2t}\frac{1}{D-2+2}\{2\}$$

$$\theta_{ss} = e^{-2t}\frac{1}{D}\{2\}$$

$$\theta_{ss} = e^{-2t}(2t)$$

Example 2.14
Find the steady-state solution of the equation

$$\frac{d^2\theta}{dt^2} + 6\frac{d\theta}{dt} + 5\theta = 2e^{-t}$$

Solution 2.14
In D-form, the equation becomes

$$(D^2 + 6D + 5)\theta = 2e^{-2t}$$

$$\theta_{ss} = \frac{1}{D^2+6D+5}\{2e^{-2t}\}$$

We can use the substitution rule, hence we have

$$\theta_{ss} = e^{2t}\frac{1}{4-12+5}\{2\}$$

$$\theta_{ss} = e^{2t}[-\frac{1}{3}\{2\}] = -\frac{2}{3}e^{2t}$$

2.3.4 Sinusoidal-Driving Function

A differential equation

$$a_n \frac{d^n\theta}{dt^n} + a_{n-1}\frac{d^{n-1}\theta}{dt^{n-1}} + \cdots + a_1\frac{d\theta}{dt} + a_0\theta = k\sin\omega t$$

in D-form becomes

$$\{F(D)\}\theta = k\sin\omega t. \quad \text{Thus,} \quad \theta_{ss} = \frac{1}{F(D)}\{k\sin\omega t\}.$$

The steady-state solution will take the form

$$\theta_{ss} = \frac{k}{Z}\sin(\omega t - \phi)$$

and a method of finding Z and ϕ is required.

Now the driving function $k\sin\omega t$ is a vector which may be represented by the complex number, $ke^{j\omega t}$. Hence the complex number representing the steady-state solution $\bar\theta_{ss}$ is given by

$$\bar\theta_{ss} = \left\{\frac{1}{F(D)}\right\}ke^{j\omega t}$$

The substitution rule is now used, i.e. we substitute $j\omega$ for D, thus

$$\bar\theta_{ss} = \frac{ke^{j\omega t}}{F(j\omega)} \quad \text{But at} \quad t = 0, d\theta/dt = 0$$

$$\therefore \quad B = -\tfrac{3}{10} \quad \therefore \quad 0 = -3e^0(-\tfrac{1}{5} + B.0) + e^0(\tfrac{2}{5}.0 + 2B) - \tfrac{2}{5}$$

$$\therefore \quad 0 = \tfrac{3}{5} + 2B - \tfrac{2}{5}$$

Hence, particular solution becomes

$$\theta_{ss} = e^{-3t}\left(-\frac{\cos 2t}{5} - \frac{3\sin 2t}{10}\right) + \frac{e^{-2t}}{5}$$

2.4 Problems and Solutions

Practice Problem 2.1

Find the transient solutions of the following differential equations:

(i) $\frac{d^2\theta}{dt^2} - 6\frac{d\theta}{dt} + 13\theta = 3t^2$

(ii) $\frac{d^3\theta}{dt^3} - 6\frac{d^2\theta}{dt^2} + 11\frac{d\theta}{dt} - 60 = 7e^{-t}$

(iii) $2\frac{d^2\theta}{dt^2} + 5\frac{d\theta}{dt} + 2\theta = 0$

(iv) $\frac{d^2\theta}{dt^2} + 4\theta = 6\sin 3t$

(v) $\frac{d^2\theta}{dt^2} + 4\frac{d\theta}{dt} + 2t$

(vi) $\frac{d^2\theta}{dt^2} + 6\frac{d\theta}{dt} + 9\theta = 4$

Practice Problem 2.2

Find the steady-state solutions of the following differential equations:

(i) $\frac{d^2\theta}{dt^2} + \frac{d\theta}{dt} + 3\theta = 0$

(ii) $\frac{d^2\theta}{dt^2} + 8\frac{d^2\theta}{dt^2} + 25\theta = 50$

(iii) $\frac{d^2\theta}{dt^2} + 5\frac{d\theta}{dt} = 10$

(iv) $\frac{d\theta}{dt^2} + 6\theta = 3$

(v) $\frac{d^2\theta}{dt^2} + 2\frac{d\theta}{dt} + 24 = t$

(vi) $\frac{d^2\theta}{dt^2} + 4\theta = 4t^2$

(vii) $\frac{d^2\theta}{dt^2} + 6\frac{d\theta}{dt} + 9\theta = 50e^{2t}$

(viii) $\frac{d^2\theta}{dt^2} - 4\frac{d\theta}{dt} + 4\theta = 50e^{2t}$

(ix) $\frac{d\theta}{dt} + \theta = 10\sin 2t$

(x) $2\frac{d^2\theta}{dt^2} - 5\frac{d\theta}{dt} + 12\theta = 200\sin 4t$

Practice Problem 2.3

Find the complete particular solutions of the following equations:

(i) $\frac{d^2\theta}{dt^2} + 5\frac{d\theta}{dt} + 4\theta = 0$, *given* $\theta = 0$, $\frac{d\theta}{dt} = 2$ *at* $t = 0$

(ii) $\frac{d^2\theta}{dt^2} + 9\theta = \sin 2t$, *given* $\theta = 1$, $\frac{d\theta}{dt} = -1$ *at* $t = 0$

(iii) $\frac{d\theta}{dt} + 3\theta = e^{-2t}$, *given* $\theta = 4$ *at* $t = 0$

(iv) $\frac{d^2\theta}{dt^2} + 4\theta = t$, *given* $\theta = 0$, $\frac{d\theta}{dt} = 1$ *at* $t = 0$

(v) $\frac{d^2\theta}{dt^2} + 5\frac{d\theta}{dt} + 6\theta = 2e^{-t}$, *given* $\theta = 1$, $\frac{d\theta}{dt} = 0$ *at* $t = 0$

(vi) $\frac{d^2\theta}{dt^2} + 6\frac{d\theta}{dt} + 5\theta = 10$, *given* $\theta = 0$, $\frac{d\theta}{dt} = 0$ *at* $t = 0$

3

Laplace Transform

This chapter provides comprehensive details regarding the Laplace transform, inverse Laplace transforms covering three cases (i) when poles are real and non-repeated, (ii) when the poles are real and repeated and (iii) when the poles are complex, conversion of differential equation into Laplace transform and transfer function and problems and solutions. At the end of the chapter, relevant practice problems are given for better understanding.

3.1 Introduction

By definition, the Laplace transform of a function of time $f(t)$ is

$$F(s) = \mathcal{L}[f(t)] = \int_0^\infty f(t)e^{-st}dt \qquad (3.1)$$

where $\mathcal{L}$ indicates the Laplace transform. Note that the variable time has been integrated out of the equation and that the Laplace transform is a function of the complex variable, **s**. The inverse Laplace transform is given by

$$f(t) = \mathcal{L}^{-1}[F(s)] = \frac{1}{2\pi j} \int_{\sigma-j\infty}^{\sigma+j\infty} F(s)e^{st}ds \qquad (3.2)$$

where $\mathcal{L}^{-1}$ indicates the inverse transform and $j = \sqrt{-1}$.

Equations (3.1) and (3.2) form the Laplace transform pair. Given a function $f(t)$, we integrate (3.1) to find its Laplace transform $F(s)$. Then if this function $F(s)$ is used to evaluate (3.2), the result will be the original value of $f(t)$. The value of σ in (3.2) is determined by the singularities of $F(s)$.

We seldom use (3.2) to evaluate an inverse Laplace transform; instead, we use (3.1) to construct a table of transforms for useful time functions. Then, if possible, we can use the table to find the inverse transform rather than integrating (3.1).

As an example, we can find the Laplace transform of the exponential function e^{-at}.

From (3.1)

$$F(s) = \int_0^\infty e^{-at}e^{-st}dt = \int_0^\infty e^{-(s+a)t}dt = \frac{-e^{-(s+a)t}}{s+a}\bigg|_0^\infty$$

$$= \frac{1}{s+a} \quad Re(s+a) > 0 \tag{3.3}$$

where $Re(.)$ indicates the real part of the expression. Of course, Laplace transform tables have been derived already in many text books (references can be found at the end of this book). Therefore, we will not derive any additional transforms.

A short table of commonly required transforms is given in Table 3.1. The first two columns of this table are more complete table of Laplace transforms. From the definition of the Laplace transform (3.1), for k constant

$$\pounds[kf(t)] = k\,\pounds[f(t)] = kF(s) \tag{3.4}$$

$$\pounds[f_1(t) + f_2(t)] = \pounds[f_1(t)] + \pounds[f_2(t)] = F_2(s) \tag{3.5}$$

The use of these two relationships greatly extends the application of Table 3.1.

Examples of the Laplace transform and of the inverse Laplace transform are given for better understanding. First, however, we need to note that using the complex inversion integral (3.2) to evaluate the inverse Laplace transform results in $f(t) = 0$ for $t < 0$. Hence, to be consistent, we will always assign a value of zero to $f(t)$ for all negative time. Also, to simplify notation, we define the unit step function $u(t)$ to be

$$u(t) = \begin{cases} 0 & t < 0 \\ 1 & t \geq 0 \end{cases} \tag{3.6}$$

In Equation (3.3), the Laplace transform of e^{-at} was derived. Note that the Laplace transform of $e^{-at}\,u(t)$ is the same function. Thus for any function $f(t)$

$$\pounds[f(t)] = \pounds[f(t)u(t)] = F(s) \tag{3.7}$$

Table 3.1 Laplace transforms

Name	Time Function, $f(t)$	Laplace Transform, $F(s)$
Unit impulse	$\delta(t)$	1
Unit step	$u(t)$	$\frac{1}{s}$
Unit ramp	t	$\frac{1}{s^2}$
nth-Order ramp	t^n	$\frac{n!}{s^{n+1}}$
Exponential	e^{-at}	$\frac{1}{s+a}$
nth-Order exponential	$t^n e^{-at}$	$\frac{n!}{(s+a)^{n+1}}$
Sine	$\mathrm{Sin}bt$	$\frac{b}{s^2+b^2}$
Cosine	$\cos bt$	$\frac{s}{s^2+b^2}$
Damped sine	$e^{-at}\sin bt$	$\frac{b}{(s+a)^2+b^2}$
Damped cosine	$e^{-at}\cos bt$	$\frac{s+a}{(s+a)^2+b^2}$
Diverging sine	$t\sin bt$	$\frac{2bs}{(s^2+b^2)^2}$
Diverging cosine	$t\cos bt$	$\frac{s^2+b^2}{(s^2+b^2)^2}$

Example 3.1

The Laplace transform of the time function

$$f(t) = 5u(t) + 3e^{-2t}$$

Solution 3.1

From Table 3.1

$$\pounds[5u(t)] = 5\pounds[u(t)] = \tfrac{5}{s}$$

$$\pounds[3e^{-2t}] = 3\pounds[e^{-2t}] = \tfrac{3}{s+2}$$

From (3.5)

$$F(s) = \pounds[5u(t) + 3e^{-2t}] = \tfrac{5}{s} + \tfrac{3}{s+2}$$

This Laplace transform can also be expressed as

$$F(s) = \tfrac{5}{s} + \tfrac{3}{s+2} = \tfrac{8s+10}{s(s+2)}$$

The transforms are usually easier to manipulate in the combined form than in the sum-of-terms form.

This example illustrates an important point. As stated, we usually work with the Laplace transform expressed as a ratio of polynomials in the variables (we call this ratio of polynomials a rational function). However, the tables used to find inverse transforms contain only low-order functions. Hence a method is required for converting from a general rational function to the forms that appear in the tables. This method is called the partial-fraction expansion method. A simple example is illustrated in the relationship

$$\frac{C}{(s+a)(s+b)} = \frac{k_1}{s+a} + \frac{k_2}{s+b}$$

Given the constants a, b and c, the problem is to find the coefficients of the partial-fraction expansion k_1 and k_2. We now derive the general relationships required. Consider the general rational function

$$F(s) = \frac{b_m s^m + \ldots + b_1 s + b_0}{s^n + a_n - 1 s^{n-1} + \ldots + a_1 s + a_0} = \frac{N(s)}{D(s)} \quad m < n \qquad (3.8)$$

where $N(s)$ is the numerator polynomial and $D(s)$ is the denominator polynomial. To perform a partial-fraction expansion, first the roots of the denominator must be found. Then

$$F(s) = \frac{N(s)}{D(s)} = \frac{N(s)}{\displaystyle\prod_{i=1}^{n}(s - p_i)} = \frac{k_1}{s - p_1} + \frac{k_2}{s - p_2} + \ldots + \frac{k_n}{s - p_n} \qquad (3.9)$$

where Π indicates the product of terms. Suppose that we wish to calculate the coefficient k_j. We first multiply (3.9) by the terms $(s - p_j)$:

$$(s - p_j)F(s) = \frac{k_1(s - p_j)}{s - p_1} + \ldots + k_j + \ldots + \frac{k_n(s - p_j)}{s - p_n} \qquad (3.10)$$

If this equation is evaluated for $s = p_j$, we see then that all terms on the right side are zero except the jth term, and thus

$$k_j = (s - p_j)F(s)|_{s=pj} \quad j = 1, 2, \ldots, n \qquad (3.11)$$

In mathematics, k_j is called the residue of $F(s)$ in the pole at $s = p_j$.

If the denominator polynomial of $F(s)$ has repeated roots, $F(s)$ can be expanded as in the example

$$F(s) = \frac{N(s)}{(s - p_1)(s - p_2)^r}$$

$$= \frac{k_1}{s - p_1} + \frac{k_{21}}{s - p_2} + \frac{k_{22}}{(s - p_2)^2} + \ldots + \frac{k_{2r}}{(s - p_2)^r} \qquad (3.12)$$

where it is seen that a denominator root of multiplicity r yields r terms in the partial-fraction expansion. The coefficients of the repeated-root terms are calculated from the equation

$$k_{2j} = \frac{1}{(r-j)!} \frac{d^{r-j}}{ds^{r-j}} [(s - p_2)^r F(s)]\Big|_{s=p_2} \tag{3.13}$$

This equation is provided here without proof.

The preceding development applies to complex poles as well as real poles. Consider the case that $F(s)$ has a pair of complex poles. If we let $p_1 = a - jb$ and $p_2 = a + jb$, (3.9) can be written as

$$F(s) = \frac{k_1}{s - a + jb} + \frac{k_2}{s - a - jb} + \frac{k_3}{s - p_3} + \ldots + \frac{k_n}{s - p_n} \tag{3.14}$$

The coefficients k_1 and k_2 can be evaluated using (3.11) as before. It will be found, however, that these coefficients are complex valued, and that k_2 is the conjugate of k_1. In order to achieve a convenient form for the inverse transform, we will use the following approach. From (3.11)

$$k_1 = (s - a + jb)F(s)|_{s=a-jb} = Re^{j\theta}$$
$$k_2 = (s - a - jb)F(s)|_{s=a+jb} = Re^{-j\theta} = k_1^* \tag{3.15}$$

where the asterisk indicates the conjugate of the complex number. Define $f_1(t)$ as the inverse transform of the first two terms of (3.14). Hence, by Euler's identity

$$f_1(t) = Re^{j\theta} e^{(a-jb)t} + Re^{-j\theta} e^{(a+jb)t}$$

$$= 2Re^{at} \left[\frac{e^{j(bt-\theta)} + e^{-j(bt-\theta)}}{2} \right] \tag{3.16}$$

$$= 2Re^{at} \cos(bt - \theta)$$

This approach expresses the inverse transform in a convenient form and the calculations are relatively simple. The damped sinusoid has an amplitude of 2R and a phase angle θ, where R and θ are defined in (3.16). Three examples of finding the inverse Laplace are given next.

Example 3.2

Find the inverse Laplace transform of the following transfer function.

$$F(s) = \frac{5}{s^2 + 3s + 2} = \frac{5}{(s+1)(s+2)}$$

Solution 3.2
First the partial fractional expansion is derived:

$$F(s) = \frac{5}{(s+1)(s+2)} = \frac{k_1}{s+1} + \frac{k_2}{s+2}$$

The coefficients in the partial-fraction expansion are calculated from (3.11):

$$k_1 = (s+1)F(s)|_{s=-1} = \frac{5}{s+2}\bigg|_{s=-1} = 5$$

$$k_2 = (s+2)F(s)|_{s=-2} = \frac{5}{s+1}\bigg|_{s=-2} = -5$$

Thus the partial-fraction expansion is

$$\frac{5}{(s+1)(s+2)} = \frac{5}{s+1} + \frac{-5}{s+2}$$

This expansion can be verified by recombining the terms on the right side to yield the left side of the equation. The inverse transform of $F(s)$ is then

$$\pounds^{-1}[F(s)] = (5e^{-t} - 5e^{-2t})u(t)$$

The function $u(t)$ is often omitted, but we must then understand that the inverse transform can be non-zero only for positive time and must be zero for negative time.

Example 3.3
Find the inverse Laplace transform of the following function:

$$F(s) = \frac{2s+3}{s^3+2s^2+s} = \frac{2s+3}{s(s+1)^2} = \frac{k_1}{s} + \frac{k_{21}}{s+1} + \frac{k_{22}}{(s+1)^2}$$

Solution 3.3
The coefficients k_1, k_{21} and k_{22} can easily be evaluated:

$$k_1 = sF(s)|_{s=0} = \frac{2s+3}{(s+1)^2}\bigg|_{s=0} = 3$$

$$k_{22} = (s+1)^2 F(s)|_{s=-1} = \frac{2s+3}{s}\bigg|_{s=-1} = -1$$

We use (3.13) to find k_{21}:

$$k_{21} = \frac{1}{(2-1)!}\frac{d}{ds}[(s+1)^2 F(s)]|_{s=-1} = \frac{d}{ds}\left[\frac{2s+3}{s}\right]\Big|_{s=-1}$$

$$= \frac{s(2)-(2s+3)(1)}{s^2}\Big|_{s=-1} = \frac{-2-1}{1} = -3$$

Thus, the partial-fraction expansion yields

$$F(s) = \frac{2s+3}{s(s+1)^2} = \frac{3}{s} + \frac{-3}{s+1} + \frac{-1}{(s+1)^2}$$

Then, from Table 3.1

$$f(t) = 3 - 3e^{-t} - te^{-t}$$

Example 3.4
Find the inverse transform of a function having complex poles

$$F(s) = \frac{10}{s^3 + 4s^2 + 9s + 10}$$

Solution 3.4

$$F(s) = \frac{10}{s^3 + 4s^2 + 9s + 10} = \frac{10}{(s+2)(s^2+2s+5)}$$

$$= \frac{10}{(s+2)[(s+1)^2 + 2^2]}$$

$$F(s) = \frac{k_1}{s+2} + \frac{k_2}{s+1+j2} + \frac{k_2^*}{s+1-j2}$$

Evaluating the coefficient k_1 as before

$$k_1 = (s+2)F(s)|_{s=-2} = \frac{10}{(s+1)^2 + 4}\Big|_{s=-2} = \frac{10}{5} = 2$$

Coefficient k_2 is calculated from (3.15):

$$k_2 = (s+1+j2)F(s) = \frac{10}{(s+2)(s+1-j2)}\Big|_{s=-1-j2}$$

$$= \frac{10}{(-1-j2+2)(-1-j2+1-j2)} = \frac{10}{(1-j2)(-j4)}$$

$$= \frac{10}{(2.236 \angle -63.4^0)(4\angle -90^0)} = 1.118\angle 153.4^0 = R\angle\theta$$

Therefore, using (3.16)

$$f(t) = 2e^{-2t} + 2.236e^{-t}\cos(2t - 153.4^0)$$

3.2 Theorems of Laplace Transform

For the analysis and design of control systems, however, we require several theorems of the Laplace transform. As an example of the theorems, we derive the final-value theorem. We shall see later that this theorem is very useful in control system analysis and design. Suppose that we wish to calculate the final value of *f(t)*, that is, $\lim_{t \to \infty} f(t)$. However, we wish to calculate this final value directly from the Laplace transform *F(s)* without finding the inverse Laplace transform. The final-value theorem allows us to do this. To derive this theorem, it is first necessary to find the Laplace transform of the derivative of a general function *f(t)*.

$$\pounds\left[\frac{df}{dt}\right] = \int_0^\infty e^{-st}\frac{df}{dt}dt \tag{3.17}$$

This expression can be integrated by parts, with

$$u = e^{-st} \quad dv\frac{df}{dt}dt$$

Thus

$$\pounds\left[\frac{df}{dt}\right] = uv\Big|_0^\infty - \int_0^\infty vdu = f(t)e^{-st}\Big|_0^\infty + s\int_0^\infty e^{-st}f(t)\,dt$$

$$= 0 - f(0) + sF(s) = sF(s) - f(0) \tag{3.18}$$

To be mathematically correct, the initial-condition term should be $f(0^+)$, where

$$f(0^+) = \lim_{t \to 0} f(t) \quad t > 0 \tag{3.19}$$

However, we will use the notation *f(0)*.

Now the final-value theorem can be derived. From (3.19)

$$\lim_{s \to 0}\left[\pounds\left(\frac{df}{dt}\right)\right] = \lim_{s \to 0}\int_0^\infty e^{-st}\frac{df}{dt}dt$$

$$= \int_0^\infty \frac{df}{dt}dt = \lim_{t \to \infty} f(t) - f(0) \tag{3.20}$$

Then, from (3.19) and (3.20)

$$\lim_{t \to \infty} f(t) - f(0) = \lim_{s \to 0}[sF(s) - f(0)] \qquad (3.21)$$

$$\lim_{t \to \infty} f(t) = \lim_{s \to 0} sF(s) \qquad (3.22)$$

provided that the limit on the left side of this relationship exists. The right-side limit may exist without the existence of the left-side limit. For this case, the right side of (3.21) gives the incorrect value for the final value of *f(t)*.

Table 3.2 lists several useful theorems of the Laplace transform. No further proofs of these theorems are given here. An example of the use of these theorems is given next.

Example 3.5

Find the Laplace transform of the time function, x(t) = u(t − 2).

Solution 3.5

$$X(s) = \mathcal{L}[x(t)] = \mathcal{L}[u(t - 2)]$$

$$X(s) = \left(\int_{-\infty}^{\infty} u(t - 2)e^{-st}dt \right) = \left(\int_{2}^{\infty} e^{-st}dt \right)$$

Table 3.2 Laplace transform theorems

Name	Theorem
Derivative	$\mathcal{L}\left[\frac{df}{dt}\right] = sF(s) - f(0^+)$
nth-Order derivative	$\mathcal{L}\left[\frac{d^n f}{dt^n}\right] = s^n F(s) - s^{n-1}f(0^+)$
	$- \ldots f^n - 1(0^+)$
Integral	$\mathcal{L}\left[\int_0^t f(\tau)d\tau\right] = \frac{F(s)}{s}$
Shifting	$\mathcal{L}[f(t - t_0)u(t - t_0)] = e^{-t_0 s}F(s)$
Initial value	$\lim_{t \to 0} f(t) = \lim_{s \to \infty} sF(s)$
Final value	$\lim_{t \to \infty} f(t) = \lim_{s \to 0} sF(s)$
Frequency shift	$\mathcal{L}[e^{-at}f(t)] = F(s + a)$
Convolution integral	$\mathcal{L}^{-1}[F_1(s)F_2(s)] = \int_0^1 f_1(t - \tau)f_2(\tau)d\tau$
	$= \int_0^1 f_1(\tau)f_2(t - \tau)d\tau$

$$X(s) = \frac{-1}{s} e^{-st} \Big|_2^\infty = [0 + \frac{e^{-2s}}{s}]$$

$$X(s) = \frac{e^{-2s}}{s}$$

Example 3.6

Find the Laplace transform of the time function cos *at*.

Solution 3.6

$$F(s) = \pounds[f(t)] = \pounds[\cos at] = \frac{s}{s^2 + a^2}$$

Then, from Table 3.2

$$\pounds \left[\frac{df}{dt}\right] = \pounds[-a \sin at] = sF(s) - f(0) = \frac{s^2}{s^2 + a^2} - 1 = \frac{-a^2}{s^2 + a^2}$$

which agrees with the transform from Table 3.1. Also

$$\pounds \left(\int_0^t f(\tau) d\tau\right) = \pounds \left(\frac{\sin at}{a}\right) = \frac{F(s)}{s} = \frac{1}{s^2 + a^2}$$

which also agrees with Table 3.1. The initial value of *f(t)* is

$$f(0) = \lim_{s \to \infty} sF(s) = \lim_{s \to \infty} \left[\frac{s^2}{s^2 + a^2}\right] = 1$$

which, of course, is correct. If we carelessly apply the final-value theorem, we obtain

$$\lim_{s \to \infty} f(s) = \lim_{s \to 0} sF(s) = \lim_{s \to 0} \left[\frac{s^2}{s^2 + a^2}\right] = 0$$

which is incorrect, since cos *at* does not have a final value; the function continues to vary between 1 and −1 as time increases without bound. This last exercise emphasizes the point that the final-value theorem does not apply to functions that have no final value.

Example 3.7

As a second example, given the time function $f(t) = e^{-0.5t}$, which is then delayed by 4 s. Thus, the function that we consider is

$$f_1(t) = f(t - 4)u(t - 4) = e^{-0.5(t-4)}u(t - 4)$$

Solution 3.7

Note that *f(t)* is delayed by 4 s and that the value of the delayed function is zero for time < 4 s (the amount of the delay). Both of these conditions are necessary in order to apply the shifting theorem of Table 3.2. From this theorem

$$\mathcal{L}[f(t - t_0)u(t - t_0)] = e^{-t_0 s}F(s) \quad F(s) = \mathcal{L}[f(t)]$$

For this example, the unshifted function is $e^{-0.5t}$, and thus $F(s) = 1/(s+0.5)$. Hence

$$\mathcal{L}[e^{-0.5(t-4)}u(t - 4)] = \frac{e^{-4s}}{s+0.5}$$

Note that the case that the times function is delayed; the Laplace transform is not a ratio of polynomials in *s* but contains the exponential function.

3.3 Differential Equations and Transfer Functions

In control system analysis and design, the Laplace transform is used to transform constant-coefficient linear differential equations into algebraic equations. The algebraic equations are much easier to manipulate and analyse, simplifying the analysis of the differential equations. We generally model analogue physical systems with linear differential equations with constant coefficients when possible (when the system can be accurately modelled by these equations). Thus the Laplace transform simplifies the analysis and design of analogue linear systems.

An example of a linear differential equation modelling a physical phenomenon is Newton's law:

$$M\frac{d^2x(t)}{dt^2} = f(t) \tag{3.23}$$

where *f(t)* is the force applied to a mass *M*, with the resulting displacement *x(t)*. It is assumed that the units in (3.23) are consistent. Assume that we know the mass *M* and the applied force *f(t)*. The Laplace transform of (3.23) is, from Table 3.2

$$M[s^2X(s) - sx(0) - \dot{x}(0)] = F(s) \tag{3.24}$$

where $\dot{x}(t)$ denotes the derivative of x(t). Thus to solve for the displacement of the mass, we must know the applied force, the initial displacement, *x(0)* and the initial velocity, $\dot{x}(0)$. Then we can solve this equation for *X(s)* and take the inverse Laplace transform to find the displacement, *x(t)*. We now solve for *X(s)*:

$$X(s) = \frac{F(s)}{Ms^2} + \frac{x(0)}{s} + \frac{\dot{x}(0)}{s^2} \qquad (3.25)$$

For example, suppose that the applied force *f(t)* is zero. Then, the inverse transform of (3.25) is

$$x(t) = x(0) + \dot{x}(0)t \quad t \geq 0 \qquad (3.26)$$

If the initial velocity, $\dot{x}(t)$ is also zero, the mass will remain at its initial position *x(0)*.

If the initial velocity is not zero, then the displacement of the mass will increase at a constant rate equal to that initial velocity.

Note that if the initial conditions are all zero, (3.25) becomes

$$X(s) = \frac{1}{Ms^2} F(s) \qquad (3.27)$$

Consider a physical phenomenon (system) that can be modelled by a linear differential equation with constant coefficients. The Laplace transform of the response (output) of this system can be expressed as the product of the Laplace transform of the forcing function (input) times a function of s (provided all initial conditions are zero), which we call the transfer function. We usually denote the transfer function by *G(s)* where

$$G(s) = \frac{1}{Ms^2} \qquad (3.28)$$

Example 3.8
The relationship between the input x(t) and the output y(t) of a causal system is described by the differential equation. Determine impulse response of the system:

$$5\frac{dy(t)}{dt} + 10y(t) = 2x(t)$$

Solution 3.8
In this equation, *x(t)* is the forcing function, or the input, and y(t) is the response function (output). If we take the Laplace transform of this equation, we have

$$5[sY(s) - y(0)] + 10[Y(s)] = 2X(s)$$

Solving this equation for the response Y(s):

$$Y(s) = \frac{2X(s) + 5y(0)}{5s + 10}$$

The transfer function is obtained by setting the initial conditions to zero:

$$H(s) = \frac{Y(s)}{X(s)} = \frac{2}{5s + 10}$$

Suppose that we wish to find the impulse response with no initial conditions:

$$H(s) = \left[\frac{2}{5(s + 2)}\right]$$

The inverse transform of this expression is then

$$h(t) = \frac{2}{5}e^{-2t} \quad t \geq 0$$

Example 3.9
Suppose that a system is modelled by the differential equation, find x(t).

$$\frac{d^2x(t)}{dt^2} + 3\frac{dx(t)}{dt} + 2x(t) = 2f(t)$$

Solution 3.9
In this equation, *f(t)* is the forcing function, or the input, and x(t) is the response function (output). If we take the Laplace transform of this equation, we have

$$s^2X(s) - sx(0) - \dot{x}(0) + 3\left[sX(s) - x(0)\right] + 2X(s) = 2F(s)$$

Solving this equation for the response X(s)

$$X(s) = \frac{2F(s) + (s + 3)x(0) + \dot{x}(0)}{s^2 + 3s + 2}$$

The transfer function is obtained by setting the initial conditions to zero.

$$G(s) = \frac{X(s)}{F(s)} = \frac{2}{s^2 + 3s + 2}$$

Suppose that we wish to find the response with no initial conditions and with the system input equal to a unit step function. Then $F(s) = 1/s$, and

$$X(s) = G(s)F(s) = \left[\frac{2}{s^2 + 3s + 2}\right]\left[\frac{1}{s}\right]$$

or

$$X(s) = \frac{2}{s(s+1)(s+2)} = \frac{1}{s} + \frac{-2}{s+1} + \frac{1}{s+2}$$

by partial-fraction expansion. The inverse transform of this expression is then

$$x(t) = 1 - 2e^{-t} + e^{-2t} \quad t \ge 0$$

Note that $x(0)$ and $\dot{x}(0)$ are zero, as assumed, and that after a very long time, *x(t)* is approximately equal to unity.

In Example 3.9, the response X(s) can be expressed as

$$X(s) = G(s)F(s) + \frac{(s+3)x(0) + \dot{x}(0)}{s^2 + 3s + 2} = X_f(s) + X_{ic}(s) \quad (3.29)$$

The term $X_f(s)$ is the forced (also called the zero-state) response, and the term $X_{ic}(s)$ is the initial-condition (zero-input) response. This result is general. We see then that the total response is the sum of two terms. The forcing-function term is independent of the initial conditions, and the initial-condition term is independent of the forcing function. The concept of a transfer function is basic to the study of feedback control systems. To generalize the results, let a system having an output c(t) and an input r(t) be described by the *n*th-order differential equation:

$$\frac{d^n c}{dt^n} + a_{n-1}\frac{d^{n-1} c}{dt^{n-1}} + \ldots + \frac{dc}{dt} + a_0 c$$

$$= bm\frac{d^m r}{dt^m} + b_{m-1}\frac{d^{m-1} r}{dt^{m-1}} + \ldots + b_1\frac{dr}{dt} + b_0 r \quad (3.30)$$

If we ignore all initial conditions, the Laplace transform of (3.30) yields

$$\left(s^n + a_{n-1}s^{n-1} + \ldots + a_1 s + a_0\right)C(s)$$

$$= \left(b_m s^m + b_{m-1}s^{m-1} + \ldots + b_1 s + b_0\right)R(s) \quad (3.31)$$

Ignoring the initial conditions allows us to solve for *C(s)/R(s)* as a rational function of s, namely

$$\frac{C(s)}{R(s)} = \frac{b_m s^m + b_{m-1}s^{m-1} + \ldots + b_1 s + b_0}{s^n + a_{n-1}s^{n-1} + \ldots + a_1 s + a_0} \quad (3.32)$$

Note that the denominator polynomial of (3.32) is the coefficient of C(s) in (3.31). The reader will recall from studying classical methods for solving linear differential equations that this same polynomial set equal to zero is the characteristic equation of the differential Equation (3.30).

3.4 Problems and Solutions

Practice Problem 3.1
Use the defining integral of the Laplace transform to find the transforms of

(a) $f(t) = u(t-1)$

(b) $f(t) = e^{-5t}$

(c) $f(t) = t$.

Practice Problem 3.2
Use the Laplace transform tables to find the transforms of:

(a) $f(t) = 5e^{2t}$

(b) $f(t) = 7e^{-0.5t}\cos 3t$

(c) $f(t) = 2t\sin 5t$

(d) $f(t) = 5\cos(4t + 30^0)$

(e) $f(t) = te^{-3t}$

(f) $f(t) = 5e^{-3t}\cos(4t - 90^0)$

Practice Problem 3.3
Find the inverse Laplace transform of each function given.

(a) $F(s) = \frac{3}{s-4}$ (c) $F(s) = \frac{3}{s(s-2)}$

(b) $F(s) = \frac{3}{s+4}$ (d) $F(s) = \frac{2(s+1)}{s(s+2)}$

Practice Problem 3.4
Find the inverse Laplace transform of each function given.

(a) $F(s) = \frac{5}{s^2(s+1)}$ (c) $F(s) = \frac{s+30}{s(s^2+4s+29)}$

(b) $F(s) = \frac{5}{s(s+1)^2}$ (d) $F(s) = \frac{2s+1}{s^2 2s+10}$

Practice Problem 3.5

Consider the Laplace transform

$$F(s) = \frac{2s - 1}{s^2 + 4s + 13}$$

(a) Express the inverse transform as the sum of two complex exponentials.
(b) Using Euler's identity, manipulate the result of (a) into the form

$$f(t) = Be^{-at}\sin(bt + \theta)$$

(c) Express the inverse transform as $f(t) = Ae^{-at}\cos(bt + \theta)$.

Practice Problem 3.6

Repeat (c) and (d) of Practice Problem 3.5 for the functions

(a) $F(s) = \frac{s}{s^2 + 4}$

(b) $F(s) = \frac{2}{s^2 + 4}$

(c) $F(s) = \frac{s+2}{s^2 + 4}$

Practice Problem 3.7

The hyperbolic sine and cosine functions are defined as

$$\sinh bt = \frac{e^{bt} - e^{-bt}}{2} \qquad \cosh bt = \frac{e^{bt} + e^{-bt}}{2}$$

Find the Laplace transforms of these two functions, expressing the transforms as rational functions.

Practice Problem 3.8

(a) Plot $f(t)$ if its Laplace transform is given by

$$F(s) = \frac{e^{-t_1 s} - e^{-t_2 s}}{2} \qquad t_2 > t_1$$

(b) Repeat (a) for $t_1 > t_2$.
(c) The time function in (a) is a rectangular pulse. Find the Laplace transform of the triangular pulse shown in Figure 3.1.

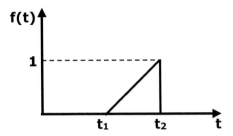

Figure 3.1 Triangular pulse.

Practice Problem 3.9

The Laplace transform of a time function is

$$F(s) = \frac{2s + 5}{s^2 + 3s + 2}$$

(a) Without first solving for *f(t)*, find $df(t)/dt$.

(b) Without first solving for f(t), find $\int_0^1 f(\tau)d\tau$.

(c) Verify the results of (a) and (b) by first solving for f(t) and then performing the operations indicated.

Practice Problem 3.10

Given $f_1(t)$, find $df(t)/dt$

(a) Find $\mathcal{L}[f_1(t)]\mathcal{L}[f_2(t)] = F_1(t)F_2(t)$

(b) Find $\mathcal{L}[f_1(t)f_2(t)] = F_{12}(s)$

(c) Is $\mathcal{L}[f_1(t)]\mathcal{L}[f_2(t)] = \mathcal{L}[f_1(t)f_2(t)]$

(d) Use the convolution integral of Table 3.2 to find the inverse transform of the results in (a).

Practice Problem 3.11

For the functions of Practice Problem 3.3.

(a) Which inverse Laplace transforms do not have final values; that is, for which of the inverse transforms does the $\lim_{t\to\infty} f(t)$ not exist?

(b) Find the final values for those functions that have final values.

Practice Problem 3.12

For the functions of Practice Problem 3.3.

(a) Which inverse Laplace transforms do not have final values? that is, for which of the inverse transforms does the $\lim_{t \to \infty} f(t)$ not exist?
(b) Find the final values for those functions that have final values.

Practice Problem 3.13

Given the differential equation

$$\frac{dx}{dt} + 3x = 5 \quad t \geq 0$$

(a) Find *x(t)* for the case the initial condition is zero.
(b) Show that your solution is correct by substituting the solution in (a) into the differential equation.
(c) Find *x(t)* for the case that x(0) = −2.
(d) Show that your solution is correct by substituting the solution in (c) into the differential equation. Also, show that the solution has the correct initial condition.

Practice Problem 3.14

Given the differential equation

$$\frac{d^2x}{dt^2} + 4\frac{dx}{dt} + 3x = 10u(t)$$

(a) Find *x(t)* for the case that the initial conditions are zero.
(b) Show that your solution is correct by substituting the solution in (a) into the differential equations.
(c) Find *x(t)* for the case that x(0) = 1 and $\dot{x}(0) = -1$.
(d) Show that your solution is correct by substituting the solution in (c) into the differential equation. Also, show that the solution has the correct initial condition.

Practice Problem 3.15

Given the differential equation

$$\frac{d^2x}{dt^2} + 2\frac{dx}{dt} + x = 5 \quad t \geq 0$$

(a) Find x(t) for the case that the initial conditions are zero.

(b) Show that your solution is correct by substituting the solution in (a) into the differential equation.

(c) Find x(t) for the case that $x(0) = 0$ and $\dot{x}(0) = 2$.

(d) Show that your solution is correct by substituting the solution in (c) into the differential equation. Also, show that the solution has the correct initial condition.

Practice Problem 3.16

Find the transfer function $C(s)/R(s)$ for each of the systems described by the differential equations.

(a) $\dot{c}(t) + c(t) = r(t)$

(b) $\ddot{c}(t) + 2\dot{c}(t) + 5c(t) = 2\dot{r}(t) + r(t)$

(c) $\ddot{c}(t) + 2\dot{c}(t) + 5c(t) = 2\dot{r}(t) + r(t - t_0)u(t - t_0)$

Practice Problem 3.17

For each of the transfer functions given, find the system differential equation if

$$G(s) = C(s)/R(s).$$

(a) $G(s) = \frac{5}{s^2(s+1)}$

(b) $G(s) = \frac{5}{s(s+1)^2}$

(c) $G(s) = \frac{s+30}{s(s^2+4s+29)}$

(d) $G(s) = \frac{2s+1}{s^2+2s+10}$

Practice Problem 3.18

(a) Give the system characteristic equation of each of the systems of Practice Problem 3.16.

(b) Give the system characteristic equation of each of the systems of Practice Problem 3.17.

Practice Problem 3.19

Equations (3.4) and (3.5) illustrate the linear properties of the Laplace transform.

This problem illustrates that these linear properties do not carry over to non-linear operations.

(a) Given $f_1(t) = e^{-t}$, $F_2(s) = \mathcal{L}[f_1(t)]$ and $\mathcal{L}[f_1^2(t)]$.

(b) In (a), is $\mathcal{L}[f_1^2(t)].[F(s)]^2$? that is, is the Laplace transform of a squared time function equal to the square of the Laplace transform of that function?

(c) Given $f_1(t) = e^{-t}$ and $f_2(t) = e^{-2t}$, find $F_1(s)$, and $\mathcal{L}[f_1(t)/f_2(t)]$.

(d) In (c), is $\mathcal{L}[f_1(t)/f_2(t)] = F_1(s)/F_2(s)$?

4

System Description

This chapter provides comprehensive details regarding the systems, continuous-time systems and its properties, discrete-time system (DTS) and its properties, block diagram representation of continuous time systems, mathematical modelling of the electrical, mechanical and electromechanical systems. At the end of the chapter, relevant practice problems are given for better understanding.

4.1 System

A system may be defined as a set of elements or functional block which are connected together and produce an output in response to the input signal. The response or output of the system depends upon transfer function of the system's mathematical functional relationship between input and output and may be written as y(t) = f(x,t).

4.2 Properties of Continuous-time System

The properties of continuous-time system are defined without regard to the mathematical models of system. However, the definitions allow us to test a particular mathematical model to determine the properties of that system.

In the following, x(t) denotes the input signal and y(t) denotes the output signal of a system. We show this relation symbolically by

$$x(t) \rightarrow y(t) \tag{4.1}$$

We read this notation as 'x(t) produces y(t)';

$$y(t) = T[x(t)] \tag{4.2}$$

4.2.1 Systems with Memory

First we define a system that has memory.

Memory
A system has memory if its output at time t_0, $y(t_0)$, depends on input values other than $x(t_0)$. Otherwise, the system is memoryless.

A system with memory is also called a dynamic system. An example of a system with memory is an integrating amplifier, described by

$$y(t) = K \int_{-\infty}^{t} x(\tau)d\tau \tag{4.3}$$

The output voltage y(t) depends on all past values of the input voltages x(t). A capacitor also has memory, if its current is defined to be the input and its voltage the output:

$$v(t) = \frac{1}{C} \int_{-\infty}^{t} i(\tau)d\tau = \frac{q(t)}{C} \tag{4.4}$$

In this equation, q(t) is the charge on the capacitor. Thus the charge on a capacitor, $q(t_0)$, is determined by the current i(t) for all time $t \leq t_0$.

A memoryless system is also called a static system. An example of a memoryless system is the ideal amplifier defined earlier. With x(t) as its input and y(t) as its output, the model of an ideal amplifier with (constant) gain K is given by

$$y(t) = Kx(t) \tag{4.5}$$

for all t. A second example is resistance, for which v(t) = R i(t). A third example is a squaring circuit, such that

$$y(t) = x^2(t) \tag{4.6}$$

4.2.2 Invertibility

A system is said to be invertible by first definition like that distinct inputs result in distinct outputs. A second definition of invertibility is that for an invertible system, the system input can be determined from its output. As an example, consider the squaring circuit mentioned earlier, which is described by

$$y(t) = x^2(t) \Rightarrow x(t) = \pm\sqrt{y(t)} \tag{4.7}$$

Suppose that the output of this circuit is constant at 4 V. The input could be either +2 V or −2 V. Hence, this system is not invertible.

An example of an invertible system is an ideal amplifier of gain K:

$$y(t) = Kx(t) \Rightarrow x(t) = \frac{1}{K}y(t) \qquad (4.8)$$

Inverse of a System

A definition related to invertibility is also the inverse of a system. An example of an identity system is an ideal amplifier with a gain of unity. The inverse of a system (denoted by T) is a second system (denoted by T_i) that, when cascaded with the system T, yields the identity system.

The notation for an inverse transformation is then

$$y(t) = T[x(t)] \Rightarrow x(t) = T_i[y(t)] \qquad (4.9)$$

Hence, T [.] denotes the inverse transformation. For an invertible system, in (4.9) we can find the unique x(t) for each y(t). We illustrate an invertible system in Figure 4.1. In Figure 4.1, for $T_2(.) = T_1(.)$

$$z(t) = T_2[y(t)] = T_2(T_1[x(t)]) = x(t) \qquad (4.10)$$

A simple example of the inverse of a system is an ideal amplifier with gain 5.

Then

$$y(t) = T[x(t)] = 5x(t) \Rightarrow x(t) = T_i[y(t)] = 0.2y(t) \qquad (4.11)$$

The inverse system is an ideal amplifier with gain 0.2.

A transducer is a physical device used in the measurement of physical variables. For example, a thermistor (a temperature-sensitive resistor) is one device used to measure temperature. To determine the temperature, we measure the resistance of a thermistor, and use the known temperature–resistance characteristics of that thermistor to determine the temperature.

In measuring physical variables (signals), we measure the effect of the physical variable on the transducer. We must be able to determine the input to the transducer (the physical variable) by measuring the transducer's output (the effect of the physical variable). This cause-and-effect relationship must

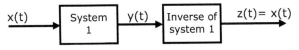

Figure 4.1   Inverse system.

be invertible. A sensor is a transducer followed by its inverse system, and is illustrated in Figure 4.2.

A glass-bulb thermometer is a second example of a transducer. The glass bulb is the system, and the scale attached to the bulb is the inverse system. A change in temperature causes a change in the density of the liquid in the bulb. As a result, the level in the column of the liquid changes. The calibrated scale then converts the liquid level of units of temperature.

From the last example, we see that the output signal of the inverse system seldom has the same units as the system input signal; however, the amplitudes of the two signals are equal.

4.2.3 Causality

A system is causal if the output at any time t_0 is dependent on the input only for $t \leq t_0$.

A causal system is also called a non-anticipatory system. All physical systems are causal.

A filter is a physical device (system) for removing certain unwanted components from a signal. We can design better filters for a signal if both all past values and all future values of the signal are available. In real time (as the signal occurs in the physical system), we never know the future values of a signal. However, if we record a signal and then filter it, the 'future' values of the signal are available. Thus we can design better filters if the filters are to operate only on recorded signal; of course, the filtering is not performed in real time.

A system described by, with t in seconds

$$y(t) = x(t - 2) \tag{4.12}$$

is causal since the present output is equal to the input of 2 s ago. For example, we can realize this system by recording the signal x(t) on magnetic tape. The playback head is then placed 2 s downstream of the tape from the recording

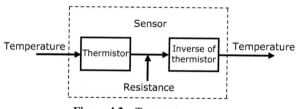

Figure 4.2   Temperature sensor.

head. A system described by (4.12) is called an ideal time delay. The form of the signal is not altered; the signal is simply delayed.

A system described by

$$y(t) = x(t + 2) \tag{4.13}$$

is not causal, since, for example, the output is t = 0 is equal to the input at t = 2 s. This system is an ideal time advance, which is not physically realizable.

Another example of a causal system is illustrated in Figure 4.3. In this system a time delay of 30 s is followed by a time advance of 25 s. Hence the total system is causal and can be realized physically. However, the time-advance part of the system is not causal, but it can be realized if preceded by a time delay of at least 25 s. An example of this type of system is the non-real-time filtering described earlier.

4.2.4 Stability

Many definitions exist for the stability of a system; we now give the bounded-input bounded-output (BIBO) definition.

BIBO Stability

A system is stable if the output remains bounded for any bounded input. By definition, a signal x(t) is bounded if there exists a number M such that

$$|x(t)| \leq M \quad \text{for all } t \tag{4.14}$$

Hence a system is BIBO stable if, for a number R

$$|y(t)| \leq R \quad \text{for all } t \tag{4.15}$$

for all x(t) such that (4.14) is satisfied. Bounded x(t) and y(t) are illustrated in Figure 4.4. To determine the BIBO stability for a given system, given any

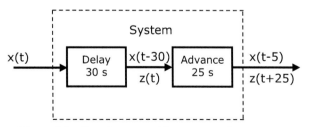

Figure 4.3   Causal system with a non-causal component.

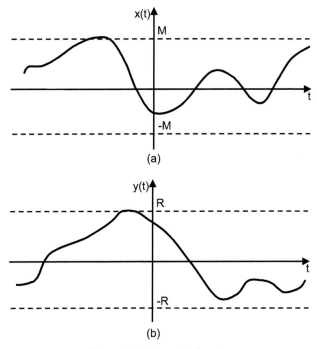

Figure 4.4   Bounded functions.

value M in (4.15), a value R (in general a function of M) must be found such that (4.15) is satisfied.

The identity system y(t) = x(t) is stable, since in (4.14) and (4.15), R = M. The integrating amplifier

$$y(t) = K \int_0^t x(\tau)d\tau \qquad (4.16)$$

is not stable. If the input x(t) is the unit step u(t), the output is the ramp function y(t) = Kt, which is unbounded. In (4.14), $|x(t)| \leq 1$, and in (4.15)

$$\lim_{t \to \infty} y(t) = \lim_{t \to \infty} Kt = \infty$$

$$y(t) = x^2(t) \qquad (4.17)$$

is stable, since y(t) is bounded for any bounded input. In (4.16) and (4.17), R = M^2. Stability is a basic property required of almost all physical systems. Generally, a system that is not stable cannot be controlled and is of no value.

An example of an unstable system is a public address system that has broken into oscillation; the output of this system is unrelated to its input.

A second example of an unstable system has been seen several times in television news segments: the first stage of a space booster or a missile that went out of control (unstable) and had to be destroyed.

4.2.5 Time Invariance

A system is said to be time invariant if a time shift in the input signal results only in the same time shift in the output signal. For a time-invariant system for which the input x(t) produces the output y(t) [x(t) $\rightarrow$ y(t)], then x(t $-$ t$_0$) produces y(t $-$ t$_0$), that is

$$x(t - t_0) \rightarrow y(t - t_0) \tag{4.18}$$

for all t$_0$. A time-invariant system is also called a fixed system. A test for time invariance is given by

$$x(t)\,|_{t-t_0} = y(t)|_{x(t-t_0)} \tag{4.19}$$

provided that y(t) is expressed as an explicit function of x(t). This test is illustrated in Figure 4.5. The signal y(t $-$ t$_0$) is obtained by delaying y(t) by t$_0$ seconds. Define y$_d$(t) as the system output for the delayed input x(t $-$ t$_0$), such that

$$x(t - t_0) \rightarrow y_d(t) \tag{4.20}$$

The system in Figure 4.47 is time invariant provided that

$$y(t - t_0) = y_d(t) \tag{4.21}$$

The left side of this expression is evaluated with each t replaced with (t $-$ t$_0$), and the right side is evaluated with each x(t) replaced with x(t $-$ t$_0$). A system that is not time invariant is time varying.

As an example of time invariance, consider the following system:

$$y(t) = \sin[x(t)]$$

In (4.19) and (4.21)

$$y_d(t) = y(t)\,\big|_{x(t-t_0)} = \sin[x(t - t_0)] = y(t)\big|_{t-t_0}$$

and the system is time invariant.

Consider next the system

$$y(t) = e^{-t}x(t)$$

In (4.19) and (4.21)

$$y_d(t) = y(t)|_{x(t-t_0)} = e^{-t}x(t - t_0)$$

and

$$y(t)|_{t-t_0} = e^{-(t-t_0)}x(t - t_0)$$

The last two expressions are not equal; therefore, (4.21) is not satisfied and the system is time varying.

An example of a time-varying physical system is the booster stage of the National Aeronautics and Space Administration (NASA) shuttle. Newton's second law applied to the shuttle yields

$$f_z(t) = M(t)\frac{d^2 z(t)}{dt^2} \tag{4.22}$$

where the force $f_z(t)$ is the engine thrust in the z-axis direction developed by burning fuel, $M(t)$ is the mass of the shuttle and $z(t)$ is the position along the z-axis. The mass decreases as the fuel burns, and hence is a function of time.

A time-varying system has characteristics that vary with time. The manner in which this type of system responds to a particular input depends on the time that the input is applied.

In designing controllers for automatic control systems, we sometimes intentionally use time-varying gains to improve the system characteristics. Of course, time-varying gains result in a time-varying system. For example, the control system that automatically lands aircraft on US Navy aircraft

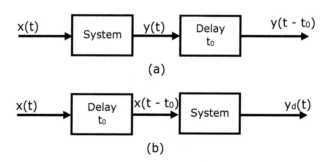

(a)

(b)

Figure 4.5   Test for time invariance.

carriers uses time-varying gains. The time-varying gains are based on time-to-touchdown, and the gains increase as the time-to-touchdown decreases. This increase in gain results in a decreased time-of-response in correcting errors in the plane's flight path.

4.2.6 Linearity

The property of linearity is one of the most important properties that we consider. Once again we define the system input signal to be x(t) and the output signal to be y(t).

Linear System
A system is linear if it meets the following two criteria:

1. Additivity: If $x_1(t) \rightarrow y_1(t)$ *and* $x_2(t) \rightarrow y_2(t)$, then

$$x_1(t) + x_2(t) \rightarrow y_1(t) + y_2(t) \tag{4.23}$$

2. Homogeneity: If $x_1(t) \rightarrow y_1(t)$, then

$$ax_1(t) \rightarrow ay_1(t) \tag{4.24}$$

where a is a constant. These criteria must apply for all $x_1(t)$ and $x_2(t)$ and for all a. These two criteria can be combined to understand the principle of superposition. A system satisfies the principle of superposition if, with the inputs and outputs as defined earlier

$$a_1 x_1(t) + a_2 x_2(t) \rightarrow a_1 y_1(t) + a_2 y_2(t) \tag{4.25}$$

where a_1 and a_2 are constants. A system is linear if it satisfies the principle of superposition. No physical system is linear under all operating conditions. However, a physical system can be tested using (4.26) to determine ranges of operation for which the system may be approximately linear.

An example of a linear system is an ideal amplifier, described by y(t) = Kx(t). An example of a non-linear system is the squaring circuit mentioned earlier:

$$y(t) = x^2(t)$$

For inputs of $x_1(t)$ and $x_2(t)$, the outputs are

$$x_1(t) \rightarrow x_1^2(t) = y_1(t) \tag{4.26}$$
$$x_2(t) \rightarrow x_2^2(t) = y_2(t)$$

However, the input $[x_1(t) + x_2(t)]$ produces the output

$$x_1(t) + x_2(t) \rightarrow [x_1(t) + x_2(t)]^2 = x_1^2(t) + 2x_1(t)x_2(t) \quad (4.27)$$
$$+x_2^2(t) = y_1(t) + y_2(t) + 2x_1(t)x_2(t)$$

and $[x_1(t) + x_2(t)]$ does not produce $[y_1(t) + y_2(t)]$. Hence the squaring circuit is nonlinear.

A linear time-invariant (LTI) system is a linear system that is also time invariant. The LTI system, for both continuous-time and DTS, is the type that is emphasized in this book.

An important class of continuous-time LTI systems is that which is modelled by linear differential equations with constant coefficients. An example of this type of system is the RL circuit, which is modelled as follows:

$$L\frac{di(t)}{dt} + Ri(t) = v(t)$$

Example 4.7
Testing for properties for a particular system
The characteristics for the system

$$y(t) = \sin 2t\, x(t)$$

are now investigated.

Solution 4.7
Note that this system can be considered to be an amplifier with a time-varying gain that varies between -1 and 1, that is, with the gain $K(t) = \sin 2t$ and $y(t) = K(t)\, x(t)$.

1. This system is memoryless, since the output is a function of the input at only the present time.
2. The system is not invertible, since, for example, $y(\pi) = 0$, regardless of the value of the input. Hence, the system has no inverse.
3. The system is causal, since the output does not depend on the input at a future time.
4. The system is stable, since the output is bounded for all bounded inputs. If $|x(t) \leq M, |y(t) \leq M$ also.
5. The system is time varying. From (4.26) and (4.27)

$$y_d(t) = y(t)|_{x(t-t_0)} = \sin 2t\, x(t - t_0)$$

and $\qquad y(t)|_{t-t_0} = \sin 2(t - t_0)x(t - t_0).$

6. The system is linear, since

$$a_1x_1(t) + a_2x_2(t) \rightarrow \sin 2t[a_1x_1(t) + a_2x_2(t)]$$
$$= a_1 \sin 2tx_1(t) + a_2 \sin 2tx_2(t)$$
$$= a_1y_1(t) + a_2y(t)$$

Example 4.8
Consider the system described by the equation y(t) = bx(t).

Solution 4.8
This system is easily shown to be linear using superposition. However, the system y(t) = [bx(t) + c] is nonlinear. By superposition

$$y(t) = b[a_1x_1(t) + a_2x_2(t)] + c \neq y_1(t) + y_2(t)$$

The most important general system properties are linearity and time invariance, since the analysis and design procedures for LTI systems are simplest and easiest to apply.

4.3 Discrete-Time Systems

A DTS shown in Figure 4.6 is a device that operates on discrete-time signals (input), according to some rules, to produce another discrete-time signal (output or response) of the systems.

4.3.1 System's Representation

A DTS can be represented by two ways:

(a) Difference Equations
(b) Block diagram

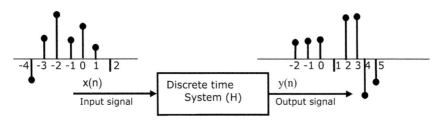

Figure 4.6 Representation of a discrete time system

4.4 Symbol Used to Represent DTS

The operation of a DTS may be described simply by drawing a block schematic. We use different building block to form a complete schematic.

(a) Adder
(b) Constant Multiplier
(c) Signal Multiplier
(d) Unit delay
(e) Unit advance

4.4.1 An Adder

Addition operation shown in Figure 4.7 is memoryless.

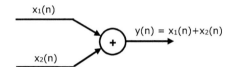

Figure 4.7   Graphical representation of adder

4.4.2 A Constant Multiplier

A constant multiplier is simply represented by applying a scale factor on input shown in Figures 4.8 and 4.9. It is also memoryless.

$$x(n) \quad a \quad y(n) = ax(n)$$

Figure 4.8   Graphical representation of a constant multiplier

4.4.3 A Signal Multiplier

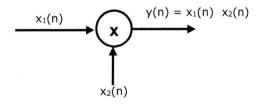

Figure 4.9   Graphical representation of a multiplier

4.4.4 Unit Delay Element

It is the system that simply delays the signal passing through it by one sample and is shown in Figure 4.10. If input signal is x(n), then the output is x(n−1). The sample x(n−1) is stored in memory at discrete time n-1 and can be recalled at time n to form $y(n) = x(n-1)$.

The symbol of unit delay $= z^{-1}$

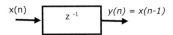

Figure 4.10   Unit delay element

4.4.5 Unit Advanced Element

A unit advance moves the input x(n) ahead by one sample in time to yield x(n+1). This advancement is shown in Figure 4.11 denoted by z.

x(n) → [z] → y(n) = x(n+1)

Figure 4.11   Unit advanced element

Example 4.9

Sketch the block diagram

$y(n) = \frac{1}{4}\,y(n-1) + \frac{1}{2}\,x(n) + \frac{1}{2}\,x(n-1)$ where x(n) is the input y(n) is the output.

Solution 4.9

$y(n) = 0.5x(n) + 0.5x(n-1) + 0.25y(n-1)$

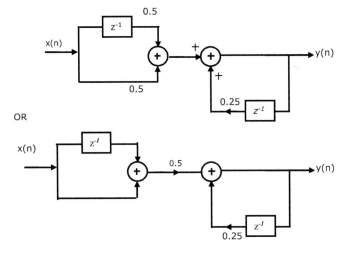

Example 4.10

A DTS is realized by the structure shown in the figure provided. Determine a realized for its inverse system, that is, the system which produces x(n) as an output when y(n) is used as an input.

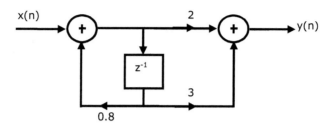

Solution 4.10

The inverse system is characterized by the difference equation:

$$x(n) = -1.5x(n-1) + \tfrac{1}{2}y(n) - 0.4y(n-1)$$

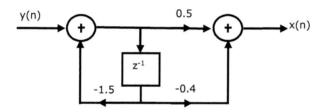

4.5 Properties of DTS

In designing the systems, the general characteristic of the systems has to be considered. A number of properties or categories that can be used to develop general characteristics of the system are now explained.

4.5.1 Static Versus Dynamic Systems

A DTS is called static or memoryless if its output at any instant *n* depends at most on the input sample at the same time, but not on past or future samples of the input. In any other case, the system is said to be dynamic or to have memory. If the output of a system at time *n* is completely determined by the input samples in the interval from n − N to n (N ≥ 0), the system is said to have memory of duration N. If N = 0, the system is static. If 0<N<∞, the system is said to have finite memory, whereas if $N = \infty$, the system is said to have infinite memory.

The systems described by the following input–output equations

$$y(n) = a\,x(n) \tag{4.28}$$

$$y(n) = n\,x(n) + bx^3(n) \tag{4.29}$$

are both static or memoryless. Note that there is no need to have stored any of the past inputs or outputs in order to compute the present output. On the other hand, the systems described by the following input–output relations are dynamic systems or systems with memory.

$$y(n) = x(n) + 3x(n-1) \tag{4.30}$$

$$y(n) = \sum_{k=0}^{n} x(n-k) \tag{4.31}$$

$$y(n) = \sum_{k=0}^{\infty} x(n-k) \tag{4.32}$$

The systems described by (4.30) and (4.31) have finite memory, whereas the system described by (4.32) has infinite memory.

4.5.2 Time Invariant Versus Time-Variant System

The system is said to be time invariant if its input–output characteristics do not change with time.

Method to Work Out for Time Invariant and Time-Variant System
Consider the following system

$$y(n) = H[x(n)] \tag{4.33}$$

If the input signal is delayed by k unit in the function only and again delayed by k unit in the overall system irrespective that it is a function or not. If $y(n,k) = y(n - k)$ it means that the characteristics of system do not change with time, and the system is time invariant.

$$x(n) \xrightarrow{H} y(n) \quad \text{relaxed}$$

$$x(n-k) \xrightarrow{H} y(n-k) \quad k \text{ shift} \tag{4.34}$$

y(n,k) means the delay is to be given in function only, while y(n−k) means that wherever n is existing it has to be replaced by n−k unit. If $y(n, k) = y(n − k)$, then the system is called time invariant.

Example 4.11
Determine if the shown in Figure 4.12 are time invariant or time variant.

Solution 4.11

(a) This system is described by the input–output equations

$$y(n) = H[x(n)] = x(n)\, x(n − 1) \tag{1}$$

Now if the input is delayed by k units in time and applied to the system, it is clear from the block diagram that the output will be

$$y(n, k) = x(n − k)\, x(n − k − 1) \tag{2}$$

On the other hand, from (1) we note that if we delay y(n) by k units in time, we obtain

$$y(n − k) = x(n − k)\, x(n − k − 1) \tag{3}$$

Since the right-hand sides of (2) and (3) are identical, it follows that y(n,k) = y(n−k). Therefore, the system is time invariant.

(b) The input–out equation for this system is

$$y(n) = H[x(n)] = nx(n) \tag{4}$$

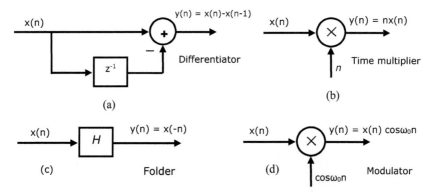

Figure 4.12   Time-invariant (a) and some time-variant systems (b)−(c)−(d)

The response of this system to x(n−k) is

$$y(n, k) = nx(n - k) \tag{5}$$

Now if we delay y(n) in (4) by k units in time, we obtain

$$y(n - k) = (n - k)x(n - k) = nx(n - k)kx(n - k) \tag{6}$$

This system is time variant, since y(n,k) ≠ y(n−k).

(c) This system is described by the input–out relation

$$y(n) = H[x(n)] = x(-n) \tag{7}$$

The response of this system to $x(n - k)$ is $y(n, k) = H[x(n - k)]$

$$= x(-n - k) \tag{8}$$

Now, if we delay the output y(n), as given by (7), by k units in time, the result will be

$$y(n - k) = x(-n + k) \tag{9}$$

Since y(n, k) ≠ y(n−k), the system is time variant.

(d) The input–output equation for this system is

$$y(n) = x(n) \cos \omega_0 n \tag{10}$$

The response of this system to $x(n - k)$ is $y(n, k) = x(n - k)$

$$\cos \omega_0 n \tag{11}$$

If the expression in (10) is delayed by k units and the result is compared to (11), it is evident that the system is time variant.

Example 4.12

Check the DTS for time invariance which is described by the following difference equation.

$$y(n) = 4nx(n)$$

Solution 4.12
The response to a delayed input is

$$y(n, k) = 4nx(n - k)$$

The delayed response will be

$$y(n - k) = 4(n - k)x(n - k)$$

it is clear that both responses are not equal, i.e.

$$y(n, k) \# y(n - k)$$

Therefore, the given DTS y(n)=4 nx(n) is not time invariant. It is a time-varying system.

Example 4.13
Check the DTS for time invariance which is described by the following difference equation:

$$y(n) = ax(n - 1) + bx(n - 2)$$

Solution 4.13
The response to a delayed input is

$$y(n, k) = ax(n - k - 1) + bx(n - k - 2)$$

The delayed response will be

$$y(n - k) = ax((n - k) - 1) + bx((n - k) - 2)$$

It is clear that both responses are not equal, i.e.

$$y(n, k) \# y(n - k)$$

Therefore, the given DTS $y(n) = a\, x(n-1) + b\, x(n-2)$ is time variant.

4.5.3 Linear Versus Non-Linear System

4.5.3.1 Linear system
A linear system is that which satisfies the properties of superposition theorem. The response of a system to a weighted sum of signal is equal to the corresponding weighted sum of the response of each individual input signals.

$$H[a_1 x_1(n) + a_2 x_2(n)] = a_1 H[x_1(n)] + a_2 H[x_2(n)] \qquad (4.35)$$

4.5.3.2 Non-linear system

If a system produces a non-zero output with a zero input, the system may be either non-relaxed or non-linear. If a relaxed system does not satisfy the superposition principle it is non-linear.

Example 4.14

Determine if the systems described by the following input–output equations are linear or non-linear.

(a) $y(n) = nx(n)$ (b) $y(n) = x(n^2)$ (c) $y(n) = x^2(n)$

(d) $y(n) = Ax(n) + B$ (e) $y(n) = ex(n)$

Solution 4.14

(a) For two input sequences $x_1(n)$ and $x_2(n)$ as shown in Figure 4.13, the corresponding outputs are

$$y_1(n) = nx_1(n)$$
$$y_2(n) = nx_2(n)$$

(1)

A linear combination of the two input sequence results in the following output

$$y_3(n) = H[a_1x_1(n) + a_2x_2(n)] = n[a_1x_1(n) + a_2x_2(n)]$$
$$= a_1nx_1(n) + a_2nx_2(n)$$

(2)

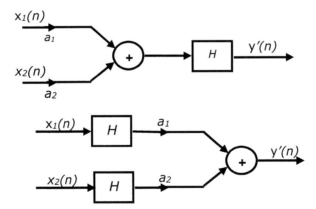

Figure 4.13 Graphical representation of the superposition principle H is linear if and only if $y(n) = y'(n)$.

On the other hand, a linear combination of the two outputs in (2) results in the output

$$a_1 y_1(n) + a_2 y_2(n) = a_1 n x_1(n) + a_2 n x_2(n) \tag{3}$$

Since the right-hand sides of (2) and (3) are identical, the system is linear.

(b) As in part (a), we find the response of the system to two separate input signals $x_1(n)$ and $x_2(n)$. The result is

$$y_1(n) = x_1(n^2)$$
$$y_2(n) = x_2(n^2) \tag{4}$$

The output of the system to a linear combination of $x_1(n)$ and $x_2(n)$ is

$$y_3(n) = H[a_1 x_1(n) + a_2 x_2(n)] = a_1 x_1(n^2) + a_2 x_2(n^2) \tag{5}$$

Finally, a linear combination of the two outputs in (2.21) yields

$$a_1 y_1(n) + a_2 y_2(n) = a_1 x_1(n^2) + a_2 x_2(n^2) \tag{6}$$

By comparing (5) with (6), we conclude that the system is linear.

(c) The output of the system is the square of the input. (Electronic devices that have such an input–output characteristic are called square-law devices.). From our earlier discussion it is clear that such a system is memoryless.

To illustrate that this system is non-linear, the responses of the system to two separate input signals are

$$y_1(n) = x_1^2(n) y_2(n) = x_2^2(n) \tag{7}$$

The response of the system to a linear combination of these two input signals is

$$y_3(n) = H[a_1 x_1(n) + a_2 x_2(n)]$$
$$= [a_1 x_1(n) + a_2 x_2(n)]^2 \tag{8}$$
$$= a_1^2 x_1^2(n) + 2 a_1 a_2 x_1(n) x_2(n) + a_2^2 x_2^2(n)$$

On the other hand, if the system is linear, it would produce a linear combination of the two outputs in (7), namely

$$a_1 y_1(n) + a_2 y_2(n) = a_1 x_1^2(n) + a_2 x_2^2(n) \tag{9}$$

Since the actual output of the system, as given by (8), is not equal to (9), the system is non-linear.

(d) Assuming that the system is excited by $x_1(n)$ and $x_2(n)$ separately, we obtain the corresponding outputs

$$y_1(n) = Ax_1(n) + B$$
$$y_2(n) = Ax_2(n) + B \tag{10}$$

A linear combination of $x_1(n)$ and $x_2(n)$ produces the output

$$y_3(n) = H[a_1 x_1(n) + a_2 x_2(n)]$$
$$= A[a_1 x_1(n) + a_2 x_2(n)] + B \tag{11}$$
$$= Aa_1 x_1(n) + a_2 Ax_2(n) + B$$

On the other hand, if the system were linear, its output to the linear combination of $x_1(n)$ and $x_2(n)$ would be a linear combination of $y_1(n)$ and $y_2(n)$, that is

$$a_1 y_1(n) + a_2 y_2(n) = a_1 Ax_1(n) + a_1 B + a_2 Ax_2(n) + a_2 B \tag{12}$$

Clearly, (11) and (12) are different and hence the system fails to satisfy the linearity test.

The reason that this system fails to satisfy the linearity test is not that the system is non-linear (in fact, the system is described by a linear equation) but it is the presence of the constant B. Consequently, the output depends on both the input excitation and on the parameter $B \neq 0$. Hence, for $B \neq 0$, the system is not relaxed. If we set $B = 0$, the system is now relaxed and the linearity test is satisfied.

(e) Note that the system described by the input–output equation

$$y(n) = e^{x(n)}$$

is relaxed. If $x(n) = 0$, we find that $y(n) = 1$. This is an indication that the system is non-linear. This, in fact, is the conclusion reached when the linearity test, as described earlier, is applied.

Example 4.15
Determine if the system described by the following input–output equations is linear or non-linear.

$$y(n) = x^2(n) - x(n-1)x(n+1)$$

Solution 4.15

For two input sequences $x_1(n)$ and $x_2(n)$, the corresponding outputs are $y_1(n)$ and $y_2(n)$:

$$y_1(n) = x_1^2(n) - x_1(n-1)x_1(n+1)$$
$$y_2(n) = x_2^2(n) - x_2(n-1)x_2(n+1) \tag{1}$$

$$a_1y_1(n) + a_2y_2(n) = [a_1x_1^2(n) - x_1(n-1)a_1x_1(n+1)]$$
$$+ [a_2x_2^2(n) - x_2(n-1)a_2x_2(n+1)]$$

A linear combination of the two input sequence results in the output

$$y_3(n) = H[a_1x_1(n) + a_2x_2(n)]$$
$$= [(a_1x_1(n) + a_2x_2(n)]^2 - [(a_1x_1(n-1)a_2x_2(n-1)]$$
$$+[(a_1x_1(n+1)a_2x_2(n+1)]$$
$$= a_1^2[(x_1^2(n) + x_1(n-1)x_1(n+1)] \tag{2}$$
$$+a_2^2[x_2^2(n) - x_2(n-1)x_2(n+1)]$$
$$+a_1a_2[2x_1(n)x_2(n) - x_1(n-1)x_2(n+1) - x_1(n+1)x_2(n-1)]$$

Since the right-hand sides of (1) and (2) are not identical, the system is non-linear.

Example 4.16

A DTS is represented by the following difference equation in which x(n) is input and y(n) is output. Determine if the system described by the input–output equations is linear or non-linear.

$$y(n) = 3y^2(n-1) - nx(n) + 4x(n-1) - 2x(n+1)$$

Solution 4.16

The given expression is $y(n) = 3\,y^2\,(n-1) - n\,x(n) + 4\,x(n-1) - 2\,x(n+1)$.

It may be noted that the real condition for linearity is $H[ax(n)] = a\,H[x(n)]$.

$$H[ax(n)] = ay(n) = 3a^2y^2(n-1) - anx(n) + 4ax(n-1) - 2ax(n+1)$$
$$aH[x(n)] = a[y(n)] = 3ay^2(n-1) - anx(n) + 4ax(n-1) - 2ax(n+1)$$

From the above it is clear that $H[ax(n)] \# a\,H[x(n)]$. The system is non-linear.

4.5.4 Causal vs Non-Causal System

A system is said to be causal if the output of the system at any time n depends only on present and past inputs [i.e. x(n), x(n−1), x(n−2) ......], but does not depends on future inputs [x (n +1), x (n +2) ...].

$$y(n) = F[x(n), x(n-1), x(n-2)...] \tag{4.36}$$

If a system does not satisfy this definition, it is called non-causal.

Example 4.17

Determine if the systems described by the following input–output equations are causal or non-causal.

(a) $y(n) = x(n) - x(n-1)$

(b) $y(n) = ax(n)$

(c) $y(n) = x(n) + 3x(n+4)$

(d) $y(n) = x(n^2)$

(e) $y(n) = x(2n)$

(f) $y(n) = x(-n)$

Solution 4.17

The systems described in parts (a) and (b) are clearly causal, since the output depends only on the present and past inputs.

On the other hand, the systems in parts (c), (d) and (e) are clearly non-casual, since the output depends on future values of the input.

The system in (f) is also non-causal, as we note by selecting, for example, n = −1, which yields y(−1) = x(1). Thus the output at n = −1 depends on the input at n = 1, which is two units of time into the future.

Example 4.18

A DTS is represented by the following difference equation in which x(n) is input and y(n) is output. Determine if the system described by the input–output equations is causal or non-causal.

$$y(n) = 3y^2(n-1) - nx(n) + 4x(n-1) - 2x(n+1)$$

Solution 4.18

It may be noted that the required condition for causality is that the output of a causal system must be dependent only on the present and past values of the input. From the given equation, it is obvious that output y(n) is dependent on future sample value x(n+1).

4.5.5 Stable Versus Unstable System

An arbitrary relaxed system is said to be BIBO stable if and only if every bounded input produces a bounded output.

By definition, a signal x(n) is bounded if there exists a member M such that

$$|x(n)| \le M \text{ for all } n \tag{4.37}$$

Hence a system is BIBO stable if, for a number R

$$|y(n)| \le R \text{ for all } x(n) \tag{4.38}$$

If for some bounded input sequence x(n), the output is unbounded (infinite), the system is unstable. A system is stable if the output remains bounded for any bounded input.

Example 4.19
Check whether the following systems are BIBO stable or not.

(a) $y(n) = ax^2(n)$

(b) $y(n) = ax(n) + b$

Solution 4.19

(a) The given expression is $y(n) = ax^2(n)$
If $x(n) = \delta(n)$, *then* $y(n) = h(n)$. Thus the impulse response is given by $h(n) = a\delta^2(n)$.
Now
when n = 0, h(0) = a $\delta^2(0)$ = a
when n=1, h(1) = a $\delta^2(1)$ = 0
In general, we have h(n) = a when n = 0, h(n) = 0 when n ≠ 0.
We know the necessary and sufficient condition for BIBO stability is expressed as

$$\sum_{n=0}^{\infty} |h(k)| < \infty.$$

Here, we have

$$\sum_{n=0}^{\infty} |h(k)| = |h(0)| + |h(1)| + |h(2)| + \ldots |h(k)| + \ldots = |\alpha|$$

Therefore, we conclude that the given system is BIBO stable only if $\alpha < \infty$

(b) The given system *is* $y(n) = ax(n) + b$.
If $x(n) = \delta(n)$, *then* $y(n) = h(n)$. Thus the impulse response is given
by $h(n) = a\delta(n) + b$
when n=0, h(0)=a δ(0)+b=a+b; when n=1, h(1)=a δ(1)+b=b.
Now, h(1) $=$ h(2) $=$ .... $=$ h(k) $=$ b. Therefore, we have
h(n)=a+b when n=0, h(n)=b when n$\neq$0. Also, we know the necessary and sufficient condition for the BIBO stability is expressed as
$\sum_{n=0}^{\infty} |h(k)| < \infty$ Here we have

$$\sum_{n=0}^{\infty} |h(k)| = |h(0)| + |h(1)| + |h(2)| + ... |h(k)| + ...$$

$$\sum_{n=0}^{\infty} |h(k)| = |a + b| + |b| + |b| + ... |b| + ...$$

Therefore, we conclude from the above expression that this series
never converges since the ration between the successive terms is one.
Therefore, the given system is BIBO unstable.

4.6 Systems' Mathematical Model

4.6.1 Electrical Systems

The variables involved in electrical systems are usually current i and voltage
v. Charge q is the fundamental quantity but usually does not appear in
engineering systems, as it is related to current by the equation:

$$i = \frac{Cdv}{dt}$$

The basic components in electrical engineering systems are the resistor,
the inductor L and the capacitor C. From the systems' view point these
components are described by the relationship between the variable (voltage
and current) at their terminals.

4.6.1.1 The resistor R

$$v = iR$$

$$I = \frac{V}{R}$$

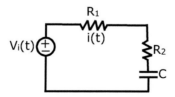

4.6.1.2 The inductor L

$$v = LDi$$

$$i = \frac{I}{LD}v$$

$$I = \frac{I}{sL}V$$

4.6.1.3 The capacitor C

$$V = \frac{I}{CD}i$$

$$i = CD\,v$$

$$I = Cs\,V$$

Example 4.20

Determine the differential equation and its Laplace transformed form for the electrical systems shown in figure.

Solution 4.20

$$R_1 i(t) + R_2 i(t) + \frac{1}{C}\int i\tau\,dt = v_i(t) \quad R_1 I(s) + R_2 I(s) + \frac{1}{sC}I(s) = V_1(s)$$

Example 4.21

Determine the differential equation relating v_o to v_I for the electrical systems shown in figure.

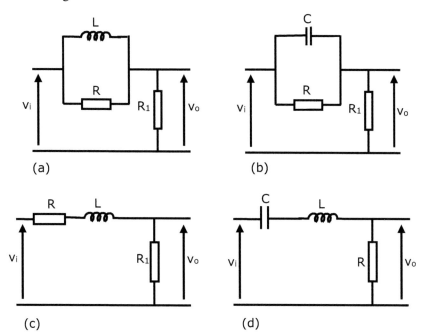

(a)   (b)

(c)   (d)

Solution 4.21

With reference to Figure (a), let the currents in components L, R and R_1 be i_1, i_2 and i_3, respectively, all flowing towards the node v_0. Then

$$i_1 + i_2 + i_3 = 0 \tag{1}$$

$$v_1 + v_0 = L\frac{di_1}{dt}, \quad \frac{di_1}{dt} = \frac{v_1 - v_0}{L}$$

$$\frac{v_1 - v_0}{R} = i_2, \quad -\frac{v_0}{R_1} = i_3$$

Differentiating Equation (1) and substituting for di_1/dt, di_2/dt and di_3/dt

$$\frac{v_1 - v_0}{R} + \frac{1}{R}\frac{dv_1}{dt} - \frac{1}{R}\frac{dv_0}{dt} - \frac{1}{R_1}\frac{dv_0}{dt} = 0$$

$$T_1\frac{dv_0}{dt} + v_0 = T_2\frac{dv_1}{dt} + v_1$$

where $T_1 = L\left[\frac{1}{R_1} + \frac{1}{R_2}\right]$,　$T_2 = \frac{L}{R}$

With reference to Figure (b) let the currents in the components C, R and R_1 be i_1, i_2 and i_3, respectively, all flowing towards the node v_0. Then

$$i_1 = C\frac{d(v_1 - v_0)}{dt}, \quad i_2 = \frac{v_1 - v_0}{R}, \quad i_3 = -\frac{v_0}{R_1}$$

$$i_1 + i_2 + i_3 = 0$$

Substituting for i_1, i_2 and i_3

$$C\frac{dv_1}{dt} - C\frac{dv_0}{dt} + \frac{v_1}{R} - \frac{v_0}{R} - \frac{v_0}{R} = 0$$

$$T\frac{dv_0}{dt} + v_0 = T\frac{dv_1}{dt} + kv_1$$

where $T = \frac{CR_1R_2}{R_1+R_2}, = k = \frac{R_1}{R_1+R_2}$

Referring to Figure (c) and applying Kirchhoff's law with mesh current i, we have

$$v_1 = iR + L\frac{di}{dt} + v_0$$

But $i = C\frac{dv_0}{dt}, \frac{di}{dt} = Ci = C\frac{d^2v_0}{dt^2}$

Substituting $LC\frac{d^2v_0}{dt^2} + RC\frac{dv_0}{dt} + v_0 = v_1$

Referring to Figure (d), let the mesh current be i and the voltage across the capacitor be v_0. Then

$$i = C\frac{dv_0}{dt}, \quad v_c = \frac{1}{C}\int i\,dt$$

Applying Kirchhoff's voltage law

$$v_1 = \frac{1}{C}\int i\,dt + L\frac{di}{dt} + v_0$$

Differentiating

$$\frac{dv_i}{dt} = \frac{1}{C}i + L\frac{d^2i}{dt^2} + \frac{dv_0}{dt}$$

But

$$i = \frac{v_0}{R}, \frac{d^2i}{dt^2} = \frac{1}{R}\frac{d^2v_0}{dt^2}; \quad \text{substituting} \quad \frac{L}{R}\frac{d^2v_0}{dt^2} + \frac{v_0}{dt} + \frac{1}{RC}v_0 = \frac{dv_1}{dt}$$

Example 4.22

The circuit shown in the figure provided is that of an oscilloscope probe. R_I and C_I represent the input resistance and capacitance of the Y amplifier. R and C represent the resistance and capacitance of the robe (C is adjustable).

Obtain the differential equation relating v_o to v_I. If the probe capacitance C is adjusted such that $CR = C_iR_I$ show that a solution of this differential equation is $v_o = R_iv_i/(R_i + R)$. What is the practical implication of this result?

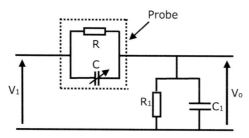

Probe

Solution 4.22

Referring to the figure and denoting the currents through R, C, R_1 and C_1 by i_1, i_2, i_3 and i_4, respectively (all currents flow towards the node at v), then

$$i_1 + i_2 + i_3 + i_4 = 0$$

$$i_1 = \frac{v_i - v_0}{R}, \quad i_2 = C\frac{d(v_i - v_0)}{dt}$$

$$i_3 = \frac{v_0}{R_1}, \quad i_4 = C_1\frac{dv_0}{dt}$$

Substituting

$$\frac{v_i - v_0}{R} + C\frac{dv_i}{dt} + C\frac{dv_0}{dt} - \frac{v_0}{R_1} - C_i\frac{dv_0}{dt} = 0$$

$$\frac{(C + C_1)RR_i}{R + R_i}\frac{dv_0}{dt} + v_0 = \frac{CRR_i}{R + R_1}\frac{dv_i}{dt} + \frac{R_i}{R + R_1}v_i$$

Writing $CR = C_iR_1 = T$ this reduces to

$T\frac{dv_0}{dt} + v_0 = \frac{R_i}{R+R_1}\left[T\frac{dv_i}{dt} + v_i\right]$, and direct substitution shows that a solution

$$v_0 = \frac{R_i}{R + R_1}v_i$$

fits the equation. The practical implication is that if the probe is adjusted such that $C_iR_i = CR$ then the relationship between input and output is an algebraic (as opposed to a differential) equation. The effect of the input capacitance of the oscilloscope has been eliminated. This advantage has been obtained at the expense of additional attenuation by a factor $R_i/(R+R_i)$.

4.6.2 Mechanical Translational Systems

Mechanical systems are in general concerned with the motion of a mechanical assembly in three-dimensional spaces. Usually the motion will combine the translation and rotational motion with reference to certain framework. However, for the purpose of this book motion will be either considered as translational along a fixed direction or rotational about a fixed co-ordinate axis. The signals (variable) involved are force, displacement, velocity and acceleration for the translational along a fixed direction or rotational about a fixed co-ordinate axis. The signals (variables) involved are force, displacement, velocity and acceleration for the translational system and torque, angular displacement, angular velocity and acceleration for the rotational system. It is convenient to consider separately the elements involved in the translational and rotational motion.

4.6.2.1 The mass element

$$f = MD^2x = Mx^{\bullet\bullet}$$

4.6.2.2 The damper element

$$f = BD$$
$$f = B(x_1^{\bullet} - x_2^{\bullet})$$

4.6.2.3 The spring element

$$f = K(x_1 - x_2)$$

It should be noted that mass and force, wherever it is connected in the system to that node, the other end of the mass and force is connected to ground (keeping in mind the concepts of acceleration due to gravity treated to the ground).

Example 4.23

The following figure shows three mechanical systems. Derive the differential equation relating applied force f(t) to displacement x(t) for the system shown in Figure (a).

For the system shown in Figures (b) and (c) derive the relationship between input displacement $x_I(t)$ and the displacement x(t).

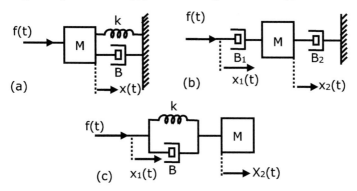

Solution 4.23

Referring to Figure (a)

$$f(t) = Mx^{\bullet\bullet} + Bx^{\bullet} + Kx$$

Referring to Figure (b)

$$f(t) = B_1(x_1^{\bullet} - x_2^{\bullet})$$
$$0 = Mx_2^{\bullet\bullet} + B_2x_2^{\bullet} + B_1(x_2^{\bullet} - x_1^{\bullet})$$

Referring to Figure (c)

$$f(t) = B_{(}x_1^{\bullet} - x_2^{\bullet}) + K_1(x_1 - x_2)$$
$$0 = Mx_2^{\bullet\bullet} + B(x_2^{\bullet} - x_1^{\bullet}) + K(x_2 - x_1)$$

Example 4.24

The following figure shows a simplified version of an accelerometer. A mass M is suspended by two springs (stiffness each k/2) within the body of the accelerometer. Damping is provided giving a damping force B per unit velocity (mass relative to body). The accelerometer output z is the difference between mass displacement and body displacement.

$$z = y - x$$

Derive the differential equation relating accelerometer output z to body acceleration $d^2 \, x/dt^2$.

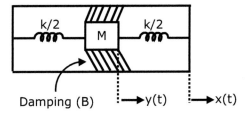

Solution 4.24

Referring to the figure

$$f = Mx^{\bullet\bullet} + 2\frac{k}{2}(x - y) + Bx^{\bullet}$$

$$= -kz$$

Total force due to damper $= B\frac{d}{dt}(x - y) = -B\frac{dz}{dt}$

$$-kz + -B\frac{dz}{dt} = M\frac{d^2y}{dt^2}$$

$$= M\frac{d^2z}{dt^2} + M\frac{d^2y}{dt^2}$$

$$M\frac{d^2z}{dt^2} + B\frac{dz}{dt} + kz = -M\frac{d^2x}{dt^2}$$

d^2x/d^2 is body acceleration, the minus sign occurs because the output has been taken as $(y - x)$.

4.6.3 Mechanical Rotational System

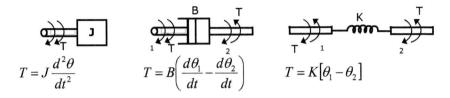

$$T = J\frac{d^2\theta}{dt^2}$$

$$T = B\left(\frac{d\theta_1}{dt} - \frac{d\theta_2}{dt}\right)$$

$$T = K[\theta_1 - \theta_2]$$

Example 4.25

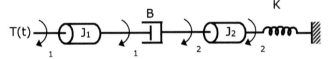

Draw the mechanical network, and write the differential equation of performance.

Solution 4.25

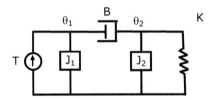

Two nodes, therefore, two equations are to be written

$$T = J_1 s^2 \theta_1 + B(s\theta_1 - s\theta_2)$$

$$0 = K\theta_2 + J_2 s^2 \theta_2 + B(s\theta_2 - s\theta_1).$$

Example 4.26
Draw the mechanical network, and write the differential equation of performance.

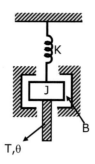

Solution 4.26

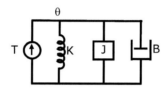

One node one equation is to be written

$$T = Js^2\theta + Bs\theta + K\theta \qquad \frac{\theta(s)}{T(s)} = \frac{1}{Js^2 + Bs + k}$$

4.6.4 Electromechanical Systems

Electromechanical systems are a combination of electrical and mechanical systems. Two mainly used systems are described here.

4.6.4.1 DC generator

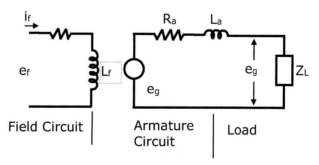

$$E_f = (R_f + sL_f)I_f$$
$$E_g = (R_a + sL_a)I_a + E_a$$
$$E_g = (R_a + sL_a) + i_a Z_L$$
$$e_q = K\phi \frac{d\theta}{dt}, \frac{d\theta}{dt}$$

$\frac{d\theta}{dt}$ (speed) is assumed constant. $e_g = k_g i_f$

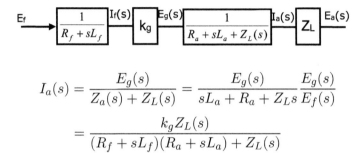

$$I_a(s) = \frac{E_g(s)}{Z_a(s) + Z_L(s)} = \frac{E_g(s)}{sL_a + R_a + Z_L s} \frac{E_g(s)}{E_f(s)}$$
$$= \frac{k_g Z_L(s)}{(R_f + sL_f)(R_a + sL_a) + Z_L(s)}$$

4.6.4.2 Servo motor

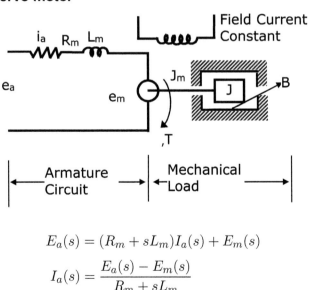

$$E_a(s) = (R_m + sL_m)I_a(s) + E_m(s)$$
$$I_a(s) = \frac{E_a(s) - E_m(s)}{R_m + sL_m}$$
$$e_m = K\phi \frac{d\theta}{dt} \quad \text{K motor parameter}$$
$$\phi \text{ is the field flex}$$

Assume ϕ remains constant

$$Em(s) = k_m s\theta(s)$$

$$T(t)\alpha\phi i_a \quad T(s) = k_\tau I_a(s)$$

$$T(s) = (Js^2 + Bs)\theta(s) \quad \theta(s) = \frac{T(s)}{Js^2 + Bs}$$

$$T = \frac{Jd^2\theta}{dt^2} + B\frac{d\theta}{dt}$$

$$T(s) = k_T I_a(s)$$

$$I_a(s) = \frac{E_a(s) - E_m(s)}{R_m + sL_m}$$

$$\theta = \frac{T}{Js^2 + B_s}; E_m = k_m s\theta(s)$$

$$G(s) = \frac{k_T}{JL_m s^3 + (BL_m + JR_m)s^2 + (BR_m + k_\tau k_m)s}$$

$$G(s) = \frac{kT}{JR_m s^2 + (BR_m + k_\tau k_m)s}$$

Example 4.27
The following figure shows an armature controlled d.c. motor that is part of a position control system. The torque of the motor is related to the armature current by the constant K_T and the back electromotive force (emf) e_g is related to the speed by the constant K_v. Using the state variables θ, ω and i_a derive a state variable description of the system.

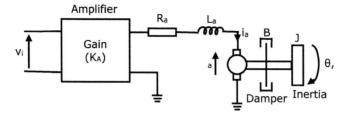

Solution 4.27

The system is governed by the following equations:

Electrical

$$K_A v_i - e_g = i_a R_a + L\frac{di_a}{dt}$$

Electromechanical

$$e_g = K_v \omega, \quad T = K_T i_a$$

Mechanical

$$T - B\omega = J\frac{d\omega}{dt}, \quad \omega = \frac{d\theta}{dt}$$

Substituting for e_g and T gives the equations

$$K_A v_i - K_v \omega = i_a R_a + L\frac{di_a}{dt} \quad K_T i_a - B\omega = J\frac{d\omega}{dt} \quad \omega^\bullet = \frac{d\omega}{dt}$$

4.7 Problems and Solutions

Problem 4.1

Two DTSs H_1 and H_2 are connected in cascade to form a new system H as shown in the following figure. Prove or disprove the following statements:

(a) If H_1 and H_2 are linear, then H is linear (i.e., the cascade connection of two linear systems is linear).

(b) If H_1 and H_2 are time invariant, then H is time invariant.

(c) If H_1 and H_2 are causal, then H is causal.

(d) If H_1 and H_2 are linear and time invariant, the same holds for H.

(e) If H_1 and H_2 are linear and time invariant, then interchanging their order does not change the system H.

(f) As in part (e) except that H_1, H_2 are now time varying. (Hint: use an example.)

(g) If H_1 and H_2 are non-linear, then H is non-linear.

(h) If H_1 and H_2 are stable, then H is stable.

(i) Show by an example that the inverse of parts (c) and (h) do not hold in general.

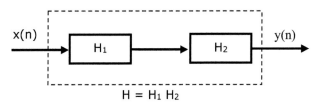

$$H = H_1 H_2$$

Solution 4.1

(a) True

If $v_1(n) = H_1[x_1(n)]$ and $v_2(n) = H_1[x_2(n)]$, then $a_1x_1(n) + a_2x_2(n)$ yields

$a_1v_1(n) + a_2v_2(n$ by the linearity property of H).

Similarly, if $y_1(n) = H_2[v_1(n)]$ and $y_2(n) = H_2[v_2(n)]$,

Then $b_1v_1(n) + b_2v_2(n) \rightarrow y(n) = b_1y_1(n) + b_2y_2(n)$ by the linearity property of H_2.

Since $v_1(n) = H_1[x_1(n)]$ and $v_2(n) = H_2[x_2(n)]$, it follows that $A_1x_1(n) + A_2x_2(n)$ yields the output $A_1H[x_1(n)] + A_2H[x_1(n)]$, where $H = H_1H_2$. Hence, H is linear.

(b) True

For H_1, if $x(n) \rightarrow v(n)$ and $x(n-k) \rightarrow v(n-k)$ for H_2, if $v(n) \rightarrow y(n)$.

Then $v(n-k) \rightarrow y(n-k)$. Hence, for H_1H_2, $x(n) \rightarrow y(n)$ and $x(n-k) \rightarrow y(n-k)$.

Therefore, $H = H_1H_2$ is time invariant.

(c) True

H_1 is causal $\Rightarrow v(n)$ depends only on $x(k)$ for $k \leq n$

H_2 is causal $\Rightarrow y(n)$ depends only on $v(k)$ for $k \leq n$

Therefore, $y(n)$ depends only on $x(k)$ for $k \leq n$. Hence, H is causal.

(d) True as we combine (a) and (b)

(e) True as this follows from $h_1(n)$. $h_2(n) = h_2(n)$. $h_1(n)$

(f) False

For example, consider H_1: $y(n) = n\,x(n)$ and H_2: $y(n) = n\,x(n+1)$. Then

$$H_2[H_1[\delta(n)]] = H_2(0) = 0$$
$$H_1[H_2[\delta(n)]] = H_1[\delta(n+1)] = -\delta(n+1) \neq 0$$

(g) False

For example, consider H_1: $y(n) = x(n) + b$ and H_2: $y(n) = x(n) - b$

where b $\neq$ 0. Then

H[x(n)] = H$_2$[H$_1$ (x(n))]] = H$_2$[x(n)+b] = x(n). Hence, H is linear.

(h) True

H$_1$ is stable $\Rightarrow$ v(n) is bounded if x(n) is bounded H$_2$ is stable $\Rightarrow$ y(n) is bounded. Hence, y(n) is bounded if x(n) is bounded $\Rightarrow$ H = H$_1$H$_2$ is stable.

(i) Inverse of (c)

H$_1$ and for H$_2$ are non-causal $\Rightarrow$ H is non-causal

For example, H$_1$: y(n) = x(n + 1), H$_2$: y(n) = x(n − 2) $\Rightarrow$ H: y(n) = x(n − 1),

which is causal. Hence, the inverse of (c) is false.

Inverse of (h)

H$_1$ and for H$_2$ is unstable, implies H is unstable.

For example, H$_1$: y(n) = e$^{x(n)}$, stable and H$_2$: y(n) = ln [x(n)], which is unstable.

But H: y(n) = x(n), which is stable. Hence, the inverse of (h) is false.

Problem 4.2

The DTS

$$y(n) = ny(n-1) + x(n), n \geq 0 \text{ is at rest [i.e., } y(-1) = 0].$$

Check if the system is linear, time invariant and BIBO stable.

Solution 4.2

If $H[a_1y_1(n) + a_2y_2(n)] = a_1H[y_1(n)] + a_2H[y_2(n)]$, the system is linear.
$H[a_1y_1(n) + a_2y_2(n)] = ny_1(n-1) + x_1(n) + ny_2(n-1) + x_2(n)$ and
$a_1H[y_1(n)] + a_2 H[y_2(n)]$ produces n $y_1(n-1) + x_1(n) + ny_2(n-1) + x_2(n)$.

Hence, the system is linear.

$$y(n, k) = (n)y(n - k - 1) + x(n - k)$$
$$y(n - k) = (n - k)y(n - k - 1) + x(n - k)$$
$$y(n, k) \neq y(n - k).$$

Hence, the system is time variant.

If x(n) = u(n)[an initial step function] then lx(n)l $\leq$ 1. But for this bounded input, the output is

$$y(0) = 1,$$
$$y(1) = 1 + 1 = 2,$$
$$y(2) = 2 \times 2 + 1 = 5, \ldots \text{ which is unbounded.}$$

Hence, the system is unstable.

Problem 4.3

In the following problem x(t) and x(n) denote system input signals and y(t) and y(n) denote the corresponding output signals. Each part of the problem gives a system description relating these signals. For each system determine whether or not it is:

(a) Linear
(b) Time invariant
(c) Instantaneous
(d) Causal
(e) Stable

$\quad$ (i) $y(t) = 2x(t) + x(t-1)$

$\quad$ (ii) $y(n) = 2x(n) + [x(n)]^2$

$\quad$ (iii) $y(n) = x(n) + x(n+1)$

$\quad$ (iv) $y(t) = \int_{-2}^{+2} x(\tau)d\tau$

$\quad$ (vi) $y(t) = x(t)\cos\omega t$

Solution 4.3

$\quad$ (i) Linear, time invariant, memory, causal, stable
$\quad$ (ii) Non-linear, time invariant, instantaneous, causal, stable
$\quad$ (iii) Linear, time invariant, memory, Non-causal, stable
$\quad$ (iv) Linear, time invariant, memory, causal, stable
$\quad$ (v) Linear, time invariant, memory, Non-causal, stable
$\quad$ (vi) Linear, non-time invariant, instantaneous, causal, stable

Problem 4.4

Comment on the properties of linearity and time invariance in relation to the following systems:

(a) An amplifier that limits such that its input/output relationship is described by the equation

$$v_0 = Kv_i \quad -V \leq v_i \leq +V$$
$$v_0 = KV \quad v_i > +V$$
$$v_0 = -KV \quad v_i < -V$$

(b) A space vehicle on take-off has position x that is related to the motor thrust F by the equation

$$F = M\frac{d^2x}{dt^2}$$

M is the total mass of the vehicle, and contributing to this is the mass of the fuel which is burnt. The mass therefore decreases with time.

(c) In the space vehicle of (b) a transducer relates external air pressure to a voltage v according to the relationship

$$v = kp^2$$

The value of k is, however, governed by the time into flight and is switched in value at time 20 s after launch.

Solution 4.4

(a) Provided the input signal is such that $v_1 \leq |V|$ the system is linear. For $v_1 > |V|$, the system is non-linear. The system is time invariant.
(b) Because the mass M depends upon time the system is not time invariant. The system is linear.
(c) The system is non-linear because of the p^2 term. As k depends upon time, the system is not time invariant.

Problem 4.5

(a) The signal shown in Figure (a), a unit step, is applied as the input to a LTI system. It produces as output y(t) as shown in Figure (b).

Determine the system outputs if the signals $x_1(t)$ and $x_2(t)$ as shown in Figure (c) are applied in turn to the same system.

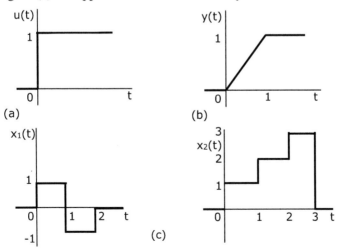

(b) The discrete signal x(n) shown in Figure (d) is applied to a LTI system and it produces a response as shown in Figure (e).

   Determine the system output if the signals $x_1(n)$ and $x_2(n)$ as shown in Figure (f) are applied in turn to the same system.

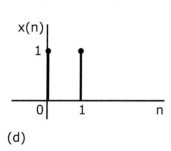

(d)

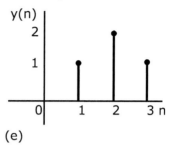

(e)

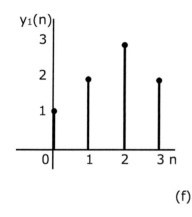

(f)

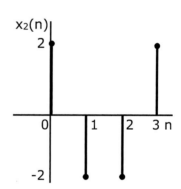

Solution 4.5

(a) The signals $x_1(t)$ and $x_2(t)$ can be expressed as the sum of scaled, time shifted versions of the signal u(t).

$$x_1(t) = u(t) - 2u(t-1) + u(t-2)$$
$$x_2(t) = u(t) - u(t-1) + u(t-2) - 3u(t-3)$$

Because the system is linear and time invariant the resulting outputs are

$$y_1(t) = y(t) - 2y(t-1) + y(t-2)$$
$$y_2(t) = y(t) + y(t-1) + y(t-2) - 3y(t-3)$$

The plots are as shown:

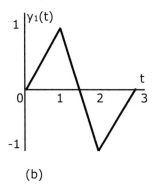

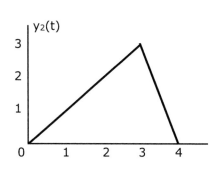

(b)

(b) As in part (a) the signals $x_1(n)$ and $x_2(n)$ can be represented by combinations of linear time shifted versions of $x(n)$.

$$x_1(n) = x(n) + x(n-1) + 2x(n-2)$$
$$x_2(t) = 2x(n) - 4x(n-1) + 2x(n-2)$$

These give output signals

$$y_1(n) = y(n) + y(n-1) + 2y(n-2)$$
$$y_2(t) = 2y(n) - 4y(n-1) + 2y(n-2)$$

The waveforms are as shown hereunder:

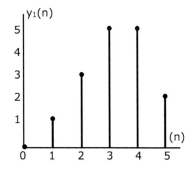

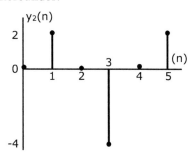

Problem 4.6

The following Figures (a) and (b) show two systems using operational amplifiers. Use the virtual earth principle to obtain the differential equation relating $v_o(t)$ to $v_I(t)$ in each case.

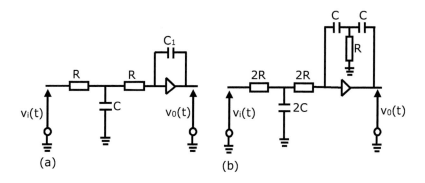

(a)   (b)

Solution 4.6

With reference to Figure (a), denoting the voltage at the node formed by the junction of R, R and C by v_x, then nodal analysis at this point gives

$$\frac{v_1 - v_2}{R} = \frac{v_x}{R} + C\frac{dv_x}{dt} \tag{1}$$

where the voltage at the virtual earth point has been taken as zero. At the virtual earth point

$$\frac{v_x}{R} = -C\frac{dv_0}{dt}$$

$$v_x = -CR\frac{dv_0}{dt}$$

$$\frac{dv_x}{dt} = -CR\frac{d^2v_0}{dt^2}$$

Substituting into (1)

$$\frac{v_i}{R} = -\frac{2}{R}CR\frac{dv_0}{dt} - C^2R\frac{d^2v_0}{dt^2}$$

$$(CR)^2\frac{d^2v_0}{dt^2} + 2CR\frac{dv_0}{dt} = -v_i$$

With reference to Figure (b), denoting the voltage at the node formed by the junction of 2R, 2R and 2C by v_x, and that at the node formed by C, C and R by v_y then nodal analysis for v_x gives

$$\frac{v_1 - v_x}{2R} = 2C\frac{dv_x}{dt} + \frac{v_x}{2R} \tag{1}$$

$$v_1 = 4CR\frac{dv_x}{dt} + 2v_x$$

Nodal analysis for v_y gives

$$C\left[\frac{dv_0}{dt} - \frac{dv_y}{dt}\right] = \frac{v_y}{R} + C\frac{dv_y}{dt}$$

$$CR\frac{dv_0}{dt} = v_y + 2CR\frac{dv_y}{dt}$$

(2)

Nodal analysis at the virtual earth point gives

$$\frac{v_x}{2R} = -C\frac{dv_y}{dt}$$

(3)

$$v_x = -2CR\frac{dv_y}{dt}$$

$$\frac{dv_x}{dt} = -2CR\frac{d^2v_y}{dt^2}$$

(4)

Substituting from (3) and (4) into (1) gives

$$v_i = -4CR\left[2CR\frac{d^2v_y}{dt^2} + \frac{dv_y}{dt}\right]$$

(5)

Differentiation of (2) yields

$$CR\frac{d^2v_0}{dt^2} = 2CR\frac{d^2v_y}{dt^2} + \frac{dv_y}{dt}$$

(6)

Comparison of (5) and (6) gives

$$4(CR)^2\frac{d^2v_0}{dt^2} = -v_i$$

Problem 4.7
The figure provided shows a large industrial vibrator. The constants of the system are as follows:

R = resistance of coil = 100 Ω

L = inductance of coil = 10 H

e = back e.m.f. in coil

$$e = K_v = \frac{dx}{dt}; \quad K_v = 15\,\text{V/ms}^{-1}$$

f = mechanical force produced

$$f = K_T i; \quad K_T = 15\text{N/A}$$

M = mass of table plus load = 1000 kg
K = stiffness of table suspension = 150 000 N/m
B = damper constant = 15 000 N/ms^{-1}

Show that the system is described by the following differential equation.

$$v_i = iR + L\frac{di}{dt} + K_v\frac{dx}{dt}$$

$$K_T i = M\frac{d^2x}{dt^2} + B\frac{dx}{dt} + kx$$

Using the values given for the system constants, draw an analogue computer diagram to represent the system. Use integrator time constants of unity and show all amplifier gains and potentiometer settings.

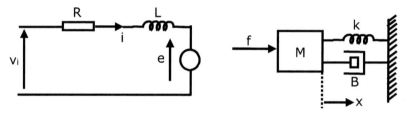

Solution 4.7
Considering the electrical circuit, Kirchhoff's mesh law gives

$$v_i - e = iR + L\frac{di}{dt}$$

But $e = K_v\frac{dx}{dt}$, hence

$$v_i = iR + L\frac{di}{dt} + K_v\frac{dx}{dt} \qquad (1)$$

Considering the mechanical system
Net force = Mass × acceleration

$$f - kx - B\frac{dx}{dt} = M\frac{d^2x}{dt^2}$$

But $f = K_T i$, hence

$$K_T i = M\frac{d^2x}{dt^2} + B\frac{dx}{dt} + kx \qquad (2)$$

Inserting figures, Equations (1) and (2) can be written as

$$\frac{di}{dt} = -\left[10i + 1.5\frac{dx}{dt} = 0.1v_i\right]$$

$$\frac{d^2x}{dt^2} = -\left[15\frac{dx}{dt} + 150x - 0.015i\right]$$

These equations can be represented by the analogue computer diagram shown on the following diagram.

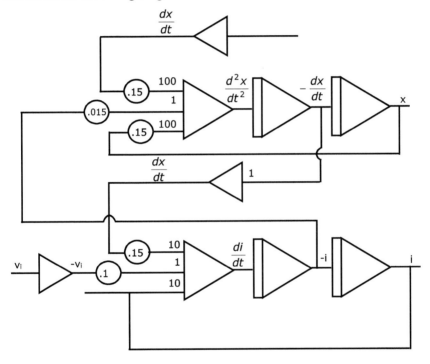

Problem 4.8

A system is described by the differential equation

$$\frac{d^2y}{dt^2} + 2.4\frac{dy}{dt} + 4y = 4x$$

The response of this system to a unit step input is given by

$$y(t) = 1 - 1.56e^{-1.2t}\sin(1.6t + 0.9272)$$

For values of T equal to 0.5, 0.1 and 0.05 obtain numerically the solution of the difference equation for a unit step input. Compare these responses with the true response given.

Note: These solutions need the use of a computer or a programmable calculator.

Solution 4.8

The approximations required are

$$\frac{dy}{dt} \approx \frac{y(n) - y(n-1)}{T}$$

$$\frac{d^2y}{dt^2} \approx \frac{y(n) - 2y(n-1) + y(n-2)}{T^2}$$

where T is the sampling time. Substituting these expressions into the given differential equation gives

$$[1 + 2.4T + 4T^2]\, y(n) - [2 + 2.4T]\, y(n-1) + y(n-2) = 4T^2 x(n)$$

This equation can be rewritten as

$$y(n) = b_0 x(n) + a_1 y(n-1) + a_2 y(n-2)$$

where

$$b_0 = \frac{4T^2}{1+2.4T+4T^2}, \quad a_0 = \frac{2+2.4T}{1+2.4T+4T^2}$$

$$a_2 = \frac{1}{1+2.4T+4T^2}$$

For the sampling times given the coefficients are as follows

T	b_0	a_1	a_2
0.5	0.3125	1.0000	−0.3125
0.1	0.0313	1.7500	−0.7813
0.05	0.0088	1.8761	−0.8849

The following table shows points on the true response and the corresponding points on the discrete approximation for the sampling times shown.

	True	Discrete Approximation		
Time s	Response	T = 0.5s	T = 0.1s	T = 001s
0	0.0000	0.3125	0.0312	0.0088
0.5	0.3224	0.6250	0.4168	0.3727
1.0	0.7830	0.8398	0.8045	0.7952
1.5	1.0381	0.9570	1.0097	1.0027
2.0	1.0945	1.0071	1.0648	1.0783
2.5	1.0608	1.0205	1.0503	1.0553
3.0	1.0180	1.0183	1.0223	1.0207
3.5	0.9955	1.0119	1.0036	1.0001
4.0	0.9911	1.0062	0.9965	0.9940

Problem 4.9
Give the difference equations that describe the two discrete realizations shown in the figure provided.

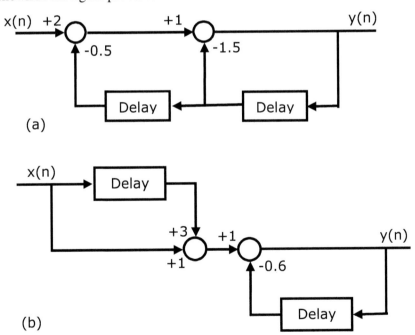

(a)

(b)

Solution 4.10

$$y(n) = -1.5y(n-1) - 0.5y(n-2) + 2x(n)$$
$$y(n) + 1.5y(n-1) - 0.5y(n-2) = 2x(n)$$

$$y(n) = -0.6y(n-1) + x(n) + 3x(n-1)$$
$$y(n) + 6.6y(n-1) = x(n) + 3x(n-1)$$

Problem 4.11

Obtain direct realization representations of the following difference equations

$$32y(n) - 16y(n-1) - 10y(n-2) + 3y(n-3) = 8x(n)$$
$$4y(n) - y(n-2) = 2x(n) + 3x(n-1)$$

Solution 4.11

Rewriting the equations such that y(n) is on the left-hand side with a coefficient of unity gives

$$y(n) = 0.5y(n-1) + 0.3125y(n-2) - 0.0938y(n-3) + 0.25x(n)$$
$$y(n) = 0.25y(n-2) + 0.5x(n) + 0.75x(n-1)$$

The direct realizations are as shown

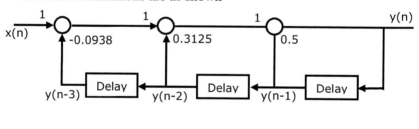

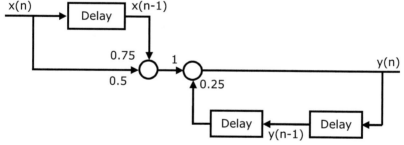

Practice Problem 4.12

Consider the circuit of the following figure:

(a) Find the transfer function $V_2(s)/V_1(s)$.
(b) A constant input voltage of 10 V is applied to the circuit. Using the final-value theorem of the Laplace transform (see Appendix C), find the steady-state values of the output voltage for the circuit of (a). Explain this final value using the properties of the circuit elements.

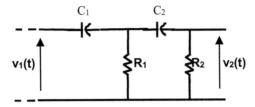

Practice Problem 4.13

(a) Write the differential equations for the mechanical system shown in Figure (a).
(b) A force $f(t)$ is applied downward to the mass M. Find the transfer function from the applied force to the displacement, $x_1(t)$, of the mass; that is, find $X_1(a)/F(s)$.
(c) Repeat (a) for the system of Figure (b).
(d) A force $f(t)$ is applied downward to the mass M in Figure (b). Find the transfer function $Y(s)/F(s)$.

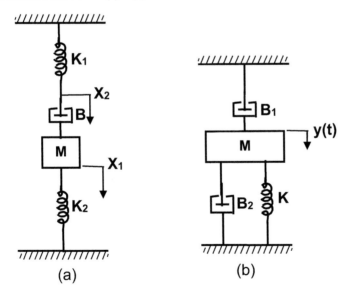

(a)   (b)

Practice Problem 4.14
Consider the mechanical system of the following figure:
(a) Write the differential equations that describe this system.

(b) Find the transfer function from the applied force $f(t)$ to the displacement, $x_1(t)$, of the mass; that is, find $X_1(s)/F(s)$.

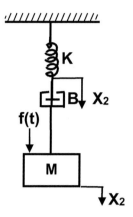

Practice Problem 4.15

Consider the mechanical system of the following figure, where the numerical values of the parameters are given. Assume that all units are consistent. The system is excited by initial conditions only.

(a) Write the differential equations that describe the motion of the system.
(b) Assume that a force $f(t)$ is applied at the connection point which has the displacement, x_2. Express the transfer function $X_2(s)/F(s)$ as a ratio of determinants. Do not evaluate the determinants.

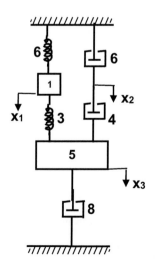

Practice Problem 4.16

(a) A mechanical translational system has the following equations, with x_1, x_2 and x_3 denoting displacements. Draw a model of the system.

$$2x_1^{\bullet\bullet} + 10x_1 - 6x_2 = 0$$
$$-6x_1 + 5x_2^{\bullet\bullet} + 7x_2^{\bullet} + 4x_3^{\bullet} = 0$$
$$-4x_2^{\bullet\bullet} + 7x_3^{\bullet\bullet} + 5x_3^{\bullet} = 0$$

In these equations, x_i denotes the derivative of x_i with respect to time, x_i the second derivative of x_i with respect to time and so on.

(b) In the equations of (a), replace x_1 with i_1, x_2 with i_2 and x_3 with i_3. Of course, x_1 will also be replaced with di_1/dt, and so forth. After these replacements, assume that the equations now model an electrical circuit. Draw the circuit diagram for these equations.

Practice Problem 4.17

For the servomotor, suppose that the shaft cannot be considered rigid but must be modelled as a torsional spring, as shown in the following figure.

(a) Write the system differential equations.
(b) Find the transfer function $\Theta_l(s)/E_a(s)$.

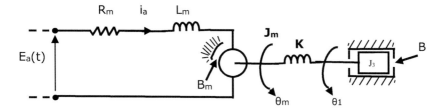

Practice Problem 4.18

(a) This problem involves writing non-linear system equations. Consider the dc generator of the following figure. A prime mover, such as a steam turbine, drives the generator. In a physical system, the prime mover cannot maintain the generator speed constant when the electrical load on the generator changes. Suppose that to a first-order approximation, the generator speed $d\theta/dt$ varies inversely with the armature current i_a;

that is

$$\frac{d\theta}{dt} = \frac{k_a}{i_a}$$

(The application of this approximation is very limited and obviously cannot be used for i_a very small.) Write the non-linear system differential equations.

(b) Label those equations in (a) that are linear and thus to which we can apply the Laplace transform. This linearity means that we can represent part of the system but not the entire system by transfer functions. In general, certain parts of non-linear systems are linear and can be represented by transfer functions.

(c) Suppose that $Z_1(s)>>(L_a s + R_a)$ over the frequency range of interest. Repeat (a) and (b) for this case, and reduce the number of equations to the minimum number possible.

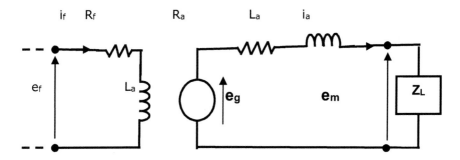

Practice Problem 4.19

Consider the circuit shown in the following figure.

(a) Find the transfer function $V_2(s)/V_1(s)$.
(b) Find the transfer function $V_2(s)/I(s)$.
(c) Evaluate the transfer functions in (a) and (b) for R1 = R2 = 10 Ω and C = 0.5 F.

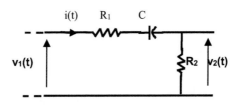

5

Control System Response

This chapter provides comprehensive details regarding the response of the first-order and second-order systems, system's step response, system's ramp response, systems' dc gain, system's frequency response and problems and solutions.

5.1 Convolution

Consider any arbitrary continuous-time input signal x(t) as shown in Figure 5.1.

This input waveform can be represented in terms of element areas as shown in Figure 5.2.

Now, if the limit $\Delta t \to 0$, nth element area can be considered as a rectangle of width Δt and height [x(n Δt)].

As an example, the shaded part in Figure 5.2 has a width of Δt and height [x(2 Δt)] and thus area of this shaded portion is 'x(2 Δt) (Δt)'.

In the limit $\Delta t \to 0$, this element area approaches a delta function of strength x(2 Δt) (Δt) which is located at t = Δt.

Figure 5.1   Arbitrary continuous-time input signal.

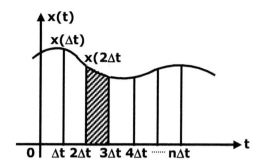

Figure 5.2   Arbitrary continuous-time signal in terms of element areas.

We can represent this delta function as

$$x(2\Delta t)(\Delta t).\delta(t - 2\Delta t)$$

Hence, the function x(t) is a continuous sum of such impulse function. Mathematically

$$x(t) = \underset{\Delta t \to 0}{Lim} \sum_{n=-\infty}^{\infty} x(n\Delta t)(\Delta t).\delta(t - n\Delta t) \qquad (5.1)$$

Now, let *h(t)* be the unit impulse response of a linear-time invariant (LTI) system. This means that *h(t)* is the response of the LTI system when input function is an unit impulse function *δ(t)* located at t = 0 and of unit strength.

Hence, obviously, the response of the LTI system for an impulse function of strength x (n Δt) (Δt) located at *t = n Δt* can be expressed as *x (n Δt) (Δt) h (t-nΔt).*

According to the superposition theorem, the response *y(t)* of the LTI system for input function *x(t)* given by Equation (5.1), may be expressed as

$$y(t) = \underset{\Delta t \to 0}{Lim} \sum_{n=-\infty}^{\infty} x(n\Delta t)h(t - n\Delta t)(\Delta t) \qquad (5.2)$$

In the Δt → 0, the summation given by (5.2) becomes an integration providing y(t).

$$y(t) = \int_{-\infty}^{\infty} x(\tau)h(t - \tau)d\tau \qquad (5.3)$$

or

$$y(t) = x(t) \otimes h(t) \qquad (5.4)$$

$$y(t) = h(t) \otimes x(t) \tag{5.5}$$

The expression in equation (5.4) is called the convolution integral for continuous-time LTI systems.

From equation (5.4) it is clear that the output of any general input may be found by convolving the given input signal x(t), and can be evaluated by equation (5.4).

5.2 Convolution Integral Formula

h(t) specifies the input–output characteristics of the system

$$y(t) = \int_{-\infty}^{\infty} x(\tau)h(t - \tau)d\tau$$

We now carry out the calculation of the output signal y(t) for any input signal.

Suppose the system's input is a unit step function u(t). Let s(t) denotes the initial step response.

$$s(t) = \int_{-\infty}^{\infty} u(\tau)h(t - \tau)d\tau \quad s(t) = \int_{0}^{\infty} h(t - \tau)d\tau$$

$$s(t) = \int_{0}^{\infty} h(t - \tau)d\tau$$

Since u(τ) is zero for $\tau < 0$. If the system is causal h h(t-τ) is zero for (t-τ) < 0 and for τ>t and $s(t) = \int_{0}^{\infty} h(\tau)d\tau$.

The unit step response is directly calculated for the unit impulse response.

$$h(t) = \frac{ds(t)}{dt}$$

Example 5.1
Obtain the convolution of the given two continuous-time functions:

$$y(t) = u(t + 1) * u(t - 2)$$

Solution 5.1

We know that for continuous-time signals, the convolution integral is expressed as

$$y(t) = x(t) \otimes h(t) = \int_{\tau=-\infty}^{\infty} x(\tau)h(t-\tau)d\tau \qquad \text{(i)}$$

$$x(t) = u(t+1) \Rightarrow u(\tau+1)$$

$$h(t) = u(t-2) \Rightarrow u(t-\tau-2) = u(-\tau+t-2)$$

Using equation (i), we get

$$y(t) = \int_{-1}^{t-2} (1)(1)d\tau = [t-2-(-1)] = t-1$$

$$y(t) = t-1$$

Example 5.2

Obtain the convolution of the given two continuous-time functions:

$$y(t) = e^{-2t}u(t+1) * u(t-2)$$

Solution 5.2

We know that for continuous-time signals, the convolution integral is expressed as

$$y(t) = x(t) \otimes h(t) = \int_{\tau=-\infty}^{\infty} x(\tau)h(t-\tau)d\tau \qquad \text{(i)}$$

$$x(t) = e^{-2t}u(t) \Rightarrow e^{-2\tau}u(\tau)$$

$$h(t) = u(t+2) \Rightarrow \quad h(t-\tau) = u(-\tau+t+2)$$

Using equation (i), we get

$$y(t) = \int_{0}^{t+2} e^{-2\tau}(1)d\tau = \frac{e^{-2\tau}}{-2} = -\frac{1}{2}[e^{-2(t+2)}-1] = \frac{1}{2}[1-e^{-2(t+2)}]$$

Example 5.3

Obtain the convolution of the given two continuous-time functions:

$$y(t) = [u(t+2) - u(t-1)] * u(-t+2)$$

Solution 5.3

We know that for continuous-time signals, the convolution integral is expressed as

$$y(t) = x(t) \otimes h(t) = \int_{\tau=-\infty}^{\infty} x(\tau)h(t-\tau)d\tau \qquad \text{(i)}$$

$$x(t) = [u(t+2) - u(t-1)] \Rightarrow [u(\tau+2) - u(\tau-1)]$$

$$h(t) = u(-t+2) \Rightarrow h(t-\tau) = u(-\tau+2-t)$$

Using equation (i), we get

$$y(t) = \int_{-2}^{2-t} (1)(1)d\tau = [2 - t + 2] = 4 - t$$

Example 5.4

Obtain the convolution of the given two continuous-time functions:

$$x(t) = e^{-t^2}$$

and $\qquad\qquad h(t) = 3t^2 \qquad\qquad$ for all values of t.

Solution 5.4

We know that for continuous-time signals, the convolution integral is expressed as

$$y(t) = x(t) \otimes h(t) = \int_{\tau=-\infty}^{\infty} x(\tau)h(t-\tau)d\tau \qquad \text{(i)}$$

Now $\qquad\qquad\qquad x(\tau) = e^{-t^2}$

and $\qquad\qquad\qquad h(t-\tau) = 3(t-\tau)^2$

Using equation (i), we get

$$y(t) = \int_{-\infty}^{\infty} e^{-\tau^2}.[3t(t-\tau)^2]d\tau = \int_{-\infty}^{\infty} e^{-\tau^2}[3t^2 - 6t\tau + 3t\tau^2]]d\tau$$

$$y(t) = 3t^2 \int_{-\infty}^{\infty} e^{-\tau^2}d\tau - 6t \int_{-\infty}^{\infty} \tau e^{-\tau^2}d\tau + 3 \int_{-\infty}^{\infty} \tau^2 e^{-\tau^2}d\tau$$

or $\quad y(t) = 3t^2\sqrt{\pi} - 0 + 1.5\sqrt{\pi}$

$\qquad y(t) = 5.31t^2 + 2.659 \quad$ for all t. $Ans.$

Example 5.5

The input signal x(t) and impulse response h(t) of a continuous-time system are described as

$$x(t) = e^{3t}u(t)$$

and

$$h(t) = u(t-1)$$

Find out the output.

Solution 5.5

We know that for continuous-time system, the output is given by the convolution integral which is expressed as

$$y(t) = x(t) \otimes h(\tau) = \int_{\tau=-\infty}^{\infty} x(\tau)h(t-\tau)d\tau \qquad \text{(i)}$$

Now

$$x(\tau) = e^{-3}\tau.u(\tau)$$

and

$$h(t-\tau) = u(t-\tau-1)$$

Therefore, using (i), we get

$$y(t) = \int_{-\infty}^{\infty} e^{-3\tau}.u(\tau).u(t-\tau-1)d\tau$$

at this point, it may be noted that, the limits of integration will be modified according to the above equations. Thus, we have

$$y(t) = \int_{0}^{t-1} e^{-3\tau}(1)d\tau$$

$$y(t) = -\frac{1}{3}[e^{-3\tau}]_0^{t-1} y(t) = \frac{1}{3}[1 - e^{-3(t-1)}]$$

is the required output.

Example 5.6

By using continuous-time convolution integral, find out the response of the system to unit step input signal. Impulse response is given as

$$h(t) = \frac{R}{L}e^{-tR/L}.u(t).$$

Solution 5.6

We know that for continuous-time systems, the convolution integral is expressed as

$$y(t) = x(t) \otimes h(t) = \int_{\tau=-\infty}^{\infty} x(\tau)h(t-\tau)d\tau \qquad \text{(i)}$$

Here, $\qquad x(\tau) = u(\tau) = 1 \quad \text{for} \quad \tau \geq 0$

and $\qquad h(t-\tau) = \frac{R}{L}e^{-(t-\tau)R/L}.u(t-\tau)$

Thus, using equation (i), we get

$$y(t) = \int_0^{\infty} 1.\frac{R}{L}e^{-(t-\tau)R/L}.u(t-\tau)d\tau$$

Now,

$$u(t-\tau) = 1 \text{ for } t \geq \tau \text{ or } \tau \leq t.$$

Hence

$$y(t) = \frac{R}{L}\int_0^t e^{-(t-\tau)R/L}d\tau = \frac{R}{L}\int_0^t e^{-tR/L}.e^{\tau R/L}d\tau$$

or $\quad y(t) = \frac{R}{L}\int_0^t e^{-tR/L}.e^{\tau R/L}d\tau = \frac{R}{L}.e^{\tau R/L}.\frac{1}{R/L}\left[e^{\tau R/L}\right]_0^t$

or $\quad y(t) = e^{-tR/L}\left[e^{tR/L} - e^0\right] = e^{-tR/L}\left[e^{tR/L} - 1\right]$

$\quad y(t) = 1 - e^{-tR/L} \text{ for } t \geq 0$

This is the required output of the system.

Example 5.7

Determine the convolution of the given two continuous-time functions:

$$x(t) = 3\cos 2t \qquad \text{for all } t$$

and $\qquad h(t) = e^{-|t|} = e^t \quad \text{for } t < 0$

$\qquad\qquad e^{-t} \qquad\qquad\quad \text{for } t \geq 0$

Solution 5.7

We know that for the continuous-time systems, the convolution integral is expressed as

$$y(t) = x(t) \otimes h(t) = \int_{\tau=-\infty}^{\infty} x(\tau)h(t-\tau)d\tau$$

Here $x(\tau) = 3 \cos 2\tau$ for all values of τ

and $h(t - \tau) = e^{-(t-\tau)}$ for $t - \tau < 0$ or $\tau > t$

               $e^{(\tau - t)}$ for $t - \tau \geq 0$ or $\tau \leq t$

Therefore, using equation (i), we get

$$y(t) = \int_{\tau=-\infty}^{\infty} x(\tau)h(t-\tau)d\tau = \int_{\tau=-\infty}^{t} (3\cos 2\tau)e^{(\tau-t)}d\tau$$

$$+ \int_{\tau=\infty}^{\infty} (3\cos 2\tau)e^{(\tau-t)}d\tau$$

or $y(t) = 3e^{-t} \int_{-\infty}^{t} e^{\tau}\cos 2\tau d\tau + 3e^{t}\int_{t}^{\infty} e^{-\tau}\cos 2\tau d\tau$

or $y(t) = 3e^{-t}\left[\dfrac{e^{\tau}.(\cos 2\tau + 2\sin 2\tau)}{5}\right]_{-\infty}^{t}$

$$+ 3e^{t}\left[e^{-\tau}(-\cos 2\tau + 2\sin 2\tau\right]_{t}^{\infty}$$

or $y(t) = \dfrac{6}{5}\cos 2t = 1.2\cos 2t$ for all values of t

Example 5.8
Given a continuous-time LTI system with unit impulse response h(t). A continuous-time signal x(t) is applied to the input of this LTI system where

$$x(t) = e^{-at}.u(t) \quad \text{for } a > 0$$
and
$$h(t) = u(t)$$

Evaluate the output y(t).

Solution 5.8
We know that the output y(t) of a continuous-time LTI system is given by the convolution of the applied input and unit impulse response as

$$y(t) = x(t) \otimes h(t) = \int_{-\infty}^{\infty} x(\tau)h(t-\tau)d\tau$$

Now the following figure shows the graphical representation of x(t), h(t) and h(t − t) for a negative value of t(t < 0) and for a positive value of t(t > 0).

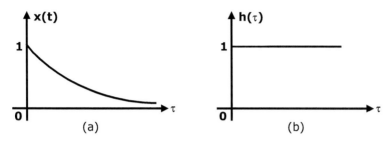

(a)   (b)

From the figure, it may be observed that for t < 0, the product of x(t) and h(t − τ) is zero and so according to equation (i), the output y(t) is zero, i.e.
Also, for t > 05555, the product of x(t) and h(t − τ) is expressed as
Therefore, the output y(t) for t > 0 is expressed as

$$y(t) = \int_0^t x(\tau)h(t - \tau)d\tau y(t)$$

$$\text{or} \quad y(t) = -\frac{1}{\alpha}[e^{-\alpha\tau}]_0^t y(t)$$

$$= \frac{1}{\alpha}(1 - e^{-\alpha t})$$

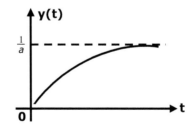

Hence, for all values of t, the output y(t) will be

$$y(t) = \frac{1}{\alpha}(1 - e^{-\alpha t})u(t)$$

The figure shows this output y(t).

Example 5.9
A continuous-time signal x(t) is applied to the input of a continuous-time LTI system with unit impulse response h(t). Find the output y(t) given that

$$x(t) = e^{2t}.u(-t)$$

and

$$h(t) = u(t - 3)$$

Evaluate the output y(t).

Solution 5.9
We know that the output y(t) of a continuous-time LTI system is given by the convolution of the applied input and unit impulse response as

$$y(t) = x(t) \otimes h(t) = \int_{-\infty}^{\infty} x(\tau)h(t - \tau)d\tau$$

The figure shows the graphical representation of signals $x(\tau)$ and $h(t - \tau)$. From the figure it may be observed that these two signals have regions of non-zero overlap irrespective of the value of t.

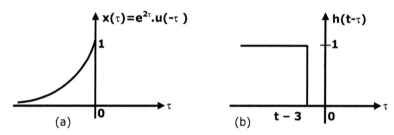

Also, when $t - 3 \leq 0$, the product $x(t) h(t - \tau)$ is non-zero for $-\infty < \tau < t - 3$ and so according to equation (i), the output y(t) will be

$$y(t) = \int_{-\infty}^{t} x(\tau)h(t - \tau)d\tau; \quad y(t) = \int_{-\infty}^{t-3} e^{2\tau}d\tau; \quad y(t) = \frac{1}{2}e^{2(t-3)}$$

Again $t - 3 \geq 0$, the product $x(\tau) h(t-\tau)$ is non-zero for $-\infty < \tau < 0$ and so according to equation (i), the output y(t) will be

$$y(t) = \int_{-\infty}^{0} x(\tau)h(t - \tau)d\tau$$

$$= \int_{-\infty}^{0} e^{2\tau}d\tau = \frac{1}{2}$$

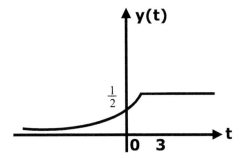

Example 5.10

In Figure x(t) represents the input to a system and h(t) represents the corresponding impulse response. Use convolution to obtain the system output. Assume all initial energy storage in the systems is zero.

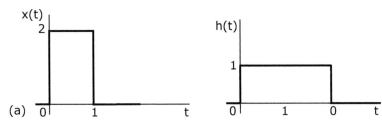

Solution 5.10

(a) The functions $h(\tau)$ and $x(t-\tau)$ are shown hereunder for different ranges of τ. The product of these functions within these ranges always results in values 2 or 0. The integral can then be performed over the appropriate range.

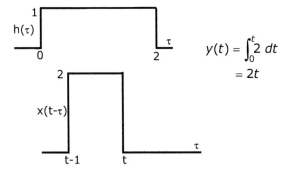

$$y(t) = \int_0^t 2 \, dt$$
$$= 2t$$

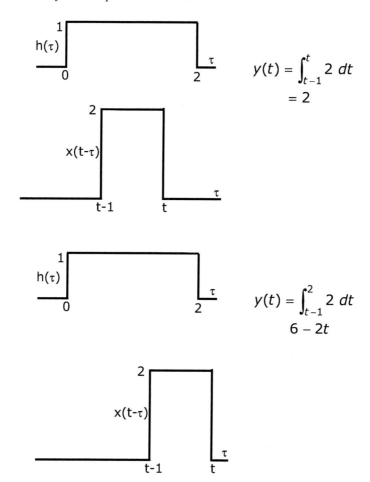

$$y(t) = \int_{t-1}^{t} 2 \, dt$$
$$= 2$$

$$y(t) = \int_{t-1}^{2} 2 \, dt$$
$$6 - 2t$$

Outside these ranges the product and the integral are zero. The function y(t) is as shown hereunder.

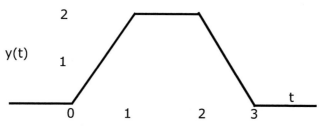

5.3 Time Response of First-Order Systems

The transfer function of a general first-order system can be written as

$$G(s) = \frac{C(s)}{R(s)} = \frac{b_0}{s + a_0} \tag{5.6}$$

where R(s) is the input function and C(s) is the output function. A more common notation for the first-order transfer function is

$$G(s) = \frac{C(s)}{R(s)} = \frac{K}{\tau s + 1} \tag{5.7}$$

since physical meaning can be given to both K. We call (5.7) the standard form of the first-order system.

$$a_0 = \frac{1}{\tau} = b_0 = \frac{K}{\tau} \tag{5.8}$$

Initial conditions are ignored in the calculation of transfer functions. We now show how the initial condition can be included in (5.7). First, we rewrite (5.7) as

$$\left(s + \frac{1}{\tau}\right)C(s) = \frac{K}{\tau}R(s) \tag{5.9}$$

the inverse Laplace transform of (5.8).

$$c(t) + \frac{1}{\tau}c(t) = \frac{K}{\tau}r(t) \tag{5.10}$$

Now we take the Laplace transform of (5.9) and include the initial condition term.

$$sC(s) + c(0) + \frac{1}{\tau}C(s) = \frac{K}{\tau}R(s) \tag{5.11}$$

Solving for C(s) yields

$$C(s) = \frac{c(0)}{s + (1/\tau)} + \frac{(K/\tau)R(s)}{s + (1/\tau)} \tag{5.12}$$

and we have that effect of c(0) on C(s). This equation can be represented in a block diagram form, as shown in Figure 5.3(a). However, initial conditions usually are not shown as inputs on the system block diagram. Note that the initial condition as an input has a Laplace transform of c(0), which is a constant. The inverse Laplace transform of a constant is an impulse function.

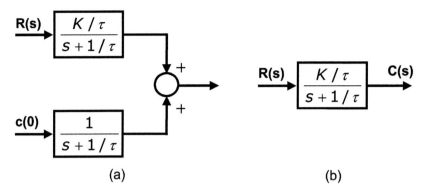

Figure 5.3   First-order system.

Hence, for this representation, the initial condition as an input appears as the impulse function c(0) δ(t). Here, we can see that the impulse function has a practical meaning, even though the impulse function is not a physically realizable signal. Since we usually ignore initial conditions in block diagrams, we normally use the system block diagram as shown in Figure 5.3(b).

5.3.1 System Step Response

We now find the unit step response for the standard first-order system of (5.11). For the case, R(s) = 1/s and, from (5.11)

$$C(s) = \frac{K/\tau}{s[s + (1/\tau)]} = \frac{K}{s} + \frac{-K}{s + (1/\tau)} \tag{5.13}$$

The inverse Laplace trnasfrom of this expression yields

$$c(t) = K(1 - e^{-t/\tau}), t > 0 \tag{5.14}$$

The response can also be expressed as

$$c(t) = K(1 - e^{-t/\tau})u(t)$$

The first term in the response (5.11) originates in the pole of the input R(s) and is called the forced response; since this term does not go to zero with increasing time, it is also called the steady-state response. The second term in (5.11) originates in the pole of the transfer function G(s) and is called the natural response; since this term goes to zero with increasing time, it is also called the transient response.

The step response of a first-order system as given in (5.11) is plotted in Figure 5.4. The two components of the response are plotted separately, along with the complete response.

Note that the exponentially decaying term has an initial slope of K/τ; that is

$$\frac{d}{dt}(-Ke^{-t/\tau})_{t=0} = \frac{K}{\tau}e^{-t/\tau}\bigg|_{t=0} = \frac{K}{\tau} \tag{5.15}$$

Mathematically, the exponential term does not decay to zero in a finite length of time. However, if the term continued to decay at its initial rate, it would reach a value of zero in τ seconds. The parameter τ is called the time constant and has the units of seconds.

The decay of an exponential function is illustrated in Table 5.1 as a function of the time constant, τ. It is seen from this table that the exponential function has decayed to less than 2% of its initial value in four time constants

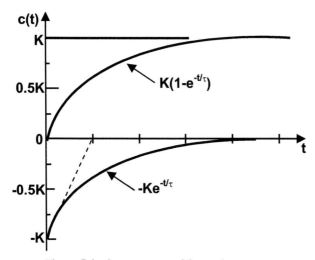

Figure 5.4 Step response of first-order system.

Table 5.1 Decay of exponential function.

t	$e^{-t/\tau}$
0	1
τ	0.3679
2τ	0.1353
3τ	0.0498
4τ	0.0183
5	0.0067

and to less than 1% of its initial value in five time constants. In a practical sense, we often consider an exponential term to have decayed to zero in four to five time constants. We somewhat arbitrarily consider an exponential term to be zero after four time constants. Other authors sometimes use other definitions. The output of a physical system that is modelled by the first-order transfer function will not continue to vary for all finite time. Thus it is reasonable to make the approximation that the response becomes constant after a number of time constants. In addition, the identification of time constants with exponential terms allows us to compare the decay of one term relative to another. For example, if we were designing a position-control system for the pen of a plotter for a digital computer, we know immediately that a time constant of 1 s for the control system is much too slow, whereas values in the region of 0.1 s and faster (smaller) are more reasonable.

Consider again the response of (5.14) as plotted in Figure 5.4. Note that

$$\lim_{t \to \infty} c(t) = K \tag{5.16}$$

We call this limit the final value, or the steady-state value, of the response. Hence in the general first-order transfer function

$$G(s) = \frac{K}{\tau s + 1}$$

where τ is the system time constant and K is the steady-state response to a unit step input. Thus both parameters in the transfer function have meaning, and this accounts for the popularity of writing the transfer function in this manner. An example is now given.

Example 5.11

We wish to find the unit step response of a system with the transfer function

$$G(s) = \frac{2.5}{0.5s + 1} = \frac{5}{s + 2}$$

Then

$$G(s) = G(s)R(s) = \frac{5}{s + 2}\frac{1}{s} = \frac{5/2}{s} + \frac{-5/2}{s + 2}$$

and

$$c(t) = (5/2)(1 - e^{-2t})$$

Hence the pole of the transfer function at $s = -2$ gives the time constant $\tau = 1/s$. The steady-state value of the response is 5/2, and since the system constant is 0.5 s, the output reaches steady state in about 2 s.

5.3.2 System dc Gain

The system dc gain is the steady-state gain to a constant input for the case that the output has a final value, and it is equal to the system transfer function evaluated at $s = 0$. A general procedure for finding the steady-state response to a unit step input for a system of any order is now developed. From Appendix B, the final-value theorem of the Laplace transform is

$$\lim_{t \to \infty} c(t) = \lim_{s \to 0} sC(s) = \lim_{s \to 0} sG(s)R(s) \qquad (5.17)$$

provided that c(t) has a final value. For the case that the input is a unit step, R(s) is equal to 1/s and

$$\lim_{t \to \infty} c(t) = \lim_{s \to 0} sG(s)\frac{1}{s} = \lim_{s \to 0} G(s) \qquad (5.18)$$

If c(t) has a final value, the right-hand limit in this equation exists. If c(t) does not have a final value, the right-hand limit may still exist but has no meaning. Assume that c(t) has a final value. Since the steady-state input is unity, then the output is also the gain in the steady state; that is, G(0) is the system steady-state gain for the input constant. This is true independent of the order of the system. This gain is of such importance that it deserves a name. Since a constant electrical signal is called a dc signal, we will call G(0) the system dc gain.

Note that for the dc gain of a system to have meaning, it is not necessary that a step function be applied to the system. It is only necessary that the input function r(t) be constant for a period longer than four time constants. This point is illustrated in Figure 5.5. The input becomes constant at a value R_c, and the output then becomes constant 4τ s later at the value C_c. Then

$$\frac{C_c}{R_c} = \lim_{s \to 0} G(s)$$

5.3.3 System Ramp Response

We now consider ramp responses. For the input equal to a unit ramp function, $r(t) = t$ and $R(s) = 1/s^2$. Then from (5.7)

$$C(s) = G(s)R(s) = \frac{K/\tau}{s^2[s + (1/\tau)]} = \frac{K}{s^2} + \frac{-K\tau}{s} + \frac{K\tau}{s + (1/\tau)} \qquad (5.19)$$

Therefore

$$c(t) = Kt - K\tau + K\tau e^{-t/\tau}$$

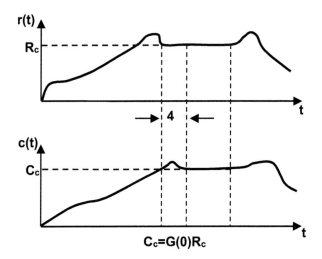

Figure 5.5   Input and output illustrating dc gain.

Thus the ramp response is composed of three terms: a ramp, a constant and an exponential.

First, the exponential has the same time constant as in the step response. Of course, this is true regardless of the input. However, the amplitude of the exponential is different in the ramp response as compared to the step response. The amplitude is different by the factor τ; if τ is large, the exponential can have a major effect on the system response. This can be a problem in designing systems to follow ramp inputs.

Next, the steady-state response $c_{ss}(t)$ is given by

$$c_{ss}(t) = Kt - K\tau \tag{5.20}$$

Here we define the steady-state response to be composed of those terms that do not approach zero as time increases. The system error for a control system is defined later; however, it is a measure of the difference between the desired system output and the actual system output. In general, we wish this error to be zero in the steady state for a control system. For this example, if we desire that the output follow a ramp, the error will include the constant term, $K\tau$. A design specification might be to alter the system such that the steady-state system error for a ramp input is zero or at least is very small. Problems such as this are covered in the chapters on design.

5.4 Time Response of Second-Order Systems

In this section we investigate the response of second-order systems to certain inputs. We assume that the system transfer function is of the form

$$G(s) = \frac{C(s)}{R(s)} = \frac{b_0}{s^2 + a_1 s + a_0} \tag{5.21}$$

However, as in the first-order case, the coefficients are generally written in a manner such that they have physical meaning. The standard form of the second-order transfer function is given by

$$G(s) = \frac{\omega_n^2}{s^2 + 2\zeta\omega_n s + \omega_n^2} \tag{5.22}$$

where ζ is defined to be the dimensionless damping ratio and ω_n is defined to be the natural frequency, or undamped natural frequency. Note also that the dc gain, $G(0)$, is unity, later we consider both the cases that the dc gain is other than unity and the case that the numerator is other than a constant. Note also that all system characteristics of the standard second-order system are functions of only ζ and ω_n since ζ and ω_n are the only parameters that appear in the transfer function (5.22).

Consider first the unit step response of this second-order system:

$$C(s) = G(s)R(s) = \frac{\omega_n^2}{s(s^2 + 2\zeta\omega_n s + \omega_n^2)} \tag{5.23}$$

The inverse Laplace transform is not derived, the result is

$$c(t) = 1 - \frac{1}{\beta} e^{-\zeta\omega_n t} \sin(\beta\zeta\omega_n t + \theta) \tag{5.24}$$

where $\beta = \sqrt{1 - \zeta^2}$ and $\theta = \tan^{-1}(\beta/\zeta)$. In this response, $\tau = 1/\zeta\omega_n$ is the time constant of the exponentially damped sinusoid in seconds (we can usually ignore this term after approximately four time constants). Also, $\beta\omega_n$ is the frequency of the damped sinusoid.

We wish now to show typical step responses for a second-order system. The step response given by (5.20) is a function of both ζ and ω_n. If we specify ζ, we still cannot plot $c(t)$ without specifying ω_n. To simplify the plots, we give $c(t)$ for a specified ζ as a function of $\omega_n t$. A family of such curves for various values of ζ is very useful and is given in Figure 5.6 and 5.7 for $0 \le \zeta \le 2$. Note that for $0 < \zeta < 1$, the response is a damped sinusoid. For $\zeta = 0$, the

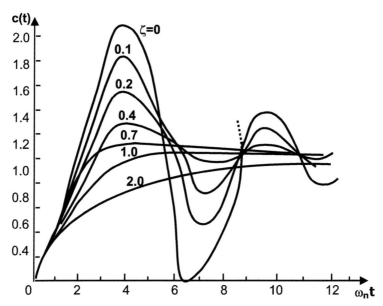

Figure 5.6   Step response for second-order system (5.24).

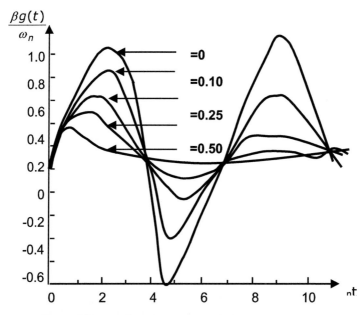

Figure 5.7   Impulse response for second-order system (5.29).

sinusoid is undamped, or of sustained amplitude. For $\zeta \geq 1$, the oscillations have ceased. It is apparent that for $\zeta < 0$, the response grows without limit.

The two poles of the transfer function G(s) occur at

$$s = -\zeta w_n \pm j w_n \sqrt{1 - \zeta^2}$$

For $\zeta > 1$, these poles are real and unequal, and the damped sinusoid portion of c(t) is replaced by the weighted sum of two exponential functions; that is

$$c(t) = 1 + k_1 e^{-t/\tau_1} + k_2 e^{-t/\tau_2} \tag{5.25}$$

where $\tau_1 = 1/(\zeta w_n + w_n \sqrt{\zeta^2 - 1})$, $\tau_2 = 1/(\zeta w_n + w_n \sqrt{\zeta^2 - 1})$ are the two system time constants. For $\zeta = 1$, the poles of G(s) are real and equal, so that

$$c(t) = 1 + k_1 e^{-t/\tau} + k_2 e^{-t/\tau}, \tau = 1/w_n$$

For $0 < \zeta < 1$, the system is said to be underdamped, and for $\zeta = 0$ it is said to be undamped. For $\zeta = 1$, the system is said to be critically damped, and for $\zeta > 1$, the system is overdamped.

For a LTI system

$$C(s) = G(s)R(s) \tag{5.26}$$

For the case that r(t) is a unit impulse function, R(s) $= 1$ and

$$c(t) = L^{-1}[G(s)] = g(t) \tag{5.27}$$

where g(t) is the unit impulse response, or weighting function, of a system with the transfer function G(s). Then, by the convolution integral for a general input r(t)

$$c(t) = \int_0^t g(\tau)r(t - \tau)d\tau \tag{5.28}$$

from (5.26). [In (5.28), τ is the variable f integration and is not related to the time constant.] Hence, all response information for a general input is contained in the impulse response g(t).

An initial condition on a first-order system can be modelled as an impulse function input. While the initial condition excitation of higher-order systems cannot be modelled as simply as that of the first-order system, the impulse response of any system does give an indication of the nature of the initial-condition response, and thus the transient response of the system. The unit impulse response of the second-order system (5.22) is given in Figure 5.7. This figure is a plot of the function.

$$c(t) = L^{-1}\left(\frac{\omega_n^2}{s^2 + 2\zeta\omega_n s + \omega_n^2}\right) = \frac{\omega_n}{\beta}e^{-\zeta\omega_n t}\sin\beta\omega_n t = g(t) \quad (5.29)$$

Compare Figure 5.6 with Figure 5.7 and note the similarity of the information. In fact, the unit impulse response of a system is the derivative of the unit step response. The impulse response of the second-order system can also be considered to be the response to certain initial conditions, with r(t) = 0.

5.5 Time Response Specifications in Design

Before a control system is designed, specifications must be developed that describe the characteristics that the system should possess. For example, some of these specifications may be written in terms of the system step response. Control system specifications in the time domain for the standard second-order system are developed in this section.

A typical unit step response of a standard second-order system is shown in Figure 5.6. Some characteristics that describe this response are now developed. The rise time of the response can be defined in different ways; we define rise time, T_r, as the time required for the response to rise from 10% of the final value to 90% of the final value, as shown in the figure. The peak value of the step response is denoted by M_{pt}, the time to reach this peak value is T_p and the per cent overshoot is defined by the equation

$$\text{percent overshoot} = \frac{M_{pt} - C_{ss}}{C_{ss}} \times 100 \quad (5.30)$$

where c_{ss} is the steady-state, or final, value of c(t). In Figure 5.8 c_{ss} = 1. The settling time, T_s, is the time required for the output to settle to within a certain per cent of its final value. Two common values used are 5% and 2%. For the second-order response of (5.28), approximately four time constants are required for c(t) to settle to within 2% of the final value. Regardless of the percentage used, the settling time will be directly proportional to the time constant τ for a standard second-order under-damped system; that is

$$T_s = k\tau = \frac{k}{\zeta\omega_n} \quad (5.31)$$

where k is determined by the defined percentage. Whereas all the preceding parameters are defined on the basis of the under-damped response, T_r, T_s and

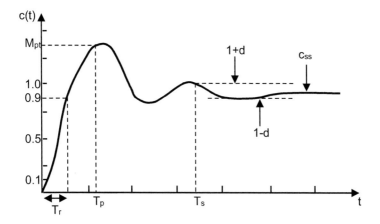

Figure 5.8 Typical step response.

c_{ss} are all equally meaningful for the overdamped and the critically damped responses. Of course, M_{pp}, T_p and per cent overshoot have no meaning for these cases.

We would like to develop analytical relationships for per cent overshoot and T_p in Figure 5.8. The impulse response of the second-order system is given in (5.29)

$$g(t) = \frac{\omega_n}{\beta} e^{-\zeta \omega_n t} \sin \beta \omega_n t \tag{5.32}$$

As just stated, this response is the derivative of the step response of Figure 5.6. Thus the first value of time greater than zero for which this expression is zero is that time for which the slope of the step response is zero, or that time T_p. Therefore

$$\beta \omega_n T_p = \pi \quad \beta = \sqrt{1 - \zeta^2}$$

or

$$T_p = \frac{\pi}{\omega_n \sqrt{1 - \zeta^2}} \tag{5.33}$$

The peak value of the step response occurs at this time, and evaluating the sinusoid in (5.31) at $t = T_p$ yields

$$\sin(\beta \omega_n t + \theta)\big|_{\beta \omega_n t = \pi} = -\sin \theta = -\sin\left(\tan^{-1} \frac{\beta}{\zeta}\right) = -\frac{\beta}{1} \tag{5.30}$$

thus from (5.20) and (5.30)

$$c(t)|_{t=T_p} = M_{pt} = 1 + e^{-\zeta\pi/\sqrt{1-\zeta^2}} \tag{5.31}$$

Therefore, since $c_{ss} = 1$, the per cent overshoot is, from (5.29)

$$\text{percent overshoot} = e^{-\zeta\pi/\sqrt{1-\zeta^2}} \times 100 \tag{5.32}$$

The per cent overshoot is thus a function only of ζ and is plotted versus ζ in Figure 5.8. We can express (5.32) as

$$\omega_n T_p = \frac{\pi}{\sqrt{1-\zeta^2}} \tag{5.33}$$

Thus the product $\omega_n T_p$ is also a function of only ζ, and this is also plotted in Figure 5.8. Since T_p is an approximate indication of the rise time, Figure 5.10 also roughly indicates rise time. An example is now given to indicate the usefulness of these curves.

Example 5.2
We will consider the design of the servo system of Figure 5.9 and the block diagram is shown in Figure 5.11. Normally, the signal out of the difference circuit that implements the summing junction does not have sufficient power to drive the motor, and a power amplifier must be used. The transfer function of the system is

$$T(s) = \frac{0.5K_a/[s(s+2)]}{1 + 0.5K_a/[s(s+2)]} = \frac{0.5K_a}{s^2 + 2s + 0.5K_a} = \frac{\omega_n^2}{s^2 + 2\zeta\omega_n s + \omega_n^2}$$

where K_a is the gain of the power amplifier. We assume that K_a is the only parameter in the system that can be varied. Hence, in designing this system, we can set only one characteristic of the system; that is, we can set only ζ,

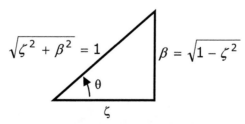

Figure 5.9   Relationship of θ, ζ, and β.

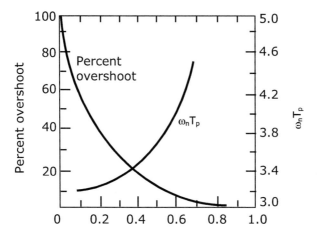

Figure 5.10 Relationship of percent overshoot, ζ, ωn, and T_p for a second-order system.

or only ω_n, or only a single function of the two. We cannot set ζ and ω_n independently by choosing the power amplifier gain, K_a.

Suppose that the servo controls the pen position of a plotter in which we can allow no overshoot for a step input. From Figure 5.12(a), $\zeta = 1$ gives the fastest response with no overshoot for a given ω_n, and we thus choose this value. From the derived transfer function

$$2\zeta\omega_n = 2(1)\omega_n = 2$$

Therefore, $\omega_n = 1$. Also, from the derived transfer function

$$\omega_n^2 = (1)^2 = 0.5K_a$$

and therefore $K_a = 0.2$. Hence setting the power amplifier gain to 2.0 results in the fastest response with no overshoot for the plotter, but in this design we have no control over the actual speed of response. Note that the settling time will be

$$T_s = \frac{4}{\zeta\omega_n} = 4s$$

This plotter would not be fast enough for most application. To improve the plotter, first we should choose a different motor, one that responds faster. Then it would probably be necessary to add a compensator in the closed-loop system, a topic that is covered in the chapters on design.

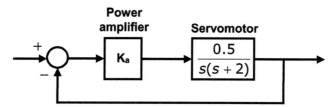

Figure 5.11   System design for servo motor.

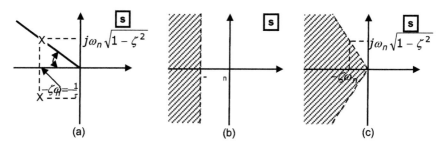

Figure 5.12   Pole location.

5.5.1 Time Response and Pole Locations

The parameters of the step response of the standard second-order system can also be related to the pole locations of the transfer function. Since the transfer function is given by

$$G(s) = \frac{\omega_n^2}{s^2 + 2\zeta\omega_n s + \omega_n^2}$$

the poles are calculated to occur at $s = -\zeta\omega \pm j\omega_n\sqrt{1 - \zeta^2}$. The poles are shown in Figure 5.12(a). The settling time T_s is related to these poles by (5.29), which is repeated here:

$$T_s = k\tau = \frac{k}{\zeta\omega_n}$$

where k is often chosen to be equal to 4. The settling time is then inversely related to the real part of the poles. If in design the settling is specified to be less than or equal to some value T_{sm}, $\zeta\omega_n \geq k/T_{sm}$, and the pole locations are then restricted to the region of the s-plane indicated in Figure 5.10(b). Hence the speed of response is increased by moving the poles to the left in the s-plane.

The angle α in Figure 5.10(a) satisfies the relationship

$$\alpha = \tan^{-1} \frac{\sqrt{1 - \zeta^2}}{\zeta} = \cos^{-1} \zeta \tag{5.34}$$

The per cent overshoot is given by

$$\text{percent overshoot} = e^{-\zeta \pi / \sqrt{1 - \zeta^2}} \times 100$$

Hence this equation can be expressed as

$$\text{percent overshoot} = e^{-\pi / \tan \alpha} \times 100$$

Decreasing the angle α reduces the per cent overshoot. Hence, specifying the per cent overshoot to be less than a particular value restricts the pole locations to the region of the s-plane, as shown in Figure 5.12(c). These relationships will now be illustrated with an example.

Example 5.3
As an example, suppose that, in the design of a second-order system, the per cent overshoot in the step response of a second-order system is limited to 4.32%, which is the value that results from $\zeta = 0.707$ (a commonly used value). Thus, in (5.34), $\alpha = \tan^{-1} 1 = \cos^{-1} 0.707 = 45$. Suppose that the design specifications require a maximum settling time of 2 s. Then $\tau \leq 0.5$, and

$$\zeta \omega_n = \frac{1}{\tau} \leq 2$$

Hence, the transfer function pole locations of the second-order system are limited to the regions of the s-plane shown in Figure 5.13. The pole locations that exactly satisfy the limits of the specifications are s $= -2 \pm j2$.

In summary, for the standard second-order system, the characteristics of the step response are related to the transfer function parameters by the following equations:

$$T_s = k\tau = \frac{k}{\zeta \omega_n}$$

where we let k = 4, and

[eq. (5.32)] percent overshoot $= e^{-\zeta \pi / \sqrt{1 - \zeta^2}} \times 100$
[eq (5.33)] $\omega_n T_p \frac{\pi}{\sqrt{1 - \zeta^2}}$

The angle α of the pole locations in the s-plane, defined in Figure 5.10, is given by

$$\alpha = \tan^{-1} \frac{\sqrt{1 - \zeta^2}}{\zeta} = \cos^{-1} \zeta$$

5.6 Frequency Response of Systems

In the preceding sections, the time responses of the first- and second-order systems were considered. In this section we give meaning to the steady-state response of systems to sinusoidal inputs, which is called the frequency response. We show later that the frequency response has meaning far beyond the calculation of the tie response to sinusoids.

Suppose that the input to a system with the transfer function G(s) is the sinusoid

$$r(t) = A \cos \omega_1 t \tag{5.39}$$

Then

$$R(s) = \frac{As}{s^2 + \omega_1^2} \tag{5.40}$$

(5.36)
and

$$C(s) = G(s)R(s) = G(s)\frac{As}{(s - j\omega_1)(s + j\omega_1)} \quad j = \sqrt{-1} \tag{5.41}$$

We can expand this expression into partial fractions of the form

$$C(s) = \frac{k_1}{s - j\omega_1} + \frac{k_2}{s + j\omega_1} + C_g(s) \tag{5.42}$$

where $C_g(s)$ is the collection of all the terms in the partial-fraction expansion that originate in the denominator of G(s). It is assumed that the system is such that the terms in $C_g(s)$ will decay to zero with increasing time. Therefore, only the first two terms in (5.38) contribute to the steady-state response. The result

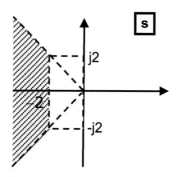

Figure 5.13   Pole for the example.

is that the steady-state response to a sinusoid is also a sinusoid of the same frequency.

$$k_1 = \frac{j\omega_1}{j2\omega_1} AG(j\omega_1) = \frac{1}{2} AG(j\omega_1) \tag{5.43}$$

$$k_2 = \frac{-j\omega_1}{-j2\omega_1} AG(-j\omega_1) = \frac{1}{2} AG(-j\omega_1) \tag{5.44}$$

and k_2 is seen to be the complex conjugate of k_1.

Since, for a given value of ω_1, $G(j\omega_1)$ is a complex number, it is convenient to express $G(j\omega_1)$ as

$$G(j\omega_1) = |G(j\omega_1)|e^{j\phi(j\omega_1)} \tag{5.45}$$

where $|G(j\omega_1)|$ is the magnitude and ϕ is the angle of the complex number. Then from (5.38) through (5.41), the sinusoidal steady-state value of $c(t)$ is

$$
\begin{aligned}
c_{ss}(t) &= k_1 e^{j\omega_1 t} + k_2 e^{-j\omega_1 t} \\
&= \frac{A}{2}|G(j\omega_1)|e^{j\phi}e^{j\omega_1 t} + \frac{A}{2}|G(j\omega_1)|e^{-j\phi}e^{-j\omega_1 t} \\
&= A|G(j\omega_1)|\frac{e^{j(\omega_1 t + \phi)} + e^{-j(\omega_1 t + \phi)}}{2} = A|G(j\omega_1)|\cos(\omega_1 t + \phi)
\end{aligned}
\tag{5.46}
$$

since

$$|G(-j\omega_1)| = |G(j\omega_1)| \tag{5.47}$$

In (5.46), $\phi = \phi$ ($j\omega_1$). We see then that the steady-state gain of a system for a sinusoidal input is the magnitude of the transfer function evaluation at $s = j\omega_1$, and the phase shift of the output sinusoid relative to the input sinusoidal is the angle of $G(j\omega_1)$. This proof is seen to be the same as that for ac circuit analysis, in which we work with the impedance $Z(j\omega_1)$. Sinusoidal steady-state response is now illustrated with an example.

Example 5.4

Consider a system with the transfer function

$$G(s) = \frac{5}{s + 2}$$

and an input of $7 \cos 3t$. Find the steady-state output.

Solution 5.4

$$G(s)|_{s=j3} = \frac{5}{2 + j3} = 1.387e^{-j56.3^0} = 1.387\angle -56.3^0$$

and the steady-state output is given by

$$c_{ss}(t) = (1.387)(7)\cos(3t - 56.3^0) = 9.709\cos(3t - 56.3^0)$$

Since the system time constant is 0.5 s, the output would reach steady-state approximately 2 s after the application of the input signal.

Recall that the sinusoidal steady-state response is determined from the system transfer function, G(s), evaluated at $s = j\omega_1$, where ω_1 is the frequency of the sinusoid. We define the frequency response function to be the function $G(j\omega)$, $0 \leq \omega < \infty$. For a given value of ω, $G(j\omega)$ is a complex number. Thus, the function $G(j\omega)$ is a complex function.

5.6.1 First-Order Systems

Consider first the frequency response of a first-order system. Now, the standard form of the first-order system is given by

$$G(s) = \frac{K}{\tau s + 1}$$

The frequency response function of this system is then

$$G(j\omega) = \frac{K}{1 + j\tau\omega} = |G(j\omega)|e^{j\phi(\omega)} \tag{5.48}$$

where the magnitude and the phase of the frequency response are given by

$$|G(j\omega)| = \frac{K}{[1 + \tau^2\omega^2]^{1/2}} \phi(\omega) = -\tan^{-1}\tau\omega \tag{5.49}$$

In this section we are interested primarily in the magnitude of the frequency response. A plot of IG(jω)I for (5.49) is given in Figure 5.14(a).

In Figure 5.14(a) the frequency ω_B denotes the frequency at which the gain is equal to $1/\sqrt{2}$ times the gain at very low frequencies; this frequency is called the system bandwidth. The factor $1/\sqrt{2}$ originated in the study of amplifiers and is the frequency at which the power output of an amplifier has decreased to one-half the maximum low-frequency value. In this book we use this definition for bandwidth. For the first-order system of (5.47), the bandwidth is forum from

$$\frac{K}{(1 + \tau^2\omega_B^2)^{1/2}} = \frac{K}{\sqrt{2}} \tag{5.50}$$

or $\omega_{\mathrm{B}} = 1/\tau$. Hence we see that the time constant τ also has meaning in the frequency domain.

It is convenient at this point to normalize frequency by the factor τ; that is, we define normalized frequency ω_{v} to be equal to $\tau\omega$. Then, the normalized frequency response $G_n(j\omega_{\mathrm{v}})$ is given by

$$G_n(j\omega_v) = G(j\omega)|_{\omega = \omega_v/\tau} = G(j\omega_v/\tau) = \frac{K}{(1 + \omega_v^2)^{1/2}} e^{j\phi} \qquad (5.51)$$

A plot of the magnitude of the normalized frequency response is given in Figure 5.14(b). The bandwidth in normalized frequency is $\omega_{\mathrm{v}} = 1$, which checks the previous calculation for the bandwidth.

In a like manner, if we normalize time by $t_{\mathrm{v}} = t/\tau$, the step response in normalized time, $c_n(\tau_{\mathrm{v}})$, is given by

$$c_n(t_v) = c(t)|_{t = \tau t v} = K(1 - e^{-t_v}) \qquad (5.52)$$

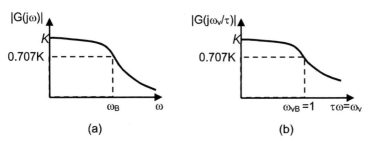

(a)   (b)

Figure 5.14   Illustration of bandwidth.

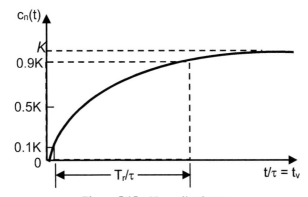

Figure 5.15   Normalized step.

from (5.9) and this normalized step response is independent of τ. A plot of this normalized step response is given in Figure 5.15.

Suppose that for a given first-order system, it is desired to decrease the rise time by a factor of 2. The new time constant required is then $\tau/2$ (Figure 5.13). Then, from (5.50), the bandwidth is increased by a factor of 2. In fact, for a first-order system, to decrease the rise time by any factor, the bandwidth must be increased by the same factor.

5.6.2 Second-Order System

We now consider second-order systems with the standard second-order transfer function

$$G(s) = \frac{\omega_n^2}{(s^2 + 2\zeta\omega_n s \omega_B^2)} = \frac{1}{(s/\omega_n)^2 + 2\zeta(s/\omega_n) + 1} \tag{5.53}$$

The frequency response is given by

$$G(j\omega) = \frac{1}{[1 - (\omega/\omega_n)^2] + j2\zeta(\omega/\omega_n)} \tag{5.54}$$

For this transfer function we define normalized frequency as $\omega_v = \omega/\omega_n$ and the magnitude of the frequency response is given by

$$|G_n(j\omega_v)| = |G(j\omega)|_{\omega=\omega_n\omega_v} = \frac{1}{((1 - \omega_v^2)^2 + (2\zeta\omega_v)^2)^{1/2}} \tag{5.55}$$

Plots of this frequency response for various values of ζ are given in Figure 5.16.

Note from Figure 5.16 that, for a given ζ, the ratio ω_B/ω_n is equal to a constant. Hence, increasing ω_n causes a bandwidth to increase by the same factor with ζ constant. As a second point, consider (5.33):

$$\omega_n T_p = \frac{\pi}{\sqrt{1 - \zeta^2}}$$

Then, for ζ constant, increasing ω_n decreases T_p, and hence the rise time T_r, by the same factor. Therefore, for ζ constant, we have the same result as for the first-order system. Increasing the bandwidth by a given factor decreases rise time by exactly the same factor for the first- and second-order systems discussed here.

The result derived in the preceding paragraph applies exactly to the two systems considered but only approximately t system in general. To increase

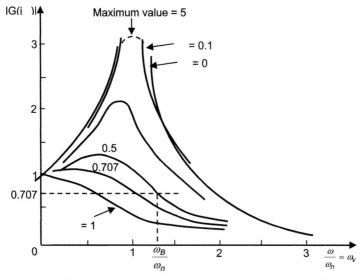

Figure 5.16   Frequency response of a second-order system (5.55).

the speed of response of a system, it is necessary to increase the system bandwidth. For a particular system, an approximate relationship can be derived, and is given by

$$\omega_B T_r \cong \text{constant} \qquad (5.56)$$

where this constant has a value in the neighbourhood of 2. This relationship can be plotted exactly for the standard second-order transfer function of (5.53) and (5.54). The rise time T_r can be obtained directly from Figure 5.4 for various values of ζ. Note that the actual rise time in seconds is $T_r = T_{rv}/\omega_n$, where T_{rv} is the normalized rise time in the units of $\omega_n t$ from Figure 5.6. The bandwidth can be obtained from Figure 5.16, and the actual bandwidth is given by $\omega_B = \omega_{Bv}\omega_n$, where ω_{Bv} is in the normalized bandwidth in the units of Figure 5.16. Hence, the bandwidth-rise time product is constant.

This product is independent of ω_n and is plotted in Figure 5.17 as a function of ζ for $0 = \zeta \leq 2$. Note that as ζ increases, the bandwidth-rise time product approaches the constant value of 2.20, which is the value for a first-order system.

It is also of value to correlate the peak in the magnitude of the frequency response (Figure 5.14), with the peak in the step response (Figure 5.6). It has been shown that the peak value in the step response is a function only of the damping ratio, ζ. The magnitude of the frequency response is given by (5.55), in the normalized frequency variable $\omega_v = \omega/\omega_n$. To find the peak value of this magnitude, we differentiate (5.55) and set this result to zero. Replacing ω_v with ω/ω_n and solving the resulting equation yields the frequency at which the peak occurs.

$$\omega_r = \omega_n\sqrt{1-\zeta^2} \quad \zeta < 0.707 \tag{5.57}$$

The maximum magnitude of the frequency response, denoted by $M_{p\omega}$, is then, from (5.55)

$$M_{p\omega} = |G(j\omega_r)| = \frac{1}{2\zeta\sqrt{1-\zeta^2}}$$

5.6.3 System dc Gain

The system dc gain is the steady-state gain to a constant input for the case that the output has a final value, and it is equal to the system transfer function evaluated at $s = 0$.

Thus the peak value of the magnitude of the frequency response is a function only of ζ, as is the peak value of the step response. These relationships are shown graphically in Figure 5.16.

The peaking in a frequency response indicates a condition called resonance, and ω_r in Figure 5.16 is called the resonant frequency. For the standard second-order system, the peak in the frequency response is directly related to the amount of overshoot in the step response (and hence in the transient response). For higher-order systems, generally the higher the peak in the magnitude of the frequency response (the more resonance present), the more the overshoot that will occur in the transient response. For this reason, we

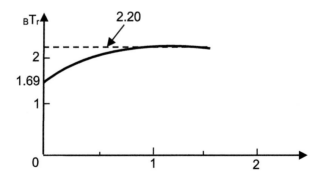

Figure 5.17   Bandwidth-rise time product versus ζ.

prefer that control systems not have significant resonances. In design, this restriction is usually specified by giving maximum allowable value for $M_{p\omega}$.

5.7 Problems and Solutions

Problem 5.1

In Figure 5.1, x(t) represents the input to a system and h(t) represents the corresponding impulse response. Use convolution to obtain the system output. Assume all initial energy storage in the systems is zero.

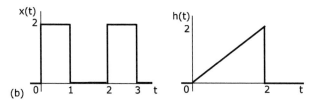

(b)

Solution 5.1

(a) Proceeding as in part (a)

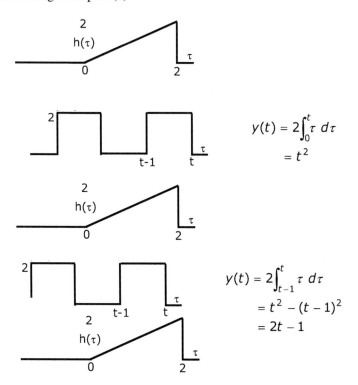

$$y(t) = 2\int_0^t \tau\, d\tau$$
$$= t^2$$

$$y(t) = 2\int_{t-1}^t \tau\, d\tau$$
$$= t^2 - (t-1)^2$$
$$= 2t - 1$$

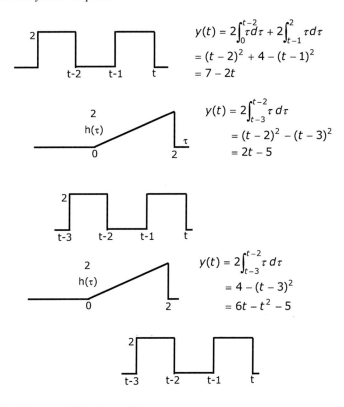

$$y(t) = 2\int_0^{t-2}\tau d\tau + 2\int_{t-1}^{2}\tau d\tau$$
$$= (t-2)^2 + 4 - (t-1)^2$$
$$= 7 - 2t$$

$$y(t) = 2\int_{t-3}^{t-2}\tau\, d\tau$$
$$= (t-2)^2 - (t-3)^2$$
$$= 2t - 5$$

$$y(t) = 2\int_{t-3}^{t-2}\tau\, d\tau$$
$$= 4 - (t-3)^2$$
$$= 6t - t^2 - 5$$

The function y(t) is shown below

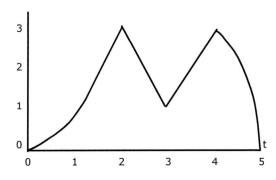

(c) In this part it is slightly easier to consider the functions x(τ) and h(t-τ):

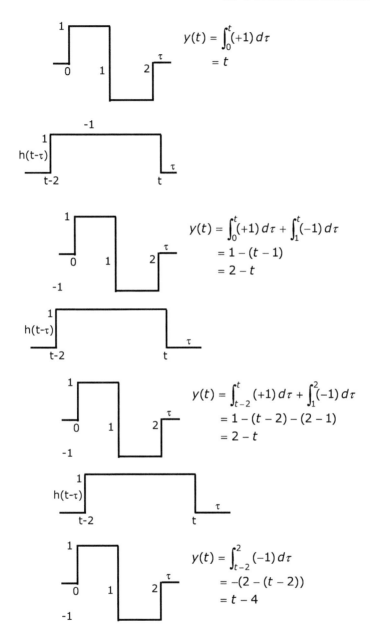

$$y(t) = \int_0^t (+1)\, d\tau$$
$$= t$$

$$y(t) = \int_0^t (+1)\, d\tau + \int_1^t (-1)\, d\tau$$
$$= 1 - (t - 1)$$
$$= 2 - t$$

$$y(t) = \int_{t-2}^t (+1)\, d\tau + \int_1^2 (-1)\, d\tau$$
$$= 1 - (t - 2) - (2 - 1)$$
$$= 2 - t$$

$$y(t) = \int_{t-2}^2 (-1)\, d\tau$$
$$= -(2 - (t - 2))$$
$$= t - 4$$

The function y(t) is as shown below

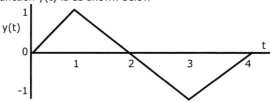

Problem 5.2
In the following figure, x(t) represents the input to a system and h(t) represents the corresponding impulse response. Use convolution to obtain the system output. Assume all initial energy storage in the systems is zero.

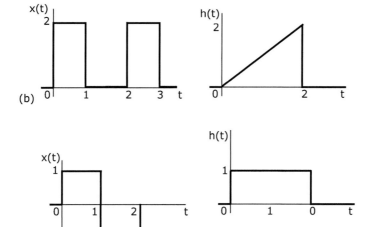

Solution 5.2

(b) Proceeding as in part (a)

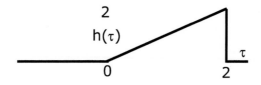

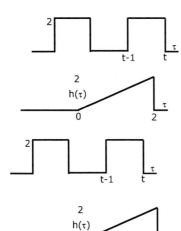

$$y(t) = 2\int_0^t \tau \, d\tau$$
$$= t^2$$

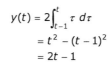

$$y(t) = 2\int_{t-1}^t \tau \, d\tau$$
$$= t^2 - (t-1)^2$$
$$= 2t - 1$$

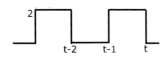

$$y(t) = 2\int_0^{t-2} \tau d\tau + 2\int_{t-1}^2 \tau d\tau$$
$$= (t-2)^2 + 4 - (t-1)^2$$
$$= 7 - 2t$$

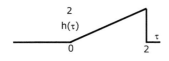

$$y(t) = 2\int_{t-3}^{t-2} \tau \, d\tau$$
$$= (t-2)^2 - (t-3)^2$$
$$= 2t - 5$$

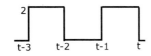

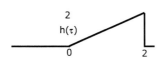

$$y(t) = 2\int_{t-3}^{t-2} \tau \, d\tau$$
$$= 4 - (t-3)^2$$
$$= 6t - t^2 - 5$$

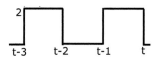

The function y(t) is shown below

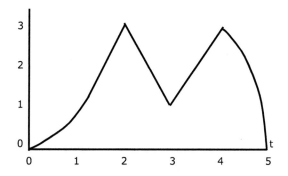

(c)　In this part it is slighty easier to consider the functions x(τ) and h(t-τ)

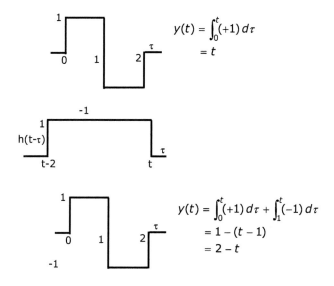

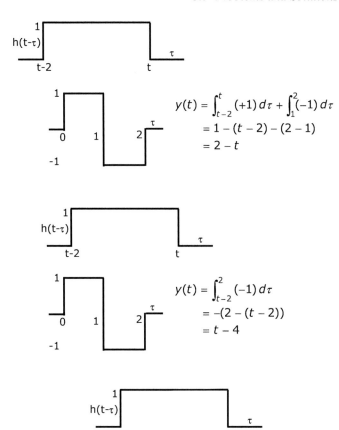

$$y(t) = \int_{t-2}^{t} (+1)\, d\tau + \int_{1}^{2} (-1)\, d\tau$$
$$= 1 - (t - 2) - (2 - 1)$$
$$= 2 - t$$

$$y(t) = \int_{t-2}^{2} (-1)\, d\tau$$
$$= -(2 - (t - 2))$$
$$= t - 4$$

The function y(t) is as shown below

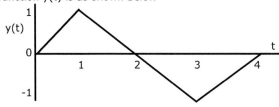

Problem 5.3

(a) Obtain the convolution of the following signals:

$$x_1(t) = e^{-at} \quad t \geq 0$$
$$= 0 \qquad t < 0$$
$$x_2(t) = e^{-bt} \quad t \geq 0$$
$$= 0 \qquad t < 0$$

Obtain an answer for the case where a = b and for case where a ≠ b.

(b) Obtain the convolution of the following signals:

$$x_1(t) = e^{-2t} \qquad t \geq 0$$
$$= 0 \qquad t < 0$$
$$x_2(t) = \sin 2\pi t \quad 0 \leq t \leq 1$$
$$= 0 \qquad \text{otherwise}$$

(Hint: Express $\sin 2\pi t$ in an exponential form and use the result of (a).)

Solution 5.3

(a) $y(t) = \int_{-\infty}^{+\infty} x_1(t - \tau) x_2(\tau) d\tau$

Because both the signals $x_1(t)$ and $x_2(t)$ are zero for $t < 0$ the convolution integral becomes

$$y(t) = \int_0^t e - a(t - \tau)e^{-b\tau} d\tau$$

$$= e^{-at} \int_0^t e^{-(b-a)\tau} d\tau$$

For a = b

$$y(t) = e^{-at} \int_0^t e^0 d\tau$$

$$= te^{-at}$$

For a $\neq$ b

$$y(t) = e^{-at} \left[\frac{e^{-(b-a)\tau}}{a - b} \right]_0^t$$

$$= \frac{e^{-bt} - e^{-at}}{(a - b)}$$

(b) $\sin 2\pi t = \frac{e^{j2\pi t} - e^{j2\pi t}}{2j}$

Using the result from part (a)

$$y(t) = \frac{1}{2j} \left[\frac{e^{-j2\pi t} - e^{-jt}}{2 + j2\pi} - \frac{e^{-j2\pi t} - e^{-2t}}{2 - j2\pi} \right]$$

The term in brackets is of the form of a complex number minus its complex conjugate, and it is equal to twice the imaginary part of the number.

$$y(t) = \frac{1}{2j} \times 2\,Im \left\{ \frac{(\cos 2\pi t - e^{-2t} + j\sin 2\pi t)(2 - j2\pi)}{(2 + j2\pi)(2 - j2\pi)} \right\}$$

$$= \frac{1}{2j} \times 2j \left\{ \frac{2\sin 2\pi t + 2\pi e^{-2t} - 2\pi \cos 2\pi t)}{4 + (2\pi)^2} \right\}$$

$$= \frac{2\sin 2\pi t + 2\pi e^{-2t} - 2\pi \cos 2\pi t)}{4 + (2\pi)^2}$$

Problem 5.4
As an impulse signal cannot be realized in practice, a short rectangular pulse can be used as a test signal to approximate the impulse. Show that if the true impulse response shows little variation over the width of the test pulse then the approximation is valid provided a scaling factor is used. What is the scaling factor?

A system has a time impulse response

$$h(t) = 10e^{-t} \quad t \geq 0$$
$$= 0 \qquad\quad t < 0$$

Use convolution to sketch the response to a finite width pulse. Using the appropriate scaling factors calculate the response to finite width pulse of 0.5, 0.1 and 0.01 s at: (a) a time equal to the pulse width and (b) time $t = 2$. Compare your answers with the true impulse response.

Solution 5.4

Consider the case where a narrow rectangular pulse is applied to an arbitrary system with impulse response h(t):

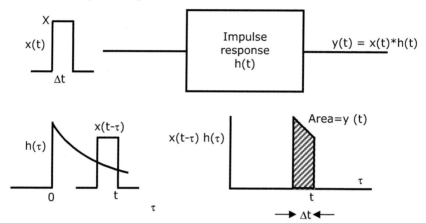

Provided $h(\tau)$ does not change appreciably over time Δt then the area y(t) is given by

$$y(t) \approx h(t)\Delta t X$$

This is the true impulse response h(t) multiplied by a scaling factor. The scaling factor is the area of the input pulse.

When the impulse response is as given, and the input is a pulse of width Δt and amplitude X, the output is given as follows:

$$t \geq \Delta t$$

$$y(t) = \int_{0-\Delta t}^{t} 10Xe^{-\tau}d\tau$$

$$0 \leq t \leq \Delta t$$

$$= 10Xe^{-t}(e^{\Delta t} - 1)$$

$$y(t) = \int_{0}^{t} 10Xe^{-\tau}d\tau$$

$$= 10X(1 - e^{-t})$$

for pulse width $\Delta t = 0.5$

(a) $t = 0.5$

Response to finite width pulse $= 10Xe^{-0.5} (e^{0.5} - 1)$

$$= 3.9347X$$

With the scaling factor of 0.5X this gives an approximation to the impulse response as 7.8694. The true impulse response at this tie is 6.0653.

(b) $t = 2$

Response to finite width pulse $= 10Xe^{-2.0} (e^{0.5} - 1)$
$$= 0.8779X$$

Proceeding as in part (a) this gives an approximate impulse response at this time of 1.7558 and the true impulse response is 1.3533.

Similar calculations apply for pulse widths of 0.1 and 0.01 s. The results are given in the following table.

Time	Impulse Response					
	$\Delta t = 0.5$		$\Delta t = 0.1$		$\Delta t = 0.01$	
	true	approx	true	approx	true	approx
Δt	6.0653	7.8695	9.0483	9.5196	9.9604	9.9502
2.0	1.3533	1.7558	1.3533	1.4233	1.3533	1.3533

Problem 5.5

Using commutative property of convolution, the order of cascaded LTI systems can be interchanged. Verify this result for the systems shown in Figure 5.5. Calculate the output signals y(t) and y(n) for the systems as shown and then with the systems interchanged.

Solution 5.5

(a) Convolution of x(t) with the impulse response of system 1 gives an intermediate result $y_1(t)$.

$$y_1(t) = t \qquad 0 \leq t \leq 1$$
$$= 2 - t \quad 1 \leq t \leq 2$$
$$= 0 \qquad \text{otherwise}$$

Convolution with the impulse response of system 2 can be considered in stages.

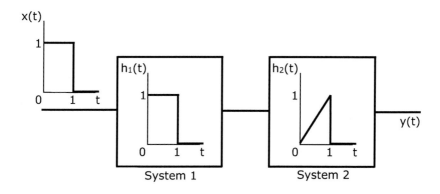

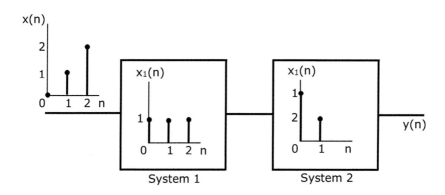

$0 \leq t \leq t$

$$y(t) = \int_0^t \tau(t - \tau)d\tau$$

$$= t\left[\frac{\tau^2}{2}\right]_0^t - \left[\frac{\tau^3}{3}\right]_0^t$$

$$= \frac{t^3}{6}$$

$1 \leq t \leq 2$

$$y(t) = \int_{t-1}^{1} \tau(t-\tau)d\tau + \int_{1}^{t} (2-\tau)(t-\tau)\,d\tau$$

$$= t\left[\frac{t\tau^2}{2} - \frac{\tau^3}{3}\right]_{t-1}^{1} + \left[2t\tau - (2+t)\frac{\tau^3}{2} + \frac{\tau^3}{3}\right]_{1}^{t}$$

$$= -\frac{t^3}{3} + t^2 - \frac{t}{2}$$

$2 \leq t \leq 3$

$$y(t) = \int_{t-1}^{2} (2-\tau)(t-\tau)d\tau$$

$$= \left[2t\tau - (2+t)\frac{\tau^2}{2} + \frac{\tau^3}{3}\right]_{t-1}^{2}$$

$$= \frac{t^3}{6} - t^2 + \frac{3t}{2}$$

Convolution f x(t) and the impulse response of system 2 gives the intermediate result $y_2(t)$.

$$y_2(t) = \int_{0}^{1} \tau d\tau = \frac{t^2}{2} \qquad 0 \leq t \leq 1$$

$$y_2(t) = \int_{t-1}^{1} \tau d\tau = \frac{2t - t^2}{2} \qquad 1 \leq t \leq 2$$

$$y_2(t) = 0 \qquad \text{otherwise}$$

Convolution with the impulse response of system 1 can be considered in stages:

$0 \leq t \leq 1$

$$y(t) = \int_{0}^{1} \frac{\tau^2}{2}d\tau = \frac{t^3}{6}$$

$1 \leq t \leq 2$

$$y(t) = \int_{t-1}^{1} \frac{\tau^2}{2}d\tau + \int_{1}^{t} \frac{(2\tau - \tau^2)}{2}d\tau$$

$$= \frac{-t^3}{3} + t^3 - \frac{t}{2}$$

$2 \le t \le 3$

$$y(t) = \int_{t-1}^{2} \frac{(2\tau - \tau^2)}{2} d\tau$$

$$= \frac{t^3}{3} - t^3 + \frac{3t}{2}$$

Convolution of the input x(n) with the impulse response of system 1 produces the intermediate variable $y_1(n)$:

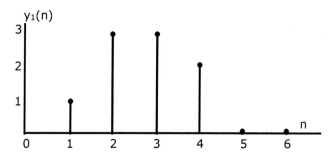

Convolution of $y_1(n)$ with the impulse response of system 2 gives

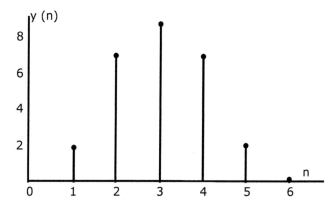

Reversing the order of the systems, convolution of the input with the impulse response of the input with the impulse response of system 2 gives the intermediate variable $y_1(n)$:

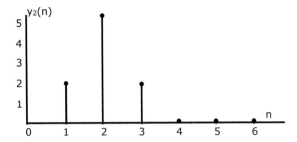

Convolution with the impulse response of system 1 gives y(n).

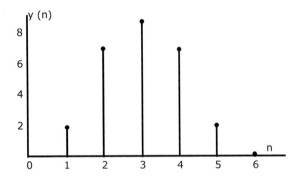

Problem 5.6

A discrete-time LTI system has the impulse response h [n] depicted in Figure P5.6(a). Use linearity and time invariance to determine the system output y[n] if the input x[n] is

(a) $x[n] = 2\delta[n] - \delta[n - i]$
(b) $x[n] = u[n] - u[n - 3]$
(c) $x[n]$ as given in Figure P5.6(b)

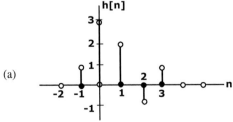

Figure P5.6(a) Channel impulse response.

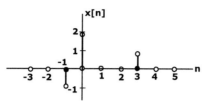

Figure P5.6(b)   LTI system input.

Solution 5.6

$$h[n] = \begin{cases} 1, & n = -1, 3 \\ -1, & n = 2 \\ 3, & n = 0 \\ 2, & n = 1 \\ 0, & \text{otherwise} \end{cases}$$

(a)

$$x[n] = 2\delta[n] - \delta[n - i]$$

$$y[n] = \sum_{K=-\infty}^{\infty} x[k]h[n - k]$$

$$\therefore \quad y[n] = 2h[n] - h[n - 1]$$

$$\therefore \quad h[n] = \begin{cases} 0, & n \le -2 \\ 2, & n = -1 \\ 5, & n = 0 \\ 1, & n = 1 \\ -4, & n = 2 \\ 1, & n = 3 \\ -1, & n = 4 \\ 0, & n \ge 5 \end{cases}$$

Hints:

For $n = 0$

$$y[n] = 2h[0] - h[0 - 1] = 2h[0] - h[-1]$$
$$= 2 \times 3 - (1) = 6 - 1 = 5$$

Similarly, all

(b)

$$x[n] = u[n] - u[n - 3]$$

graphically x[n] can be denoted as

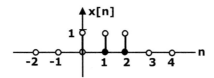

Figure P5.6(c) Discrete-time input.

This unit step function can be easily expressed in sum of impulse function:

$$i.e. \quad x[n] = \delta[n] + \delta[n - 1] + \delta[n - 2]$$

$$\therefore \quad y[n] = h[n] + h[n - 1] + h[n - 2]$$

Similarly y[n] can be expressed as :$y[n] = \begin{cases} 0, & n \leq -2 \\ 1, & n = -1 \\ 4, & n = 0 \\ 6, & n = 1 \\ 4, & n = 2 \\ 2, & n = 3 \\ 0, & n = 4 \\ 1, & n = 5 \\ 0, & n \geq 6 \end{cases}$

(c) x[n] can be represented as the sum of impulsive function of Figure P5.6(c) as

$$x[n] = -\delta[n + 1] + 2\delta[n] + \delta[n - 3]$$

$$\therefore \quad y[n] = -h[n + 1] + 2h[n] + h[n - 3]$$

Similarly y[n] can be expressed as :$y[n] = \begin{cases} 0, & n \leq -3 \\ 1, & n = -2 \\ -1, & n = -1 \\ 4, & n = 0 \\ 5, & n = 1 \\ -2, & n = 2 \\ 5, & n = 3 \\ 2, & n = 4 \\ -1, & n = 5 \\ 1, & n = 6 \\ 0, & n \geq 7 \end{cases}$

Problem 5.7

Consider the discrete-time signals depicted in the following figures. Evaluate the convolution sums indicated hereunder.

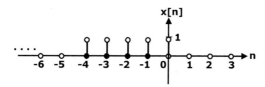

Figure P.5.7   Discrete-time signal.

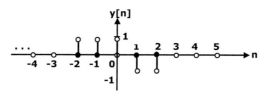

Figure P.5.7(a)   Discrete-time input

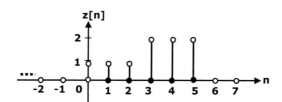

Figure P.5.7(b)   Discrete-time input

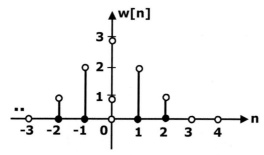

Figure P.5.7(c)   Discrete-time input.

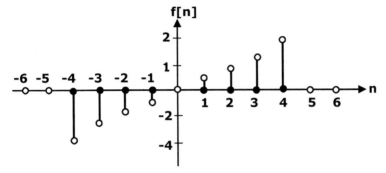

Figure P.5.7(d) Discrete-time input

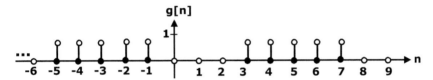

Figure P.5.7(e) Discrete-time input

(a) $m[n] = x[n] * z[n]$

(b) $m[n] = x[n] * z[n]$

(c) $m[n] = x[n] * f[n]$

(d) $m[n] = x[n] * g[n]$

(e) $m[n] = y[n] * z[n]$

(f) $m[n] = y[n] * w[n]$

(g) $m[n] = z[n] * g[n]$

(h) $m[n] = f[n] * g[n]$

Solution 5.7

(a) $m[n] = x[n] * z[n]$

$\therefore \quad m[n] = \ldots + x[-4]\delta[n + 4] + [-3]\delta[n + 3] + x[-2]\delta[n + 2]$

$\qquad + x[-1]\delta[n + 1] + x[0]\delta[n] + x[1]\delta[n-1] + x[2]\delta[n - 2] + \ldots\ldots$

$\qquad = 0 + 0 + 0 + \ldots\ldots + 0 + 0 + 0 + 0 + 1.1 + 0 + 0 + 0 + \ldots\ldots$

$\qquad = 1$

So $\therefore$ $m[n] = 1$ *Ans.*

(b) $m[n] = x[n] * y[n]$

$\because$ $m[n] = \displaystyle\sum_{K=-\infty}^{\infty} x[k]\delta[n-k]$

$\qquad k = -\infty$

$\qquad\quad = ... + x[-2]\delta[n+2] + x[-1]\delta[n+1] + x[0]\delta[n]$

$\qquad\qquad + x[1]\delta[n-1] + x[2]\delta[n-2] + ...$

$\qquad\qquad = 0 + 0 + 0 + ... + 1 \times 1 + 1 \times 1 + 1 \times 1 + 0 + 0 + 0 + 0...$

$\qquad\qquad = 1 + 1 + 1 = 3$

So, $m[n] = 3$ *Ans.*

(c) $m[n] = x[n] * f[n]$

$\because$ $m[n] = \displaystyle\sum_{K=-\infty}^{\infty} x[k]\delta[n-k]$

$\qquad\qquad = 0+0+0+......+(-4 \times 1)+(-3 \times 1)+(-2 \times 1)+(-1 \times 1)$

$\qquad\qquad\quad + 1 \times 0 + 0 + 0 + 0 + ...$

$\qquad\qquad = -4 - 3 - 2 - 1 = -10$

$\therefore m[n] = -10$ *Ans.*

(d) $m[n] = x[n] * g[n]$

$\because$ $m[n] = \displaystyle\sum_{K=-\infty}^{\infty} x[n]\delta[n-k]$

$\qquad\qquad = 0 + 0 + 0 + ...+(+1 \times 0) + 1 \times 1 + 1 \times 1 + 1 \times 1 + 1 \times 1$

$\qquad\qquad\quad + 1 \times 0 + 0 \times 0 + 0 \times 0 + 0 \times 1 + 0 \times 1 + ... + 0 + 0...$

$\qquad\qquad = 1 + 1 + 1 + 1 = 4$

$\therefore m[n] = 4$ *Ans.*

(e) $m[n] = y[n] * z[n]$

$\because \quad m[n] = \displaystyle\sum_{K=-\infty}^{\infty} y[n]\delta[n-k]$

$\qquad = 0 + 0 + 0 + \ldots + 1 \times 0 + 1 \times 0 + 1 \times 1 + 1 \times -1 + 1 \times -1$

$\qquad \quad + 2 \times 0 + 2 \times 0 + 2 \times 0 + 0 + 0 + 0 + \ldots$

$\qquad = 1 - 1 - 1 = -1$

$\therefore \ m[n] = -1 \ Ans.$

(f) $m[n] = y[n] * w[n]$

$\because \quad m[n] = \displaystyle\sum_{K=-\infty}^{\infty} y[n]\delta[n-k]$

$\qquad = 0 + 0 + 0 + \ldots + 1 \times 1 + 1 \times 2 + 1 \times 3 + (-1) \times 2 + (-1)$

$\qquad \quad \times 1 + 0 + 0 + 0 + \ldots$

$\qquad = 1 + 2 + 3 - 2 - 1 = 3$

$\therefore \ m[n] = 3 \ Ans.$

(g) $m[n] = z[n] * g[n]$

$\because \quad m[n] = \displaystyle\sum_{K=-\infty}^{\infty} z[n]\delta[n-k]$

$\qquad = 0 + 0 + 0 + \ldots + 1 \times 0 + 1 \times 0 + 1 \times 0 + 1 \times 0 + 1 \times 0 + 1$

$\qquad \quad \times 0 + 1 \times 0 + 1 \times 0 + 2 \times 1 + 2 \times 1 + 2 \times 1 + 0 + 0$

$\qquad \quad + 0 + \ldots$

$\qquad = 2 + 2 + 2 = 6$

$\therefore \ m[n] = 6 \ Ans.$

(h) $m[n] = f[n] * g[n]$

$\therefore$ $m[n] = \displaystyle\sum_{K=-\infty}^{\infty} f[n]\delta[n-k]$

$\quad\quad = 0 + 0 + 0 + \ldots + (-4) \times 1 + (-3) \times 1 + (-2) \times 1 + (-1)$
$\quad\quad\quad \times 1 + 0 + 0 + 0 + 1.5 \times 1 + 2 \times 1 + 0 + 0 + \ldots$
$\quad\quad = -4 - 3 - 2 - 1 + 1.5 + 2 = -6.5$

$\therefore$ $m[n] = -6.5$ *Ans.*

Problem 5.8

Evaluate the given discrete-time convolution sums.

(a) $y[n] = u[n] * u[n-3]$

(b) $y[n] = 2^n u[-n+2] * u[n-3]$

(c) $y[n] = \left(\frac{1}{2}\right)^n u[n-2] * u[n]$

(d) $y[n] = \cos\left(\frac{1}{2}\pi n\right) u[n] * u[n-1]$

(e) $y[n] = (u[n+10] - 2u[n+5] + u[n-6]) * u[n-2]$

(f) $y[n] = (u[n+10] - 2u[n+5] + u[n-6]) * \cos\left(\frac{1}{2}\pi n\right)$

Solution 5.8

(a) $y[n] = u[n] * u[n-3]$
$\quad\quad u[n] =$

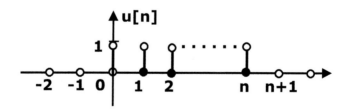

Figure P5.8(a)   Discrete-time input.

$u[n-3] =$

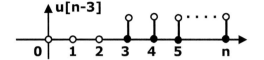

Figure P5.8(b) Discrete-time shifted input.

$$y[n] = u[n] * u[n-3]$$
$$\therefore \quad y[n] = ... + y[-4]\delta[n+4] + y[-3]\delta[n+3] + y[-2]\delta[n+2]$$
$$+ y[-1]\delta[n+1] + y[0]\delta[n] + y[1]\delta[n-1] + ...$$
$$= \sum_{K=-\infty}^{\infty} x[k]\delta[n-k]$$

$$\therefore \quad y[n] = 0 + 0 + ...0 + 1.0 + 1.0 + 1.0 + 1.1 + 1.1 + ...+$$
$$1.1 \; upto \; n$$
$$= 0 + 0 + 0 + 1 + 1 + 1 + ... + 1 \; upto \; n$$
$$y[n] = n - 2 \; Ans.$$
$$[as \; y[0] \; \delta[n] = y[1] \; \delta[n+1] = y[2] \; \delta[n-2] = 0]$$

(b) $y[n] = 2^n u[-n+2] * u[n-3]$

2n u[−n + 2] can be graphically represented as

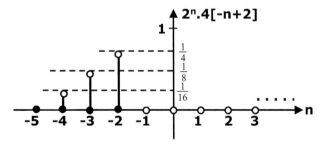

Figure P5.8(c) Discrete-time shifted input.

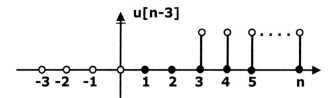

Figure P5.8(d)   Discrete-time shifted input.

$$y[n] = \sum_{K=-\infty}^{\infty} x[k]\delta[n-k] = 0+0+0+\ldots\ldots+0 \; uptp \; n$$

$$y[n] = 0 \; Ans.$$

(c) $y[n] = \left(\frac{1}{2}\right)^n u[n-2] * u[n]$

$$y[n] = \sum_{K=-\infty}^{\infty} x[k]\delta[n-k] = \left(\frac{1}{2}\right)^2 + \left(\frac{1}{2}\right)^3 + \left(\frac{1}{2}\right)^4 + \ldots\ldots + \left(\frac{1}{2}\right)^n$$

Or, $y[n] = \left(\dfrac{1-\left(\frac{1}{2}\right)^n}{\left(1-\frac{1}{2}\right)}\right) - \dfrac{1}{2} = 2\left(1-\left(\dfrac{1}{2}\right)^n\right) - \dfrac{1}{2}$ *Ans.*

$$When \; n \; is \; very-very \; l\arg e, then \; \left(\frac{1}{2}\right)^n = 0$$

$$y[n] = 2 - \frac{1}{2} = \frac{3}{2}$$

(d) $y[n] = \cos\left(\frac{1}{2}\pi n\right) u[n] * u[n-1]$

$\cos\left(\frac{1}{2}\pi n\right) \cdot u[n]$ can be represented as :

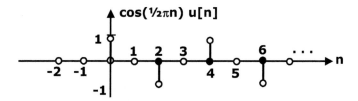

Figure P5.8(e)   Discrete-time unit step input.

and u[n − 1] can be represented as

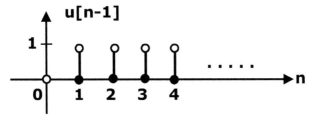

Figure P5.8(f) Discrete-time unit step shifted input.

$$y[n] = \sum_{K=-\infty}^{\infty} y[k]\delta[n-k]$$

$$y[n] = 0 + 0 + (-1) + 0 + (1) + 0 + (-1) + 0 + (1) + \dots\dots upto \; n$$

$$y[n] = 0$$

(i) $y[n] = (u[n+10] - 2u[n+5] + u[n-6]) * u[n-2]$

$(u[n+10] - 2u[n+5] + u[n-6])$ can be represented as:

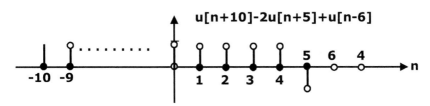

Figure P5.8 (g) Discrete-time unit step shifted input.

and $u[n-2]$ can be represented as

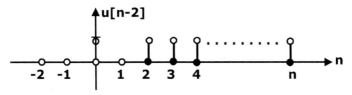

Figure P5.8 (h) Discrete-time unit step shifted input.

$$y[n] = \sum_{K=-\infty}^{\infty} x[k]\delta[n-k]$$

$$= \ldots + x[2]\delta[n-10] + x[3]\delta[n-3] + x[4]\delta[n-4]$$
$$+ x[5]\delta[n-5] + \ldots\ldots$$
$$= 0 + 0 + 0 + \ldots + 1.1 + 1.1 + 1.1 + 1.(-1) + 0 + 0 +$$
$$0 + \ldots\ldots \ upto \ n$$
$$= 1 + 1 + 1 - 1 = 2$$

$\therefore \quad y[n] = 2 \ Ans.$

(k) $\quad y[n] = (u[n+10] - 2u[n+5] + u[n-6]) * \cos\left(\frac{1}{2}\pi n\right)$

Let $x[n] = u[n+10] - 2u[n+5] + u[n-6]$ can be represented as:

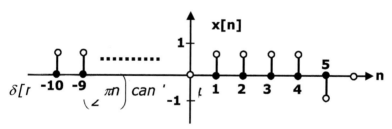

Figure P5.8 (i)   Discrete-time impulse input.

Let $\quad \delta[n] = \cos\left(\frac{1}{2}\pi n\right)$ can be represented as

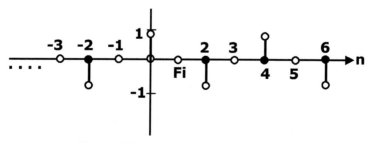

Figure P5.8 (j)   Discrete-time impulse input.

$$\therefore \quad y[n] = \sum_{K=-\infty}^{\infty} x[k]\, \delta[n-k]$$

$$= ... + x[-10]\delta[n+10] + x[-9]\delta[n+9] + ... + x[5]\delta[n-5] + ...$$

$$= 0 + 0 + 0 + -1.1 + 0 + 1.1 + 0 - 1.1 + 0 + \cdot 1.1 + 0 +$$

$$- 1.1 + 0 + 1.1 + 0 + -1.1 + 0 + 1.1 + 0 + 0 + 0 + ...$$

$$= -1 + 1 - 1 + 1 - 1 + 1 - 1 + 1$$

$$= -4 + 4 = 0$$

$$\therefore \quad y[n] = 0 \; Ans.$$

Problem 5.9

Evaluate the continuous time convolution integration given hereunder:

(a) $y(t) = u(t+1) * u(t-2)$

(b) $y(t) = e^{-2t}u(t) * u(t+2)$

(c) $y(t) = \cos(\pi t)(u(t+1) - u(t-3)) * u(t)$

(d) $y(t) = (u(t+2) - u(t-1)) * u(-t+2) = 0$

Problem 5.10

(a) The first-order plant of Figure P5.10(a) has the unit step response given in Figure P5.10(b). Find the parameters of the transfer function.

(b) Verify that your result yields the plot in Figure P5.10(b).

(c) Find the unit impulse response of the plant.

(d) The plant is connected into the closed-loop system as shown in Figure P5.10(c). Sketch the unit step response of the closed-loop system.

Solution 5.10

(a) $c(t) = K(1 - e - t/T)$

$$C(0.5) = 0.730 = K(1 - e^{-0.5/T})$$

$$C(1.0) = 1.104 = K(1 - e^{-1.0/T})$$

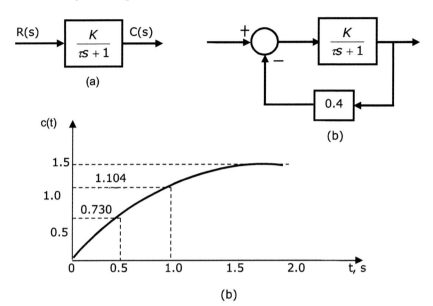

(a)

(b)

(b)

Figure P5.10   Control system response.

$$\therefore \quad \frac{1 - e^{-0.5/T}}{1 - e^{-1.0/T}} = \frac{0.730}{1.104}; \quad \therefore \quad \text{let } x = e^{-0.5/T}$$

$$\therefore \quad 1.104(1 - \gamma) = 0.730(1 - \gamma^2)$$

$$0.374 - 1.104x + 0.73x^2 = 0$$

$$x = \frac{1.104 \pm [1.21882 - 0.09208]^{1/2}}{z(0.73)} = 0.512$$

$$\therefore \quad e^{-0.5/T} = 0.512$$

$$T = \underline{0.746}$$

$$K = \frac{0.730}{1 - 0.512} = \underline{1.496}$$

(b) $C(s)\frac{1}{s}\frac{1.496}{0.746s+1} = \frac{2}{s(s+1.340)} = \frac{1.496}{s} + \frac{-1.496}{s+1.340}$

$$c(t) = 1.496(1 - e^{-1.340t})$$

$$c(0.5) = 1.46(1 - e^{-1.340(05)}) = \underline{0.730}$$

$$c(1) = 1.46(1 - e^{-1.340}) = \underline{1.104}$$

(c) impulse responce $S^{-1}\left[\dfrac{2}{s+1.340}\right] = \underline{2e^{-1.340t}}$

(d) $T(s) = \dfrac{K/(Ts+1)}{1+0.4k/(Ts+1)} = \dfrac{2}{s+1.34+0.8} = \dfrac{2}{s+2.14}$

$\quad = \dfrac{0.935}{0.467s+1},\quad \therefore \underline{k=0.935,\ T=0.467s}$

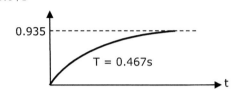

Problem 5.11

The closed-loop first-order system of Figure P5.11(a) has the unit step response given in Figure P5.11(b). Find as many of the parameters of the transfer functions as is possible from the given information (H_k is a constant gain).

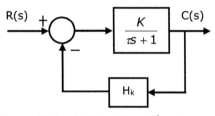

Figure P5.11   (a) Closed loop 1st order system.

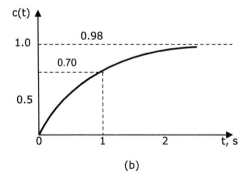

(b)

Figure P5.11   (b) Unit step response

Solution 5.11

$$T(s) = \frac{K/(Ts+1)}{1+1+KH_k/(Ts+1)} = \frac{K/(1+1+KH_k)}{\frac{T}{1+1+KH_k}s+1} = \frac{K_1}{T_1s+1}$$

$$k_1 = 0.98$$

$$k_1 = (1 - e^{-1.0/T_1}) = 0.98(1 - e^{-1.0/T_1}) = 0.70$$

$$\therefore e^{-1.0/T_1} = 0.2857, \quad T_1 = 0.798$$

$$\therefore \frac{K}{1+KH_k} = 0.98; \quad \frac{K}{1+KH_k} = 0.798$$

2 equations, 3 unknowns.

Problem 5.12

For the system shown in Figure P5.12, sketch the unit step response of the system without mathematically solving for the time response, c(t). Indicate approximate numerical values on both the amplitude axis and the time axis.

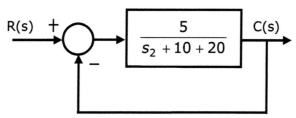

Figure P5.12   Closed loop 2nd order system

Solution 5.12

(a) $$T(s) = \frac{G(s)}{1+G(s)} = \frac{5}{s^2 + 10s + 25} = \frac{5}{(s+5)^2}$$

$$\therefore \zeta = 1, \text{ critically damped}$$

$$T = \frac{1}{5} = \underline{0.2s}, \quad dc\ gain = T(0) = \underline{0.2}$$

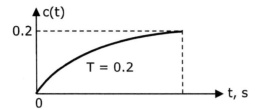

Problem 5.13

Shown in Figure P5.13 is a satellite altitude control system. The moment of inertia J of the satellite has been normalized to unity.

(a) Write the system transfer function for this system.
(b) The system is commanded to assume an altitude of $10[\theta_r(t) = 10u(t)]$. After the transients die out (the system reaches steady state), what will be the altitude angle of the satellite $[\theta_{ss}(t)]$?
(c) Design the system by finding K_v as a function of K such that the closed-loop system responds to a step input in minimum tine with no overshoot, which requires that $\zeta = 1$.
(d) Complete the system design in (c) by finding K such that the system reaches steady state approximately 6 s after a command to change the altitude angle (after the application of an input).

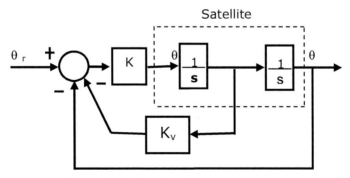

Figure P5.13 Satellite altitude control system.

(e) The rate signal is measured using a rate gyro. Suppose that the rate gyro fails, such that no signal appears in the rate path (effectively, $K_V = 0$). What is the nature of the response in this failure mode? (This failure occurred on a space lab mission of the National Aeronautics and Space Administration [NASA], with the predictable results.)

Solution 5.13

(a) $T(s) = \frac{Ks^{-2}}{1+KK\wp s^{-1}+Ks^{-2}} = \frac{K}{s^2+KK\wp s+K}$

(b) $dc\ gain = T(0) = 1$, $\quad \therefore \theta_{ss}(t) = 10^0$

(c) $critical\ damping, \zeta = 1$
$$w_n^2 = k; \quad w_n = \sqrt{K}$$

$$z\zeta w_n = 2w_n = 2\sqrt{K} = K\ K_n; \quad \therefore K_n = \frac{2}{K^{1/2}}$$

(d) $T = \underline{1.5s} = \frac{1}{\zeta w_n} = \frac{1}{\sqrt{k}}, \therefore \sqrt{K} = \frac{1}{1.5} = 0.667$

$$\therefore \underline{K = 0.444}; \quad K_n = \frac{2}{0.667} = \underline{3}$$

(e) See (g)

(f) $T(s) = T(s) = \frac{K}{s^2+K}, \quad \therefore \zeta = 0$

Response is an undamped sinusoid, with

$$w = \sqrt{K} = \underline{0.667}, \quad \therefore T = \frac{2\pi}{w} = \underline{9.42s}$$

(g) $T(s) = \frac{0.444}{s^2+1.333s+0.444}$

Problem 5.14
Consider the system of Figure P5.14.

(a) Find the range of K for which the system is
  (i) Underdamped
  (ii) Critically damped
  (iii) Overdamped

(b) Find the value of K that will result in the system having minimum settling time.

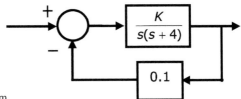

Closed loop 2nd order system

Figure P5.14

Solution 5.14

$$T(s) = \frac{K/(s^2 + 4s)}{1 + 0.1K/(s^2 + 4s)} = \frac{K}{s^2 + 4s + 0.1K} \Rightarrow W_n = \sqrt{0.1K}$$

$$2\zeta W_n = 4 = 2\zeta\sqrt{0.1K} \Rightarrow 4\zeta(0.1K) = 16 \Rightarrow K = \frac{40}{\zeta}$$

(a) (i) $\zeta < 1 \Rightarrow K > \frac{40}{1} = 40 \quad \therefore \quad 40 < k > \infty$

 (ii) $\zeta = 1 \Rightarrow K = 40 \quad \therefore \quad k = 4$

 (iii) $\zeta > 1 \Rightarrow K < 40 \quad \therefore \quad \underline{0 < k < 40}$

(b) We want T as minimum $\quad \therefore \quad$ for $\zeta \leq 1$ or $K \geq 40$
  $w_n^2 \geq 4 \Rightarrow w_n \geq 2$
  $2\zeta w_n = 4$ and $w_n = \sqrt{0.1k}, \quad \therefore \quad \zeta = \frac{4}{2\sqrt{0.1k}}$
  $\therefore \quad T = \frac{1}{\zeta w_n} = \frac{4}{\left(\frac{4}{2\sqrt{0.1k}}\right)\left(\sqrt{0.1k}\right)} = \underline{0.5}$, independent of K.
  $\therefore \quad 40 \leq k < \infty$

For $\zeta > 1$, we have two time constants with the larger one greater than 0.5s:
$$\therefore \quad T = \underline{0.5s}, \quad \text{for } K \geq 40$$

Problem 5.15

The middle-C string ($f = 256$ Hz) of a guitar is plucked, and the sound becomes inaudible after approximately 3 s.

(a) Find the approximate value of the damping ratio ζ for this string.
(b) Find the approximate value of the natural frequency for this string.
(c) Give the transfer function of the string. What is the physical input, and what is the physical output for this transfer function?
(d) List all the assumptions made in modelling the string.

Solution 5.15

(a) $w = 2\pi f = 2\pi (256) w_n \sqrt{1 - \zeta^2}$

$\therefore w_n^2 (1 - \zeta^2) = 2.567 \times 10^6$

$4T \approx 3 \Rightarrow T = \frac{1}{\zeta w_n} = 0.75 \Rightarrow \zeta = \frac{4}{3 w_n}$

$\therefore w_n^2 - w_n^2 \zeta^2 = w_n^2 - \left(\frac{4}{3}\right)^2 = 2.587 \times 10^6$

$\therefore w_n \approx 2\pi (256) rad/s, \quad f_n \approx 256 Hz$

$\zeta = \frac{1}{T w_n} \approx \frac{1}{(0.75)(2\pi [256])} = \underline{0.000829}$

(b) $f_n \approx \underline{256 \ Hz}$

(c) $G(s) = \frac{w_n^2}{s^2 + 2\zeta w_n s + w_n^2} \approx \frac{2.587 \times 10^6}{s^2 + \frac{8}{3} s + 2.587 \times 10^6}$

(d)   (i) The system can be modelled as in (c).

   (ii) The settling time is approximately equal to the time for the tone to become inaudible.

Problem 5.16

Consider the servo control system of Figure P5.6.

(a) Find the closed-loop transfer function.

(b) Find the closed-loop dc gain.

(c) If $K_v = 0$, find the closed-loop system gain (magnitude of the closed-loop frequency response) at resonance.

(d) A design specification is that the peak closed-loop gain can be no greater than 1.30 at any frequency. Design the system by finding K_v such that this specification is satisfied. Note that the damping of the system has been increased by the velocity (also called rate and derivative) feedback.

Solution 5.16

(a) $T(s) = \frac{9}{s^2 + (9k_v + 1)s + 9}$

(b) $T(0) = \frac{9}{9} = \underline{1}$

(c) $T(s) = \frac{9}{s^2 + s + 9}, \quad \therefore w_n = \sqrt{9} = \underline{3}$

$2\zeta w_n = 1, \quad \therefore \zeta = \frac{1}{2(3)} = \underline{0.1667}$

$w_n = w_n \sqrt{1 - 2\zeta^2} = 3 \left[1 - 2 \left(\frac{1}{36}\right)\right]^{1/2} = \underline{2.915 \ rad/s}$

$M_{pw} = \frac{1}{2\zeta \sqrt{1 - \zeta^2}} = \frac{1}{2(0.1667)\left[1 - \frac{1}{36}\right]^{1/2}} = \underline{3.042}$

(d) $M_{pw} = 1.30 = \frac{1}{2\zeta\sqrt{1-\zeta^2}}, \zeta \approx \underline{0.425}$,

$\therefore 2\zeta w_n = 2(0.425)(3) = 2.55 = 1 + 9kv$

$\therefore K_n = \frac{1.55}{9} = \underline{0.172}$

(e) $T(s) = \frac{9}{s^2+2.55s+9}$

Problem 5.17

Consider the control system of a satellite given in Figure P5.17.

(a) Find the closed-loop transfer function.
(b) Find the closed-loop dc gain.
(c) If $K_v = 0$, find the closed-loop system gain (magnitude of the closed-loop frequency response) at resonance.
(d) Design specifications for the system are that the peak closed-loop gain cannot be greater than 1.25 and that the system time constant $\tau = 1$ s. Design the system by finding K and K_v such that the specifications are satisfied. Note that the damping of the system has been increased by the velocity (also called rate and derivative) feedback.

Solution 5.17

(a) $T(s) = \frac{K \cdot s^{-2}}{1+KK_v s^{-1}+Ks^{-2}} = \frac{K}{s^2+KK_v s+K}$

(b) $T(0) = \frac{K}{K} = \underline{1}$

(c) $T(s) = \frac{K}{s^2+K} \Rightarrow \zeta = 0, \; w_n = \sqrt{K}, \; w_n = w_n = \underline{\sqrt{K}}$,

(d) $2\zeta w_n = KK_v = 2\zeta\sqrt{K}; \quad \therefore \zeta = \frac{K_v K}{2}, \quad w_n = \underline{\sqrt{K}}$

$$M_{pw} = \frac{1}{2\zeta\sqrt{1-\zeta^2}} = 1.25 \Rightarrow \zeta = \underline{0.44}, \text{ Figure 4.16} \quad (4.54)$$

$$\therefore T = \frac{1}{\zeta w_n} = \underline{1} \Rightarrow w_n = \frac{1}{\zeta} = \frac{1}{0.44} = \underline{2.273}$$

$$\therefore K = w_n^2 = \underline{5.165}; \quad K_v = \frac{2\zeta}{\sqrt{K}} = \frac{2(0.44)}{2.273} = \underline{0.387}$$

$$\therefore T(s) = \frac{5.165}{s^2+1.999s+5.165}$$

Problem 5.18

Consider the system, which has the transfer function

$$G(s) = \frac{8}{(s+2.5)(s^2 + 2s + 4)}$$

Both the system unit step response and the system frequency response are plotted in the figure. The final value of the unit step response is seen to be 0.8, and the initial value of the frequency response is also 0.8. Will these two values always have the same magnitude for any stable system? Give a proof for your answer.

Solution 5.18

$$G(s) = \frac{8}{(s+2.5)(s^2 + 2s + 4)}$$

DC gain $= G(0) = 0.8$
  $\therefore$ for unit step input, $C_{ss}(t) = G(0)\,(1) = G(0)$
  Also, dc gain $= G(j0) = G(0)$
  $\therefore$ final value of unit step response is always equal to the initial value of the frequency response.

Problem 5.19

Show that for a first-order system with the transfer function

$$G(s) = \frac{K}{\tau s + 1}$$

The product of bandwidth and rise time is equal to 2.197.

Solution 5.19

$$G(s) = K/(Ts+1), \quad \therefore bandwidth = w_B = 1/T$$

step response $= c(t) = K(1 - e^{-t/T})$

  Now, $0.1K = K(1 - e^{-t_1/T}) \Rightarrow e^{-t_1/T} = 0.9$

$$0.9K = K(1 - e^{-t_2/T}) \Rightarrow e^{-t_2/T} = 0.1$$

$$\therefore e^{-(t_2-t_2)/T} = 1/9 = e^{-w_B T_r}$$

$$\therefore w_B T_r = In\,9 = \underline{2.197}$$

Problem 5.20
Given the system with the transfer function

$$\frac{C(s)}{R(s)} = \frac{5}{s}$$

(a) An input

$$r(t) = 3 + 4\cos(2t - 15°)$$

is applied at $t = 0$. Find the steady-state value of $c(t)$ without finding the complete response.

(b) Find the complete response with $r(t) = 3u(t)$.

(c) Use the results of (b) to explain any unusual results in (a).

Solution 5.20

(a) $G(0) \to \infty$ $C_{ss1}(t) \to \infty$

$$G(j^2) = \frac{5}{j^2} = 2.5\angle - 90°, \quad C_{ss2}(t) = 4(2.5)\cos(2t - 105°)$$

$$= \underline{10\cos(2t - 105°)}$$

$C_{ss}(t)$ is unbounded.

(b) $C(s) = G(s)\,R(s) = \left(\frac{5}{s}\right)\left(\frac{3}{s}\right) = \frac{15}{s^2} \Rightarrow c(t) = \underline{15tu(t)}$

(c) In(b), $\displaystyle\lim_{t\to\infty} c(t) \to \infty$; this gives $C_{ss1}(t) \to \infty$ in (a).

Problem 5.21
Given the system with the transfer function

$$\frac{C(s)}{R(s)} = \frac{10}{s+4}$$

(a) An input

$$r(t) = -5 + 104\cos(5t + 30)$$

is applied at $t = 0$. Find the steady-state value of $c(t)$ without finding the complete response.

(b) Suppose that in addition to the input of (a), the initial condition is $c(0) = -3$. Find the steady-state value of $c(t)$ without solving for the complete response.

(c) The solutions in (a) and (b) apply for (approximately) $t > t_0$. Find t_0.

Solution 5.21

(a) $dc\ gain = G(0) = \frac{10}{4} = \underline{2.5}$

$$G(j^5) = \frac{10}{4 + j^5} = \frac{10}{6.403\angle 51.30^0} = \underline{1.562\angle -51.3^0}$$

$$\therefore\ C_{ss}(t) = (-5)(2.5) + (10)(1.562)\cos(5t + 30^0 - 51.30^0)$$
$$= \underline{-12.5 + 15.62\cos(5t - 21.30^0)}$$

(b) natural response $= Ke4t,\ \therefore$ goes to zero as $t \to \infty$
     $\therefore C_{ss}(t)$ same as (a).

(c) $T = \frac{1}{4} = \underline{0.25s},\quad 4T = Is,\quad \therefore t_0 \approx Is.$

Problem 5.22
For the system of Figure P4.5(a), the input r(t) = 3 cos 2t is applied at t = 0.

(a) Find the steady-state system response.
(b) Find the range of time t for which the system is in steady state.
(c) Find the steady-state response for the input r(t) = 3 cos 8 t.
(d) Why is the amplitude of the response in (a) much greater than that in (c), with the amplitudes of the input signals being equal?

Solution 5.22

(a) $G(j^2) = \frac{2}{0.5+j^2} = \frac{2}{2.062\angle 75.96^0} = \underline{0.970\angle -76^0}$
     $\therefore\ C_{ss}(t) = (3)(0.970)\cos(2t - 76^0)$
     $= \underline{2.91\cos(2t - 76^0)}$

(b) $T = \frac{1}{0.5} = 2s,\quad 4T = 8s,\quad \therefore t > 8s$

(c) $\therefore C_{ss}(t) = (3)(0.250)\cos(8t - 86.4^0)$
     $= \underline{0.750\cos(8t - 86.4^0)}$

(e) System gain at W = 2 is larger than that at W = 8.

Problem 5.23
For the system of Figure P4.5(b), the input r(t) = 3 cos 0.5t is applied at t = 0.

(a) Find the steady-state system response.
(b) Find the rage of time for which the system is in steady state.
(c) Find the steady-state response if the input is r(t) = 3 cos 10t.

(d) Find the system resonant frequency, ω_r.
(e) Find the steady-state response if the input is $r(t) = 3 \cos \omega_r t$.
(f) Why are the response amplitudes different in (a), (c) and (e)?
(g) Verify the results in (e) with a MATLAB simulation.

Solution 5.23

$$G(s) = \frac{5}{s^2 + s + 9}, w_n = \underline{3}, 2\zeta w_n = 1 \Rightarrow \zeta = \frac{1}{6}, T = \frac{1}{\zeta w_n} = \underline{0.5}$$

(a) $G(j0.5) = \frac{5}{(j0.5)^2+j0.5+9} = \frac{5}{8.75+j0.5} = \frac{5}{8.764\angle 3.29^0} = 0.571\angle 3.29^0$
$\therefore C_{ss}(t) = 3(0.571)\cos(0.5t - 3.27^0) = \underline{(1.71\cos 10.5t - 3.27^0)}$

(b) $T = 0.5s. \quad t_0 > 4T = 2s$

(c) $G(j10) = \frac{5}{(j10)^2+j10+9} = \frac{5}{-91+j10} = \frac{5}{91.6\angle 173.7^0} = 0.0546\angle - 173.7^0$
$\therefore C_{ss}(t) = 3(0.0546)\cos(10t - 173.7^0) = \underline{0.1638\cos(10t - 173.7^0)}$

(d) $G(j10) = \frac{5}{(9-w^2)+jw} \quad \therefore |G(jw)|$ varies with w.

Problem 5.24
Consider the system with the transfer function

$$\frac{C(s)}{R(s)} = \frac{K_1}{s + \alpha}$$

(a) Find the region of allowable s-plane pole locations if the system design is to yield a system settling time of less than 10 s.
(b) Solve for the ranges of K_1 and α for (a).

Solution 5.24

(a) $4T < 10 \Rightarrow T < 2.5s, \quad \therefore s = -\frac{1}{T} = -\frac{1}{25} = -0.4$

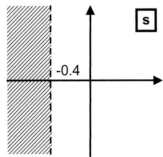

(b) $\frac{C(s)}{R(s)} = \frac{K_1}{s+0.4}$

Problem 5.25

Consider the system with the transfer function

$$\frac{C(s)}{R(s)} = \frac{w_n^2}{s^2 + 2\zeta w_n s + w_n^2}$$

(a) Find the region of allowable s-plane pole locations if the system design is to yield a system settling time of less than 2 s and an over shoot in a step response of less than 10%.

(b) Solve for the ranges of ζ and w_n for (a).

Solution 5.25

(a) For 10% overshoot, $\zeta \approx 0.57$

$$\cos \theta = \zeta = 0.57 \Rightarrow \theta < 55.2^0$$
$$T = \frac{2}{4} = 0.5, \quad \therefore Re(s) < -\frac{1}{T} = -2$$

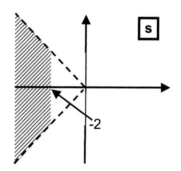

(b) $\zeta > 0.57; \quad \zeta w_n = \frac{1}{T} = 2; \quad \zeta w_n > 2$
$\therefore \zeta w_n > \frac{2}{\zeta}; \quad 0.57 < \zeta < 1$

For $\zeta > 1, \quad poles = w_n \left[-\zeta \pm \sqrt{\zeta^2 - 1} \right] < -2$

$\therefore w_n \left[\zeta - \sqrt{\zeta^2 - 1} \right] > 2$

$\therefore w_n > \dfrac{2}{\zeta - \sqrt{\zeta^2 - 1}}$

Problem 5.26

Given the third-order system of Figure P5.26.

(a) If $r(t) = 5u(t)$, find the steady-state value of $c(t)$.
(b) Approximately how many seconds is required for the system to reach steady-state response?
(c) Do you expect the transient response to be oscillatory? Justify your answer.

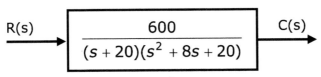

$$R(s) \quad \boxed{\dfrac{600}{(s+20)(s^2+8s+20)}} \quad C(s)$$

Figure P5.26   3rd order system.

Solution 5.26

(a) $dc = \dfrac{600}{(20)(20)} = 1.5, \quad \therefore C_{ss}(t) = (1.5)5 = \underline{7.5}$

(b) $T_1 = 1/20 = 0.05s$ $w_n = \sqrt{20} = 2\zeta w_n = 8, \quad \therefore \zeta = \dfrac{8}{2\sqrt{20}} = 0.894$
   $T_2 = 1/\zeta w_n = 2/2\zeta w_n = 2/8 = 0.25s$
   $\therefore$ setting time $= 4T_2 = 4(0.25) = \underline{/s}.$

(c) $\zeta = 0.894, \quad \therefore$ stightly oscillatory.

Problem 5.27

Given a system described by the standard second-order transfer function

$$G(s) = \dfrac{25}{s^2 + 4s + 25}$$

(a) Sketch the system unit step response. Any of the figures of this chapter may be used.
(b) Find the transfer function if this system is time scaled with $t_v = 0.01t$.
(c) From the transfer function in (b), find ζ_v and w_{nv} for the scaled system.
(d) Repeat (a) for the time-scaled system.
(e) Given the frequency scaling $s_v = as$, where the parameter a is real. Find the effect of this scaling on the time response of a standard second-order system.

Solution 5.27

(a) $w_n = \sqrt{25} = \underline{5}, \quad 2\zeta w_n = 10\zeta = 4 \Rightarrow \zeta = 0.4, \quad T = \frac{2}{4} = 0.5s.$

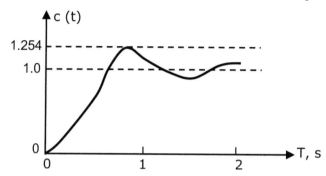

$$M_{pt} = 1 + e^{-\zeta\pi/\sqrt{1-\zeta^2}} = 1 + e^{-1.37} = \underline{1.254}$$

(b) $t_v = \frac{t}{\sigma} = 0.01t, \quad \therefore \sigma = 100$

$$\therefore G_s(s) = \frac{25 \times 10^4}{s^2 + 400s + 25 \times 10^4}$$

(c) $w_{nv} = [25 \times 10^4]^{1/2} = \underline{500}; \quad 2\zeta_v w_{nv} = 1000\zeta_v 400 \Rightarrow \zeta_v = \underline{0.4}$

$$T_v = \frac{1}{\zeta_v w_v} = \frac{1}{(0.4)(500)} = \underline{0.005s}$$

(d) as in (a), with different time axis.

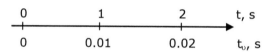

(e) $G(s_v) = \dfrac{w_n^2}{\left(\frac{s_v}{a}\right)^2 + 2\zeta w_n \left(\frac{s_v}{a}\right) + w_n^2} = \dfrac{w_n^2}{s^2 + 2\zeta(aw_n)s + (a+w_n)^2}$

$\therefore \underline{w_{nv} = aw_n}; \quad \underline{\zeta_v = \zeta}$

Problem 5.28

(a) Given a system described by the transfer function

$$G(s) = \frac{500}{s + 200}$$

Time scale this transfer function such that the time constant of the scaled system is

(b) Repeat (a) for the system transfer function

$$G(s) = \frac{0.00016}{s^2 + 0.016s + 0.000256}$$

(c) For verification, calculate the time constants directly from the scaled transfer function, for (a) and (b).

Solution 5.28

(a) $G(s) = \frac{500}{s+200}$, $\therefore T = \frac{1}{200} = \underline{0.005s}$

$t_v = 1 = \frac{t}{\sigma} \Rightarrow \sigma = T = \underline{0.005s}$

$G_s(s_v) = \frac{2.5}{s_v+1}$

(b) $G(s) = \frac{0.00016}{s^2+0.016s+0.000256} = \frac{0.00016}{(s+0.008)^2} + (.)^2$ $\therefore T = \frac{1}{0.008} = \underline{125s}$

$\therefore t_v = 1 = \frac{t}{\sigma} \Rightarrow \sigma = T = \underline{125}$

$\therefore G_s(s_v) = \frac{2.5}{s^2+2s+4}$

(c) (a) $G_s(s_v) = \frac{2.5}{s+1} = \frac{k_v}{T_v s+1}$, $\therefore T_v = \underline{1}$

(b) $G_s(s_v) = \frac{2.5}{s^2+2s+4} = \frac{2.5}{(s+1)^2+3}$ $\therefore T_v = \underline{1}$

Problem 5.29
Consider the scaling of time and frequency by the factor w_n for the standard second-order system with the transfer function

$$\frac{C(s)}{R(s)} = \frac{w_n^2}{s^2 + 2\zeta w_n s + w_n^2}$$

(a) Show that the scaling $s_v = s/w_n$ results in a transfer function that depends only on ζ.
(b) Show that the time-domain scaling $t_v = t/w_n$ in (a) results in the same transfer function.

Solution 5.29

(a) $G_s(s_v) = \frac{w_n^2}{(w_n s_v)^2 + 2\zeta w_n (w_n s_v) + w_n^2} = \frac{1}{s_v^2 + 2\zeta s_v + 1}$

(b) (5.62) with $\sigma = 1/w_n$

$$G_s(s_v) = \frac{\sigma^2 w_n^2}{s_v^2 + \sigma(2\zeta w_n s_v) + \sigma^2 w_n^2}\bigg|_{\sigma=\frac{1}{w_n}}$$

$$= \frac{1}{s_v^2 + 2\zeta s_v + w_n^2}$$

Problem 5.30

Given a system described by the third-order transfer function

$$G(s) = \frac{500}{s + 200}$$

Note that the coefficients in the numerator are not specified.

(a) Estimate the system rise time, settling time and per cent overshoot in the step response.
(b) Using a MATLAB simulation, find the step responses and compare the actual parameters of the step responses with the estimated values for the transfer-function numerator polynomials

  (i) $b_2 s^2 + b_1 s + b_o = 2s_2 + 240s + 800$
  (ii) $b_2 s^2 + b_1 s + b_o = 2s^2 + 50s + 420$
  (iii) $b_2 s^2 + b_1 s + b_o = 0.1s^2 + 402s + 820$

Solution 5.30

$$\Delta(s) = (s + 2)[(s + 10)^2 + (10)^2], \quad \therefore \text{roots} = -2, -10 \pm j^{10}$$

transient response terms: $b_1 e^{-2t}, \quad b_2 e^{-10t} \cos(10t + \theta)$

$\underline{T_1 = 0.5s}; \quad \underline{T_2 = 0.1s}$ and $\underline{w_n = 10\sqrt{2}}, \quad 2\zeta w_n = 20, \quad \zeta = 0.707$

$\therefore \underline{T_s} = \text{setting time} = 4T_1 = \underline{2s} - \zeta\pi/\sqrt{1 - \zeta^2} = \underline{4.3\%}$

$\zeta = 0.707, \%overshoot = e$

$$T_p = \frac{\pi}{w_n\sqrt{1 - \zeta^2}} = 0.314s \quad \therefore T_r \text{ somewhat less.}$$

by simulation:

	(a)	(b)	(c)
T_s	2.3s	2.3s	0.4s
% overshoot	0%	0%	3%
T_r	0.8s	1s	0.2s

Problem 5.31

A system with an unknown transfer function has a unit step response of the type shown in the figure. The peak value of this response is 1.16, the time to peak value is 0.8 s and the steady-state value of the response is unity.

(a) Can this system be modelled accurately as a first-order system? Explain your answer.

(b) If this system is to be modelled as second order, make use of the curves of Figure 4.8 to develop a second-order model for this system.

(c) Verify the transfer function of (b) using analytical techniques.

Solution 5.31

(a) No. First-order system cannot have an overshoot.

(b) 16% overshoot, $\therefore \zeta = 0.5$ Figure 4.7 and $w_n T_p = 3.6$

$$\therefore w_n = \frac{3.6}{0.8} = 4.5 \quad \therefore G(s) = \frac{w_n^2}{s^2 + 2\zeta w_n s + w_n^2} \cong \frac{20.25}{s^2 + 4.5s + 20.25}$$

(c) $0.16 = e^{-\zeta \pi / \sqrt{1-\zeta^2}}, \quad \therefore \frac{\zeta \pi}{\sqrt{1-\zeta^2}} = 1.8326$

$\therefore 9.8696 \zeta^2 = 3.3584(1 - \zeta^2) \quad \therefore \zeta = 0.504$

$w_n = \frac{\pi}{T_p \sqrt{1-\zeta^2}} = 4.54 \ rad/s$

$\therefore G(s) = \frac{20.7}{s^2 + 4.58s + 20.7}$

Problem 5.32

Repeat Problem 5.31 under the conditions that the given response is for a step input of magnitude 2; that is, $r(t) = 2u(t)$.

Solution 5.32

% overshoot = 8%

(a) Same as Problem 5.18

(b) with 8% overshoot, $\zeta \cong 0.6$ and $w_n T_p = 3.9$

$\therefore w_n = \frac{3.9}{0.8} = 4.875 \ rad/s$

$$\therefore G(s) = \frac{K w_n^2}{s^2 + 2\zeta w_n s + w_n^2} = \frac{\overbrace{0.5(23.77)}^{11.89}}{s^2 + 5.85s + 23.77}$$

(c) $0.08 = e^{-\zeta \pi / \sqrt{1-\zeta^2}}, \quad \therefore \frac{\zeta \pi}{\sqrt{1-\zeta^2}} = 2.526$

$\therefore 9.8696 \zeta^2 = 6.379(1 - \zeta^2), \quad \therefore \zeta = 0.627$

$w_n = \frac{\pi}{T_p \sqrt{1-\zeta^2}} = 5.039 \ rad/s.$

$G(s) = \frac{K w_n^2}{s^2 + 2\zeta w_n s + w_n^2} = \frac{12.70}{s^2 + 6.318s + 25.39}$

Problem 5.33

Repeat Problem 5.31 if the unit step response has a peak value of 2.34, the time to the peak value is 1.25 s and the steady-state value of the response is 1.80.

Solution 5.33

$$G(s) = \frac{K w_n^2}{s^2 + 2\zeta w_n s + w_n^2} = K = DC \ gain = G(0) = \underline{1.80}$$

(a) $\%overshoot = \frac{2.34}{1.80} * 100 - 100 = 30\% \therefore \underline{NO}$

(b) $\zeta = 0.34$ for 30% overshoot

$$w_n T_p = 3.33, \quad \therefore w_n = 3.33/1.25 = \underline{2.66 \ rads}$$

$$\therefore G(s) = \frac{1.80(7.10)}{s^2 + 1.81s + 7.10} = \frac{12.77}{s^2 + 1.81s + 7.10}$$

(c) $0.30 = e^{-\zeta \pi / \sqrt{1-\zeta^2}}, \quad \therefore \frac{\zeta \pi}{\sqrt{1-\zeta^2}} = 1.20$

$$\therefore 9.870 \zeta^2 = 1.45(1 - \zeta^2)$$

$$\therefore \zeta = \underline{0.358}, \quad w_n = \frac{\pi}{T_p \sqrt{1-\zeta^2}} = \underline{2.69 \ rad/s.}$$

$$\therefore G(s) = \frac{13.04}{s^2 + 1.926s + 7.25}$$

Problem 5.34

(a) Sketch the unit step response of the system of Figure P5.34(a), giving approximate values on the two axes.

(b) Repeat (a) for the system of Figure P5.34(b).

$$\frac{2}{s + 0.5} \quad \frac{5}{s^2 + s + 9} \quad \frac{2}{s + 0.5} \quad \frac{5}{s^2 + s + 9}$$

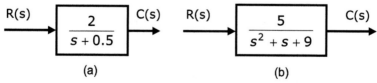

| R(s) | $\frac{2}{s+0.5}$ | C(s) | R(s) | $\frac{5}{s^2+s+9}$ | C(s) |

(a)   (b)

Figure P5.34   (a) 1st order system, (b) 2nd order system.

Solution 5.34

(a) $K = \frac{2}{0.5} = 4, \ T = 2$

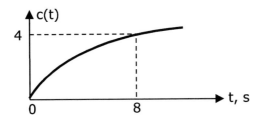

(b) $dc \ gain = \frac{5}{9} = \underline{0.556}$

$w_n = \sqrt{9} = \underline{3}; 2\zeta w_n = 6\zeta = 1, \Rightarrow \zeta = \frac{1}{6} = \underline{0.1667}$

$T = \frac{1}{\zeta w_n} = \underline{2s}; \quad \% \ overshoot \approx \underline{50\%}, \ \text{from Figure 4.8}$

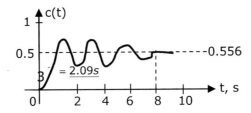

$$w_n = 3 = 2\pi/T \Rightarrow T = \frac{2\pi}{3} = \underline{2.09s}$$

Problem 5.35

Consider the servo control system of the figure, which employs rate feedback (K_v).

(a) Find the system transfer function.

(b) Design the system by finding the value of K_v such that the system is critically damped $(\zeta = 1)$.

(c) Redesign the system by finding the value of K_v such that $\zeta = 0.707$ (a commonly used value in design).

(d) Find the value of K_v such that when the input is a constant value of 10, the output is also a constant value of 10.

(e) Is the value of K_v found in (d) reasonable, considering that the feedback signal is dc(t)?

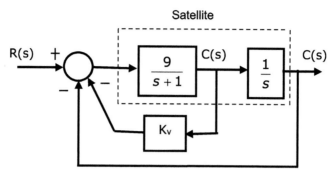

Figure P5.35   Servo control systems.

Solution 5.35

(a) By Mason's gain formula:

$$T(s) = \frac{\frac{9}{s(s+1)}}{1 + \frac{9K_\vartheta}{s+1} + \frac{9}{s(s+1)}} = \frac{9}{s^2 + (1 + 9K_\vartheta)s + 9}$$

(b) $T(s) = \frac{w_n^2}{s^2 + 2\zeta w_n s + w_n^2}$, $\therefore w_n = \sqrt{9} = \underline{3}$

$z\zeta w_n = (2)(1)(3) = 6 = 9K_v + 1 \Rightarrow K_v = \frac{5}{9} = \underline{0.5556}$

(c) $z\zeta w_n = z(0.707)3 = 4.242 \Rightarrow K_v = \frac{3.242}{9} = \underline{0.3602}$

(d) $dc\ gain = T(0) = \frac{9}{9} = 1$, $\therefore$ value in dyendent of Kv.

(e) In the study state, $dc/dt = 0$. Thus the output of the Kv block is zero, independent of the value of Kv.

(f) $\zeta = 1$, *no overshoot*, $T = \frac{1}{\zeta W_n} = \frac{1}{3} = \underline{0.333s}$ $z\zeta w_n = 6$

$\zeta = 0.707, \approx 7\%$ overshoot from Figure 4.8,

$T = \frac{1}{\zeta W_n} = \frac{1}{(0.707)(3)} = \underline{0.471s}$, $z\zeta w_n = \underline{4.242}$

Problem 5.36

The digital plotter is to be designed such that it has settling time of 0.20 s. The motor in the example cannot achieve this specification. If a motor is chosen that has the transfer function $G(s) = K/[s(\tau_m s + 1\ ]$ and is connected into the system of the figure, find the values of K, K_a and τ_m that meet the settling-time specification, along with the specification that $\zeta = 1$ to eliminate overshoot for step inputs.

Solution 5.36

$$T(s) = \frac{K\alpha K}{T_m s^2 + s + K\alpha K} = \frac{K\alpha K/T_m}{s^2 + \frac{1}{T_m}s + \frac{K\alpha K}{T_m}}$$

(a) $T_s = 4T = \frac{1}{5} \Rightarrow T = \frac{1}{20}, \quad \zeta = 1, \quad w_n^2 = \frac{K\alpha K}{T_m}$

$$\therefore 2\zeta w_n = 2w_n = 2\left(\frac{T_m}{K\alpha K}\right)^{1/2} = \frac{1}{T_m} \Rightarrow 2\sqrt{K\alpha K T_m} = 1 \quad (1)$$

$$T = \frac{1}{\zeta w_n} = \frac{1}{w_n} = \left(\frac{T_m}{K\alpha K}\right)^{1/2} = \frac{1}{20} \Rightarrow \sqrt{K\alpha K} = 20\sqrt{T_m}$$

$$\therefore (2) \; into \; (1) : 2\sqrt{K\alpha K}\sqrt{T_m} = 1 = 2(\sqrt{T_m})\sqrt{T_m} = 40T_m$$

$$\therefore T_m = 1/40 = \underline{0.025}$$

(2) $K\alpha K = (20)^2 T_m = \dfrac{400}{40} = \underline{10}$

$$w_n^2 = \frac{10}{1/40} = 7 \; w_n = \underline{20}$$

$$\therefore T(s) = \frac{40K\alpha K}{s^2 + 40s + 40K\alpha K} = \frac{400}{s^2 + 40s + 400} = \frac{400}{(s+20)^2}$$

$$T(s) = \frac{1}{20} \Rightarrow 4T = \frac{4}{20} = 0.2s; \; \zeta = 1$$

Problem 5.37
Show that the inverse Laplace transform such that

$$\mathcal{L}^{-1}\left[\frac{w_n^2}{s(s^2 + 2\zeta w_n s + w_n^2)}\right] = 1 - \frac{1}{\beta}e^{-\zeta w_n t}\sin(\beta w_n t + \theta)$$

where

$$\beta = \sqrt{1 - \zeta^2} \text{ and } \theta = \tan^{-1}(\beta/\zeta).$$

Solution 5.37

$$\zeta\left[\frac{1}{\beta}e^{-\zeta w_n t}\sin(\beta w_n t + \theta)\right]$$

$$= \zeta\left\{\frac{1}{\beta}e^{-\zeta w_n t}[\sin \beta w_n t \cos \theta + \cos \beta w_n t \sin \theta]\right\}$$

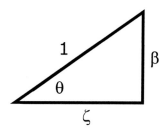

$$= \frac{1}{\beta}\left[\frac{\zeta\beta w_n}{s^2 + 2w_n s + w_n^2} + \frac{\beta(s + \zeta w_n)}{s^2 + 2\zeta w_n s + w_n^2}\right]$$

$$= \frac{s + 2\zeta w_n}{s^2 + 2\zeta w_n s + w_n^2}$$

$$\therefore \zeta\left[1 - \frac{1}{\beta}e^{-\zeta w_n t}\sin(\beta w_n t + \theta)\right]$$

$$= \frac{1}{s} - \frac{s + 2\zeta w_n}{s^2 + 2\zeta w_n s + w_n^2} = \frac{w_n^2}{s(s^2 + 2\zeta w_n s + w_n^2)}$$

Problem 5.38

(a) Show that for the system with the transfer function G(s), the unit impulse response is the derivative of the unit step response.
(b) Show that the unit step response of the system of (5.23), given in (5.30), is the derivative of the unit step response given in (5.25).
(c) Show that for a system with the transfer function G(s), the unit step response is the derivative of the unit ramp response.

Solution 5.38

(a) Impulse response: $C_g(s) = G(s)E(s) = G(s)$

   Step response: $C_u(s) = G(s)\,E(s) = \frac{1}{s}G(s)$

$$\zeta\left[\frac{dc_u(t)}{dt}\right] = sC_s(s) = s\left[\frac{1}{s}G(s)\right] = G(s) = C_g(s)$$

(b) $\zeta\left[\frac{dc_u(t)}{dt}\right] = sC_s(s) = s\left[\frac{1}{s}G(s)\right] = G(s) = C_g(s)$

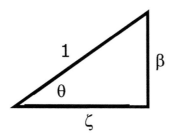

$$= \frac{\zeta w_n}{\beta} e^{-\zeta w_n t} \sin(\beta w_n t + \theta) - \frac{\beta w_n}{\beta} e^{-\zeta w_n t} \cos(\beta w_n t + \theta)$$

$$= w_n e^{-\zeta w_n t} \left[\frac{\zeta}{\beta} \sin \beta w_n t \overset{(\zeta)}{\cos\theta} + \frac{\zeta}{\beta} \cos \beta w_n t \overset{(\beta)}{\sin\theta} - \cos \beta w_n t \overset{(\zeta)}{\cos\theta} \right.$$

$$\left. + \sin w_n t \overset{(\beta)}{\sin\theta} \right]$$

$$= w_n e^{-\zeta w_n t} \left[\frac{\zeta^2}{\beta} \sin \beta w_n t + \beta \sin w_n t \right]$$

$$= w_n e^{-\zeta w_n t} \left[\frac{\zeta^2 + \beta^2}{\beta} \sin \beta w_n t \right] = \frac{w_n}{\beta} e^{-\zeta w_n t} \sin \beta w_n t$$

(c) Ramp response: $C_r(s) = G(s)E(s) = \frac{1}{s^2}G(s)$

$$\zeta \left[\frac{dc_r(t)}{dt} \right] = sC_r(s) = s \left[\frac{1}{s^2}G(s) \right] = \frac{1}{s}G(s) = C_s(s)$$

Problem 5.39
Given the standard second-order transfer function

$$\frac{C(s)}{R(s)} = \frac{w_n^2}{s^2 + 2\zeta w_n s + w_n^2}$$

Solution 5.39

$$\frac{C(s)}{R(s)} = \frac{w_n^2}{s^2 + 2\zeta w_n s + w_n^2}$$

$$\therefore (s^2 + 2\zeta W_n s + W_n^2)\,C(s) = W_n^2 R(s)$$

$$\ddot{c} + 2\zeta W_n C + W_n^2 C = W_n^2 R$$

$$\therefore s^2 C(s) - sc(0) - c(0) + 2\zeta W_n[sC(s) - C(0)] + W_n^2 C(s) = W_n^2 R(s)$$

$$\therefore C(s) = \frac{sc(0) + c(0) + 2\zeta W_n c(0)}{s^2 + 2\zeta w_n s + w_n^2} \Rightarrow \frac{w_n^2}{s^2 + 2\zeta w_n s + w_n^2}$$

$$\therefore c(0) = 0 \; ; \; c(0) = w_n^2$$

Problem 5.40

The model of a thermal stress chamber is given in Figure P5.40. The temperature of the chamber is controlled by the input voltage m(t), which controls the position of a valve in a steam line. The disturbance d(t) models the opening and closing of the door into the chamber. If the door is opened at t = 0, d(t) = 60 u(t), a step function. With the door closed, d(t) = 0.

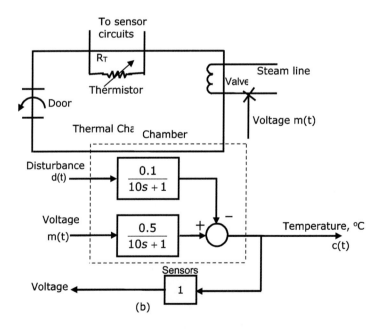

Figure P5.40 Thermal stress chamber.

(a) If the controlling voltage is m(t) = 70u(t) with the door closed, plot the chamber temperature.

(b) A tacit assumption in (a) is an initial chamber temperature of zero degrees Celsius. Assuming that the initial temperature is 20C, repeat (a).

(c) Two minutes after the application of the voltage in (b), the chamber door is opened and remains open. Add the results of the disturbance to the plot of (b).

Solution 5.40

(a) $\quad C(s) = G_p(s)\,M(s) = \frac{0.05}{s+0.1}\left(\frac{70}{s}\right) = \frac{35}{s} + \frac{-35}{s+0.1}$

$\therefore C(t) = 35(1 - e^{-0.1t}), \; t \geq 0$

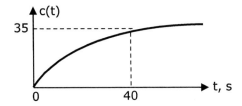

(b) $\quad C(s) = \frac{0.05}{s+0.1}\,M(s) \Rightarrow \dot{c}(t) + 0.1\,c(t) = 0.05\,m(t)$

$\therefore s\,C(s) - c(0) + 0.1\,C(0) = 0.05\,M(s) = \frac{3.5}{s}$

$C(s) = \frac{3.5}{s(s+0.1)} + \frac{20}{s+0.1}$

$= 35 - 35e^{-0.1t} + 20e^{-0.1t} = 35 - 15\,e^{-0.1t}$

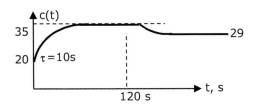

(c) $\quad C_{d1}(s) = -\frac{0.05}{s+0.1}\left(\frac{60}{s}\right) = \frac{-0.6}{s(s+0.1)} = \frac{-6}{s} + \frac{6}{s+0.1}$

$C_{d1}(t) = -6 + 6e^{-0.1t}$

$C_d(t) = C_{d1}(t - 120)\,u(t - 120)$

$= -6u(t - 120) + 6e^{-0.1(t-120)}\,u(t - 120)$

Problem 5.41

Given a system described by

$$C(s) = \frac{10}{s+4}R(s)$$

(a) Let the input signal r(t) = 0. Find the system response c(t) for the initial condition c(0) = 8.
(b) Let the input signal r(t) = 6u(t) and c(0) = 0. Find the system response c(t).
(c) Let the input signal r(t) = 6u(t) and c(0) = 8. Use the results of (a) and (b) to find the system response c(t). Sketch this response.
(d) Verify mathematically that c(t) in (c) satisfies the system differential equation.

Solution 5.41

(a) $C(s) = \frac{10}{s+4}R(s)$

$\therefore \dot{c}(t) + 4c(t) = 10r(t)$

$s\,c(t) - c(0) + 4C(s) = 10R(s)$

$C(s) = \frac{10\,R(s)}{s+4} + \frac{C(s)}{s+4}$

$R(s) = 0, \therefore C(s) = \frac{8}{s+4} \Rightarrow c(t) = 8e^{-4t}, t \geq 0$

(b) $c(s) = 0, C(s) = \frac{60}{s(s+4)} = \frac{15}{s} - \frac{15}{s+4} \Rightarrow c(t) = 15 - 15e^{-4t}, t \geq 0$

(c) $c(t) = C_a(t) + c_b(t) = 15 - 7e^{-4t}, t \geq 0$

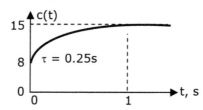

(d) $c(t) = 15 - 7e^{-4t} \Rightarrow C(0) = 15 - 7 = 8$

$\dot{c}(t) + 4c(t) = 28e^{-4t} + 60 - 28e^{-4t} = 60$

(e) $\dot{x} = -4x + 10u \quad ; \quad y = x$

Problem 5.42

Given a system described by

$$C(s) = \frac{4}{s^2 + 2.8s + 4} R(s)$$

(a) Let the input signal r(t) = 0. Find the system response c(t) for the initial conditions c(0) = 0 and c (0) = 8.

(b) Let the input signal r(t) = 6u(t) and c(0) = c (0) = 0. Find the system response c(t), using (4−20) for the inverse transform.

(c) Let the input signal r(t) = 6u(t), c(0) = 0 and c(0) = 8. Use the results of (a) and (b) to find the system response c(t).

(d) Verify that c(t) has the correct initial conditions and the correct steady-state value.

Solution 5.42

$$\ddot{c} + 2.8\dot{c} + 4c = 4\pi$$

$$s^2 C(s) - s\,c(0) - \dot{c}(0) + 2.8[sC(s) - c(0)] + 4C(s) = 4R(s)$$

$$\therefore C(s) = \frac{s\,c(0) + \dot{c}(0) + 2.8\,c(0)}{s^2 + 2.8s + 4} + \frac{4R(s)}{s^2 + 2.8s + 4}$$

(a) $R(s) = 0, C(s) = \frac{8}{s^2 + 2.8s + 4} = \frac{5.602(1.428)}{(s+1.4)^2 + (1.428)^2}$

$\therefore c(t) = 5.602e^{-1.4t} \sin 1.428t, t \geq 0$

(b) $C(s) = \frac{(4)(6)}{s(s^2 + 2.8s + 4)};$ $\therefore w_n = \sqrt{4} = 2$

                          $z\zeta\, w_n = 4\zeta = 2.8 \Rightarrow \zeta 0.7$

$\beta = \sqrt{1 - \zeta^2} = 0.714$

$c(t) = 6[1 - 1.4e^{-1.4t} \sin(1.428t + 45.57^0)]$

(c) $c(t) = C_a(t) + C_b(t)$

$$= 1 - e^{-1.4t}[1.4\sin(1.428t + 45.57) - 5.602\sin 1.428t]$$

(d) $c(0) = 6[1 - 1.4\sin(45.57^0)] = 6(1 - 1) = 0$

$\dot{c}(t) = 6[-1.4(-1.4)e^{-1.4t} \sin(1.428t + 45.57^0)]$

       $+ 6[-1.4(1.428)e^{-1.4t} \cos(1.428t + 45.57^0)]$

       $+ 5.602(-1.4)e^{-1.4t} \sin 1.428t + 5.602(1.428)e^{-1.4t} \cos 1.428t$

$\therefore \dot{c}(0) = 8.397 - 8.397 + 0 + 8 = 8$

$C_{ss}(t) = 6 - 0 + 0 = 6$

6

Control System's Stability

This chapter provides comprehensive details regarding the introduction to stability, Routh–Hurwitz stability criterion and problems and solutions.

6.1 Introduction

This chapter investigates the important topic of determining the stability of a linear time-invariant (LTI) analogue system. The bounded-input bounded-output (BIBO) definition of stability is used when investigating the stability of this type of system.

The BIBO definition of stability is as follows:

A system is BIBO stable if, for every bounded input, the output remains bounded with increasing time.

For a LTI system, this definition requires that all poles of the closed-loop transfer function (all roots of the system characteristic equation) lie in the left half of the complex plane. Hence any stability analysis requires determining if the closed-loop transfers function has any poles either on the imaginary axis or in the right half of the s-plane, except for simple poles on the imaginary axis.

The first method of stability analysis to be covered, the Routh–Hurwitz criterion, determines if any root of a polynomial lies outside the left half of the complex plane. However, this method does not find the exact location of the roots. The second procedure covered involves the calculation of the exact location of the roots. For first- and second-order systems, these roots can be found analytically. For higher-order systems a digital computer program should be used. The third procedure covered is simulation. This procedure applies to all systems; in addition, it is the only general procedure available for complex non-linear system.

All the three methods are based on the (inexact) system model. The only method to determine with certainty, the stability of a physical system, is to operate the system. Then the stability is determined only for those operating conditions tested.

Before the stability analysis techniques are presented, we consider some general properties of polynomials that will prove to be useful. We always assume that all polynomial coefficients are real. Consider the second-order polynomial

$$Q_2(s) = s^2 + a_1 s + a_0 = (s - p_1)(s - p_2) = s^2 - (p_1 + p_2)s + p_1 p_2$$

$$(6.1)$$

and the third-order polynomial

$$Q_3(s) = s^3 + a_2 s^2 + a_1 s + a_0 = (s - p_1)(s - p_2)(s - p_3)   (6.2)$$

$$= [s^2 - (p_1 + p_2)s + p_1 p_2](s - p_3)$$

$$= s^3 - (p_1 + p_2 + p_3)s^2 + (p_1 p_2 + p_1 p_3 + p_2 p_3)s - p_1 p_2 p_3$$

Extending the expansion to the nth-order polynomial

$$Q_n(s) = s^n + a_{n-1}s^{n-1} + \cdots + a_1 s + a_0   (6.3)$$

we see that the coefficients are given by

a_{n-1} = negative of the sum of all roots

a_{n-2} = sum of the products of all possible combinations of roots taken two at a time

a_{n-3} = negative of the sum of the products of all possible combinations of roots taken three at a time

$\vdots$

$a_0 = (-1)^n$ multiplied by the product of all the roots

Suppose that all roots of a polynomial are real and in the left half-plane (LHP). Then all p_1 in (6.1) and (6.2) are real and negative. Therefore, all polynomial coefficients are positive; this characteristic also applies to the general case of (6.3). The only case for which a coefficient can be negative is that there would be at least one root in the right half-plane (RHP). Note also that if all roots are in the LHP, no coefficient can be zero.

If any roots of the polynomials above are complex, the roots must appear in complex-conjugate pairs, since the polynomial coefficients are assumed real. Then, in the rules given for forming the polynomial coefficients, all imaginary parts of the products will cancel. Therefore, if all roots occur in the LHP, all coefficients of the general polynomial of (6.3) will be positive. Conversely, if not all coefficients of (6.3) are positive, the polynomial will have at least one root that is not in the LHP (that is, on the $j\omega$ axis or in the RHP). In summary, given a polynomial in (6.3)

1. If any coefficient a_i is equal to zero, then not all the roots are in the LHP.
2. If any coefficient a_i is negative, then at least one root is in the RHP.

The converse of rule 2 is not true; if all coefficients of a polynomial are positive, the roots are not necessarily confined to the LHP. We now illustrate this point with an example.

Example 6.1
For the polynomial

$$Q(s) = (s + 2)(s^2 - s + 4) = s^3 + s^2 + 2s + 8$$

all coefficients are positive the roots are at the locations

$$-2, \quad \frac{1}{2} \pm \frac{j\sqrt{15}}{2}$$

and we see that two of the three roots are in the RHP.

6.2 Routh–Hurwitz Stability Criterion

The Routh–Hurwitz criterion is an analytical procedure for determining if all roots of a polynomial have negative real parts and is used in the stability analysis of LTI systems. The criterion gives the number of roots with positive real parts, and applies to all LTI systems for which the characteristic equation is a polynomial set to zero. This requirement excludes a system that contains an ideal time delay (transport lag). For this special case, which is covered later on, the Routh–Hurwitz criterion cannot be employed.
The Routh–Hurwitz criterion applies to a polynomial of the form

$$Q(s) = a_n s^n + a_{n-1} s^n + \ldots + a_1 s + a_0 \tag{6.4}$$

where we can assume with no loss of generality that $a_0 \neq 0$. Otherwise, the polynomial can be expressed as a power of s multiplied by a polynomial in which $a_0 \neq 0$. The power of s indicates roots at the origin; hence, only the latter polynomial needs to be investigated using the Routh–Hurwitz criterion. We assume in the following developments that a_0 is not zero.

The first step in the application of the Routh–Hurwitz criterion is to form the array below, called the Routh array, where the first two rows are the coefficients of the polynomial in (6.4).

s^n	a_n	a_{n-2}	a_{n-4}	a_{n-6}	$\cdots$
s^{n-1}	a_{n-1}	a_{n-3}	a_{n-5}	a_{n-7}	$\cdots$
s^{n-2}	b_1	b_2	b_3	b_4	$\cdots$
s^{n-3}	c_1	c_2	c_3	c_4	$\cdots$
$\vdots$	$\vdots$	$\vdots$			
s^2	k_1	k_2			
s^1	l_1				
s^0	m_1				

The column with the powers of s is included as a convenient accounting method. The b row is calculated from the two rows directly above it, the c row, from the two rows directly above it, and so on. The equations for the coefficients of the array are as follows:

$$b_1 = -\frac{1}{a_{n-1}}\begin{vmatrix} a_n & a_{n-2} \\ a_{n-1} & a_{n-2} \end{vmatrix} \qquad b_2 = -\frac{1}{a_{n-1}}\begin{vmatrix} a_n & a_{n-4} \\ a_{n-1} & a_{n-5} \end{vmatrix}, \cdots$$

$$c_1 = -\frac{1}{b_1}\begin{vmatrix} a_{n-1} & a_{n-3} \\ b_1 & b_2 \end{vmatrix} \qquad c_2 = -\frac{1}{b_1}\begin{vmatrix} a_{n-1} & a_{n-5} \\ b_1 & b_3 \end{vmatrix}, \cdots \qquad (6.5)$$

Note that the determinant in the expression for the ith coefficient in a row is formed from the first column and the $(i + 1)$ column of the two preceding rows.

As an example, the Roth array for a fourth-order polynomial (which has five coefficients) is of the form

s^4	x	x	x
s^3	x	x	
s^2	x	x	
s^1	x		
s^0	x		

where each table entry is represented by the symbol x. Hence, for a general array, the final two rows of the array will have one element each, the next two rows above two elements each, the next two above three elements each and so forth.

The Routh–Hurwitz criterion may now be stated as follows:

With the Routh array calculated as previously defined, the number of polynomial roots in the RHP is equal to the number of sign changes in the first column of the array.

Example 6.2

Consider again the polynomial given in Example 6.1:

$$Q(s) = (s + 2)(s^2 + s + 4) = s^3 + s^2 + 2s + 8$$

The Routh array is

$$
\begin{array}{c|cc}
s^3 & 1 & 2 \\
s^2 & 1 & 8 \\
s^1 & -6 & \\
s^0 & 8 & \\
\end{array}
$$

where

$$
b_1 = -\frac{1}{1}\begin{vmatrix} 1 & 2 \\ 1 & 8 \end{vmatrix} = -6 \qquad c_1 = -\frac{1}{-6}\begin{vmatrix} 1 & 8 \\ -6 & 0 \end{vmatrix} = 8
$$

Since there are two sign changes on the first column (from 1 to -6 and from -6 to 8), there are two roots of the polynomial in the RHP. This was shown to be the case in Example 6.1.

This example illustrates the application of the Routh–Hurwitz criterion. Note, however, that only the stability of a system is determined. Since the locations of the roots of the characteristic equation are not found, no information about the transient response of a stable system is derived from the Routh–Hurwitz criterion. Also, the criterion gives no information about the steady-state response. We obviously need analysis techniques in addition to the Routh–Hurwitz criterion. From the equations for the calculation of the elements of the Routh array, (6.5), we see that the array cannot be completed if the first element in a row is zero. For this case, the calculation of the elements of the following row requires a division by zero. Because of this possibility, we divide the application of the criterion into three cases.

6.2.1 Case I

This case is the one just discussed. For this case none of the elements in the first column of the Routh array is zero, and no problems occur in the calculation of the array. We do not discuss this case further.

6.2.2 Case II

(a) For this case, the first element in a row is zero, with at least one non-zero element in the same row. This problem can be solved by replacing the first element of the row, which is zero, with a small number ε, which can be assumed to be either positive or negative. The calculation of the array is then continued, and some of the elements that follow that row will be a function of ε. After the array is completed, the signs of the elements in the first column are determined by allowing ε to approach zero. The number of roots of the polynomial in the RHP is then equal to the number of sign changes in this first column, as before.

An example illustrates this case.

Example 6.3

As an example of case 2, consider the polynomial

$$Q(s) = s^5 + 2s^4 + 2s^3 + 4s^2 + 11s + 10$$

The Roth array is calculated to be

s^5	1	2	11
s^4	2	4	10
s^3	ϕ^ε	6	
s^2	$-\frac{12}{\varepsilon}$	10	
s^1	6		
s^0	10		

$$b_1 = -\frac{1}{2} \begin{vmatrix} 1 & 2 \\ 2 & 4 \end{vmatrix} = 0 \qquad\qquad b_2 = -\frac{1}{2} \begin{vmatrix} 1 & 11 \\ 2 & 10 \end{vmatrix} = 6$$

$$c_1 = -\frac{1}{\varepsilon} \begin{vmatrix} 2 & 4 \\ \varepsilon & 6 \end{vmatrix} = \frac{1}{\varepsilon}(12 - 4\varepsilon) = -\frac{12}{\varepsilon} \qquad c_2 = -\frac{1}{\varepsilon} \begin{vmatrix} 2 & 10 \\ \varepsilon & 0 \end{vmatrix} = 10$$

$$d_1 = -\frac{\varepsilon}{12} \begin{vmatrix} \varepsilon & 6 \\ \frac{-12}{\varepsilon} & 10 \end{vmatrix} = \frac{\varepsilon}{12} \left[10\varepsilon + 6 \right) \left(\frac{12}{\varepsilon} \right) \right] = 6$$

$$e_1 = -\frac{1}{6} \begin{vmatrix} \frac{-12}{\varepsilon} & 10 \\ 6 & 0 \end{vmatrix} = 10$$

In the preceding calculations, the limits were taken as $\varepsilon \to 0$ at convenient points in the calculations rather than waiting until the array was complete. This procedure simplifies the calculations and the final form of the array, and the final result is the same. From the array we see that there are two sign changes in the first column whether ε is assumed positive or negative. The number of sign changes in the first column is always independent of the assumed sign of ε, which leads to the conclusion that the system that falls under case 2 is always unstable.

(b) For this case, the first element in a row is zero, with at least one non-zero element in the same row. This problem can be solved in another way if we substitute $s = 1/x$ and then reforming the characteristic equation. The calculation of the array is then continued, if a zero occurs in the first column, form up the preceding row, differentiate it with respect to s and once we differentiate we keep the coefficients of s that we get. After the array is completed, the signs of the elements in the first column are determined. The number of roots of the polynomial in the RHP is then equal to the number of sign changes in this first column, as before. An example illustrates this case.

As an example of case 2, consider the polynomial

$$Q(s) = s^5 + 2s^4 + 2s^3 + 4s^2 + 11s + 10$$

The Routh array is calculated to be

s^5	1	2	11
s^4	2	4	10
s^3	0	6	
s^2			
s^1			
s^0			

When a zero occurs in the first column, we cannot proceed further, therefore substituting $s = 1/x$, we get

$$Q(x) = 10x^5 + 11x^4 + 4x^3 + 2x^2 + 2x + 1$$

x^5	10	4	2
x^4	11	2	1
x^3	24/11	12/11	
x^2	2	1	
x^1	$-7/2$	1	
x^0	11/7		

$$b_1 = -\frac{1}{11} \begin{vmatrix} 10 & 4 \\ 11 & 2 \end{vmatrix} = \frac{24}{11} \qquad\qquad b_2 = -\frac{1}{11} \begin{vmatrix} 10 & 2 \\ 2 & 1 \end{vmatrix} = \frac{12}{11}$$

This is again an unstable case.

6.2.3 Case III

A case 3 polynomial is one for which all elements in a row of the Routh array are zero. The method described for case 2 does not give useful information in this case. A simple example illustrates case 3. Let

$$Q(s) = s^2 + 1$$

For this system the roots of the characteristic equation are on the imaginary axis, and thus the system is marginally stable. The Routh array is then

$$\begin{array}{c|cc} s^2 & 1 & 1 \\ s^1 & 0 & \\ s^0 & & \end{array}$$

and the s^1 row has no non-zero elements. The array cannot be completed because of the zero element in the first column. A second example is

$$Q(s) = (s+1)(s^2 + 2) = s^3 + s^2 + 2s + 2$$

The Routh array is

$$\begin{array}{c|cc} s^3 & 1 & 2 \\ s^2 & 1 & 2 \\ s^1 & 0 & \\ s^0 & & \end{array}$$

Again the s^1 row is zero, and the array is prematurely terminated.

A case 3 polynomial contains an even polynomial as a factor. An even polynomial is one in which the exponents of s are even integers or zero only. This even-polynomial factor is called the auxiliary polynomial and is evident in each of the examples discussed earlier. The coefficients of the auxiliary polynomial will always be the elements in the row directly above the row of zeros in the array. The exponent of the highest power in the auxiliary polynomial is the exponent that denotes the row containing its coefficients. In the first example, the row directly above the row of zeros, the s^2 row, contains the elements (1, 1). Hence the auxiliary polynomial is

$$Q_a(s) = s^2 + 1$$

For the second example, the s^1 row is all zeros and the s^2 row contains the coefficients $(1, 2)$. Thus the auxiliary equation is

$$Q_a(s) = s^2 + 2$$

The case 3 polynomial may be analysed in either of two ways. First, once the auxiliary polynomial is found, it may be factored from the characteristic equation, leaving a second polynomial. The two polynomials may then be analysed separately. However, difficulties may be encountered in applying the Routh–Hurwitz criterion to an auxiliary polynomial of high order, the second example hereunder illustrates this case. If the auxiliary polynomial is of low order, it may be factored algebraically to give the roots.

The second method for analysing case 3 polynomials is used in this book. Suppose that the row of zeros in the array is the s^i row. Then the auxiliary polynomial is differentiated with respect to s, with the coefficients of the resulting polynomial used to replace the zeros in the s^i row. The calculation of the array then continues as in the case 1 polynomial. First an example is given to illustrate the construction of the array; then the interpretation of the array is discussed.

Example 6.4

Consider the polynomial

$$Q(s) = s^4 + s^3 + 3s^2 + 2s + 2$$

The Routh array is then

s^4	1	3	2
s^3	1	2	$b_1 = -(2 - 3) = 1$
s^2	1	2	$b_2 = -(0 - 2) = 2$
s^1	$\emptyset 2$		$c_1 = -(2 - 2) = 0$
s^0	2		$d_1 = -\left(\frac{1}{2}\right)(0 - 4) = 2$

Since the s^1 row contains no non-zero elements, the auxiliary polynomial is obtained from the s^2 row and is given by

$$Q_a(s) = s^2 + 2$$

Then

$$\frac{dQ_a(s)}{ds} = 2s$$

The coefficient 2 replaces the zero in the s^1 row, and the Routh array is then completed.

The preceding example illustrates completing the array by using the derivative of the auxiliary polynomial. The array itself is interpreted in the usual way; that is, the polynomial in the example has no roots in the RHP. However, investigation of the auxiliary polynomial shows that there are roots on the imaginary axis, and thus the system is marginally stable.

The roots of an even polynomial occur in pairs that are equal in magnitude and opposite in sign. Hence, these roots can be purely imaginary, as shown in Figure 6.1(a), purely real, as shown in Figure 6.1(b), or complex, as shown in Figure 6.1(c). Since complex roots must occur in conjugate pairs, any complex roots of an even polynomial must occur in groups of four, which is apparent in Figure 6.1(c). Such roots have quadrantal symmetry: that is, the roots are symmetrical with respect to both the real and imaginary axes. For Figures 6.1(b) and 6.1(c), the Routh array will indicate the presence of roots with positive real parts. If a zero row occurs but the completed Routh array shows no sign changes, roots on the $j\omega$-axis are indicated. Thus, in any case, the presence of an auxiliary polynomial (a row of zeros in the array) indicates a non-stable system. An additional example is now given to illustrate further problems with cases 2 and 3.

Example 6.5
Consider the polynomial in Figure 6.1(c):

$$Q(s) = s^4 + 4$$

The physical systems generally do not have characteristic polynomials such as this one, but this polynomial does present a good exercise. The Routh

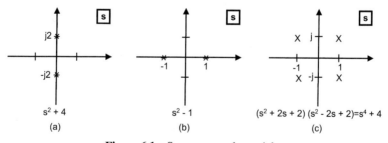

Figure 6.1 Some even polynomials.

array begins with the two rows

$$
\begin{array}{c|ccc}
s^4 & 1 & 0 & 4 \\
s^3 & 0 & 0
\end{array}
$$

and we immediately have a row of zeros. The auxiliary polynomial and its derivative are

$$
Q_a(s) = s^4 + 4 \qquad \frac{dQ_a}{ds} = 4s^3
$$

Hence, the array becomes

$$
\begin{array}{c|ccl}
s^4 & 1 & 0 & 4 \\
s^3 & \phi^4 & 0 & b_1 = -\left(\frac{1}{4}\right)(0 - 0) = 0 \\
s^2 & \phi^\varepsilon & 4 & b_2 = -\left(\frac{1}{4}\right)(0 - 16) = 4 \\
s^1 & \frac{-16}{\varepsilon} & & c_1 = -\left(\frac{1}{\varepsilon}\right)(16 - 0) = \frac{-16}{\varepsilon} \\
s^0 & 4 & & d_1 = -\left(\frac{-\varepsilon}{16}\right)\left[0 + 4\left(\frac{16}{\varepsilon}\right)\right] = 4
\end{array}
$$

The s^2 row has a non-zero element with zero for the first element; the zero is replaced with the small number. The array has two sign changes in the first column, indicating two roots with positive real parts. This result agrees with Figure 6.1(c).

This polynomial is seen to be both case 2 and case 3. The row of zeros in the array indicates the possibility of roots on the $j\omega$-axis. In this example we know that this is not the case. In general, it is necessary to factor the auxiliary equation to determine the presence of imaginary roots.

As a final point for case 3, note that if the given system is marginally stable, the auxiliary polynomial will give the frequency of oscillation of the system. This is seen from Figure 6.1(a), where the roots of the auxiliary polynomial are those poles of the closed loop transfer function that are located on the $j\omega$-axis. Thus far we have used the Routh–Hurwitz criterion only to determine the stability of systems. This criterion can also be used to aid in the design of control systems, as shown in the following two examples.

Example 6.6
The Routh–Hurwitz criterion will be employed to perform a simple design for the control system shown in Figure 6.2, in which a proportional compensator is employed. The system is type 0, and the steady-state error for a constant input of unity is

$$
e_{ss} = \frac{1}{1 + K_p} = \frac{1}{1 + K}
$$

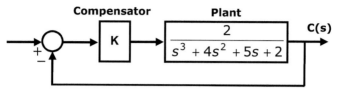

Figure 6.2   System for Example 6.6.

where the position error constant K_p is, for this example

$$K_p = \lim_{s \to 0} G_c(s)G_p(s) = \lim_{s \to 0} \frac{2K}{s^3 + 4s^2 + 5s + 2} = K$$

Suppose that a design specification is the e_{ss} must be less than 2% of a constant input. Thus, K must be greater than 49. The calculation of steady-state errors is based on the assumption of stability; thus we must ensure that the system is stable for the range of K required. The system characteristic equation is given by

$$e_{ss} = \frac{1}{1 + K_p} = \frac{1}{1 + K} < \frac{1}{50}$$

$$1 + G_c(s)G_p(s) = 1 + \frac{2K}{s^3 + 4s^2 + 5s + 2} = 0$$

or

$$Q(s) = s^3 + 4s^2 + 5s + 2 + 2K = 0$$

The Routh array for this polynomial is then

$$
\begin{array}{c|cc}
s^3 & 1 & 5 \\
s^2 & 4 & 2 + 2K \\
s^1 & \frac{18 - 2K}{4} & \\
s^0 & 2 + 2K &
\end{array}
\qquad
\begin{array}{l}
b_1 = -\frac{1}{4}(2 + 2K - 20) \\
\\
\Rightarrow \ K < 9 \\
\Rightarrow \ K > -1
\end{array}
$$

Thus the system is stable only for the compensator gain K greater than -1 but less than 9. The steady-state error criterion cannot be met with the proportional compensator; it will be necessary to use a dynamic compensation such that $G_c(s)$ is a function of s and not simply a pure gain. Of course, a PI (proportional plus integral) compensator will make the system type 1, since $G_c(s)G_p(s)$ has a pole at the origin:

$$G_c(s) = K_p + \frac{K_1}{s}$$

Hence the steady-state error for a constant input is then zero, provided that the compensated system is stable.

Example 6.7

The design in Example 6.6 is continued in this example. We replace the gain K with a PI compensator with the transfer function:

$$G_c(s) = K_p + \frac{K_1}{s} = \frac{K_{ps} + K_1}{s} = 0$$

The system characterise equation is then

$$1 + G_c(s)G_p(s) = 1 + \frac{2(K_{ps} + K_1)}{s(s^3 + 4s^2 + 5s + 2)} = 0$$

or

$$Q(s) = s^4 + 4s^3 + 5s^2 + (2 + 2K_p)s + 2K_1 = 0$$

The Routh array for this polynomial is then

$$
\begin{array}{c|ccc}
s^4 & 1 & 5 & 2K_1 \\
s^3 & 4 & 2 + 2K_p & \\
s^2 & \frac{18 - 2K_p}{4} & 2K_l & \Rightarrow \quad K_p < 9 \\
s^1 & c_1 & & \\
s^0 & 2K_1 & & \Rightarrow \quad K_1 > 0
\end{array}
$$

where

$$c_1 = \frac{-4}{18 - 2K_p}\left[8K_l - \frac{(2 - 2K_p)(18 - 2K_p)}{4} \right]$$

$$= \frac{4}{18 - 2K_p}[(1 + K_p)(9 - K_p) - 8K_l]$$

From the s^2 row, we see that $K_p < 9$ is required for stability. To simplify the design, we choose $K_p = 3$ and solve for the range of K_1 for stability. The element c_1 is then, for $K_p = 3$

$$c_1 = \frac{24 - 8K_1}{3}$$

Hence, for stability, with $K_p = 3$, the integrator gain K_1 must be less than 3.0. Also, from the s^0 row, K_1 must be greater than zero. Of course, if the value of K_p is chosen to be different, the range of K_1 for stability will also be different.

This design has considered only stability. Usually specifications in addition to stability are given in the design of a control system, and the final values of the gains K_p and K_1 are determined by all the specifications.

Example 6.8

This example is a continuation of the preceding example. If we choose the design parameters as $K_p = 3$ and $K_1 = 3$, the system is marginally stable. To show this, the Routh array is, from Example 6.7

$$\begin{array}{c|ccc} s^4 & 1 & 5 & 6 \\ s^3 & 4 & 8 \\ s^2 & 3 & 6 \\ s^1 & 0 \\ s^0 & 6 \end{array}$$

The auxiliary polynomial is the

$$Q_a(s) = 3s^2 + 6 = 3(s^2 + 2) = 3(s + j\sqrt{2})(s - j\sqrt{2})$$

This system should oscillate with the frequency $\omega = 1.14$ rad/s, and a period of $T = 2\pi/\omega = 4.44$ s.

6.3 Problems and Solutions

Problem 6.1

What conclusions, if any, can you draw concerning the root locations of the following polynomial equations, without applying the Routh–Hurwitz criterion or solving for the roots?

(a) $s^3 + 2s^2 + 3s - 1 = 0$

(b) $s^4 + s^3 + s + 2 = 0$

(c) $-s^3 - 2s^2 - 3s - 1 = 0$

(d) $s^4 + s^2 + 1 = 0$

(e) $s^3 + 2s^2 + 3s + 1 = 0$

(f) $s^4 + 1 = 0$

Solution 6.1

(a) At least one root in the RHP.

(b) Roots not all in the LHP.

(c) No contusions

(d) Quad, at least two roots not in the in RHP.

(e) No contusions

(f) Same as (d)

Problem 6.2

(a) Apply the Routh–Hurwitz criterion to each of the polynomials of Problem 6.1 to verify your conclusions.

Solution 6.2

(a)

(a)

s^3	1	3
s^2	2	-1
s^1	$7/2$	
s^0	-1	

∴ 1 root in the RHP, none on jω axis.

(b)

s^4	1	0	2
s^3	1	1	
s^2	-1	2	
s^1	3		
s^0	2		

∴ 2 roots in the RHP, none on the jω axis.

(c)

s^3	-1	-3
s^2	-2	-1
s^1	$-5/2$	
s^0	-1	

∴ All roots in the RHP.

(d)

s^4	1	1	1
s^3	ϕ^4	ϕ^2	
s^2	$1/2$	1	
s^1	-6		
s^0	1		

∴ 2 roots in the RHP, none on the jω axis.

(e)

s^3	1	3
s^2	2	1
s^1	$5/2$	
s^0	1	

∴ All roots in the LHP.

(f)

s^4	1	0	1
s^3	ϕ^4	0	
s^2	ϕ	P	1
s^1	$-4/p$		
s^0	1		

∴ 2 roots in the RHP, none on the jω axis.

Problem 6.3

Use the Routh–Hurwitz criterion to determine the number of roots in the LHP, the RHP and on the imaginary axis for the given characteristic equations:

(a) $s^4 + s^3 + 5s^2 + 2s + 4 = 0$

(b) $s^4 + 2s^2 + 1 = 0$

(c) $s^5 + 2s^4 + 4s^2 + 6s = 0$

(d) $s^4 + s^3 + s + 0.5 = 0$

(e) $s^4 + s^3 + 5s^2 + 5s + 2 = 0$

(f) $s^4 + 2s^3 + 3s^2 + 2s + 5 = 0$

Solution 6.3

(a)

s^4	1	5	4
s^3	1	2	
s^2	3	4	
s^1	2/3		
s^0	4		

$\therefore$ 4 roots in the LHP.

(b)

s^4	1	2	1
s^3	$\phi 4$	$\phi 4$	
s^2	1	1	
s^1	$\phi 4$		
s^0	1		

$\therefore$ all roots on the $j\omega$ axis.

(c)

s^5	1	5	4
s^4	2	4	
s^3	3	6	
s^2	$\phi 6$		
s^1	6		
s^0	0		

Qa $= 3s^3 + 6s$

$\therefore$ 3 roots on the $j\omega$ axis, 2 in the LHP.

(d)

s^4	1	0	0.5
s^3	1	1	
s^2	-1	1	
s^1	1.5		
s^0	0.5		

$\therefore$ 2 roots in the LHP.

(e)

s^4	1	5	2
s^3	1	5	
s^2	ϕp	2	
s^1	$-2/p$		
s^0	2		

$\therefore$ 2 roots in the RHP.

(f)

s^4	1	3	5
s^3	2	2	
s^2	2	5	
s^1	-3		
s^0	5		

$\therefore$ 2 roots in the RHP.

Problem 6.4
Consider the polynomial

$$Q(s) = s^5 + 2s^4 + 2s^3 + 4s^2 + 11s + 10$$

Apply the Routh–Hurwitz criterion to $Q(1/x)$, where $Q(1/x) = Q(s)|_{s=1/x}$,

Solution 6.4

(a) $Q(\frac{1}{x}) = 10x^5 + 11x^4 + 4x^3 + 2x^2 + 2x + 1$

s^5	10	4	2
s^4	11	2	1
s^3	24/11	12/11	
s^2	-3.5	1	
s^1	1.714		
s^0	1		

$\therefore$ 2 roots in the RHP.

Problem 6.5
Find the range of the parameter k for which all roots of the given characteristic equations are in the LHP.

(a) $s^3 + s^2 + s + k = 0$

(b) $s^3 + s^2 + ks + 1 = 0$

(c) $s^3 + ks^2 + s + 1 = 0$

(d) $ks^3 + s^2 + s + 1 = 0$

Solution 6.5

(a)

s^3	1	1
s^2	1	$\propto$
s^1	$1- \propto$	
s^0	$\propto$	

$\therefore 0 <\propto< 1$

(b)

s^3	1	$\propto$
s^2	1	1
s^1	$\propto$	-1
s^0	1	

$\therefore \propto> 1$

(c)

s^3	1	1
s^2	$\propto$	1
s^1	$(\propto -1)/\propto$	
s^0	1	

$$\therefore \propto > 1$$

(d)

s^3	$\propto$	1
s^2	1	1
s^1	$1- \propto$	
s^0	1	

$$\therefore 0 < \propto < 1$$

Problem 6.6

Consider the system of Figure P6.6.

(a) Determine the value of K (K $\neq$ 0) that will result in the system being marginally stable.
(b) Using only results in (a). Find the location of all roots of the system characteristic equation for that value of k in (a). Do not use a calculator or computer calculations.
(c) Repeat (a) and (b) for the transfer function.

$$KG_p(s) = \frac{K}{(s+4)(s+2)(s-1)} = \frac{K}{s^3 + 5s^2 + 2s - 8}$$

Two different values of K result in a marginally stable system.

Solution 6.6

Characteristics equation $\therefore 1 + G(s) = 1 + \frac{K}{s^3+3s^2+2s} = 0$,
$\therefore$ $s^3 + 3s^2 + 2s + K = 0$

(a)

s^3	1	2
s^2	3	K
s^1	$(6-k)/3$	
s^0	K	

$$\therefore 6 = K$$

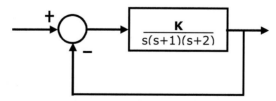

Figure P6.6   System for problem.

(b) $Q_a(s) = 3s^2 + 6 = 3(s^2 + 2)$

$$
\begin{array}{r}
s + 3 \\
s^2 + 2 \overline{\smash{\big)}\ s^3 + 3s^2 + 2s + 6} \\
\underline{s^3 \qquad\quad + 2s} \\
3s^2 \qquad + 6 \\
\underline{3s^2 \qquad + 6}
\end{array}
$$

$$\therefore Q(s) = (s + 3)(s^2 + 2)$$
roots at $s = -3, \pm j\sqrt{2}$

(c) $1 + G(s) = 1 + \dfrac{K}{s3 + 5s^2 + 2s - 8} = 0, \quad s^3 + 5s^2 + 2s + K - 8 = 0$

$$
\begin{array}{c|cc}
s^3 & 1 & 2 \\
s^2 & 5 & K - 8 \\
s^1 & (18 - k)/5 \\
s^0 & K - 8
\end{array}
$$

$$\therefore K = 8 \text{ and } K = 18$$

For $K = 8$, $Q_a = 2s$ \quad For $K = 18$, $Q_a = 2(s^2 + 2)$

Char. eg. : $s(s^2 + 5s + 2)$ \quad $Q(s) = (s + 5)(s^2 + 2) = 0$

$\therefore$ roots: $s = 0, \dfrac{-5 + \sqrt{17}}{2}, \dfrac{-5 - \sqrt{17}}{2}$ \quad $\therefore$ roots: $s = -5, \pm j\sqrt{2}$

Problem 6.7
A temperature-control system is depicted in Figure P6.7. Note the PI compensator and the rate sensor (sensor 2). Sensor 1 measures temperature and this sensor is slow compared to the rate at which the temperature can change. The sensor that measures the rate of change of temperature, sensor 2, is an operational amplifier circuit that differentiates. This system is very rare in that the sensor is slower than the plant.

(a) What is the low-frequency ($s = j\omega \approx j0$) gain of the position sensor (sensor 1)?
(b) What is the low-frequency ($s = j\omega \approx j0$) gain of the rate sensor (sensor 2)?
(c) From (a) and (b) we see that at low frequencies the system is a unity feedback system.

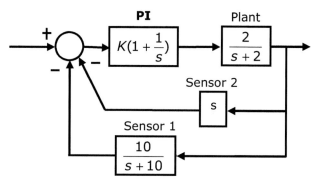

Figure P6.7   Control system for problem.

(d) From (c) we see that the steady-state error for a constant input is zero, provided that the system is stable. For what range of K is the system stable?

Solution 6.7

(a) $H_1(0) = \underline{1}$

(b) $H_2(0) = \underline{0}$

(c) $G_c(s)\, G_p(s)[H_1(s) + H_2(s)] = \left(\frac{K(s+1)}{s}\right)\left(\frac{2}{s+2}\right)\left(s + \frac{10}{s+10}\right)$

$$= \frac{2K(s+1)(s^2 + 10s + 10)}{s(s+2)(s+10)} \underline{\,\therefore +type\ 1}$$

(d) $1 + G_c(s)\, G_p(s)[H_1(s) + H_2(s)] = 0$

$\therefore \quad s^3 + 12s^2 + 20s + 2K(s^3 + 11s^2 + 20s + 10) = 0$

$\therefore \quad (1 + 2K)\, s^3 + (12 + 22K)\, s^2 + (20 + 4K)\, s + 20K = 0$

$$
\begin{array}{c|cc}
s^3 & 1 + 2K & 20 + 40K \\
s^2 & 12 + 22K & 20K \\
s^1 & \frac{840k^2 - 900k + 240}{20 + 40K} & \\
s^0 & 20 & \\
\end{array}
$$

$$\Rightarrow \underline{K > 0}$$

$\therefore$ s^0 – row requires $K > 0$. All other 1st column elements are position for $K > 0$. $\therefore$ stable for $K > 0$.

Problem 6.8

Given the closed-loop system of Figure P6.8

(a) Find the values of K that will result in the system being marginally stable.

(b) With K equal to the value found in (a), find the frequency, ω_o, of oscillation. Also, find the period $T_o = 2\pi/\omega_o$.

Solution 6.8

(a) $1 + G(s)H(s) = 1 + \left(\frac{K}{s(s+20)}\right)\left(\frac{200}{s+100}\right) = 0$

$\therefore$ $s^3 + 120s^2 + 2000s + 1600\,K = 0$

$$
\begin{array}{l|ll}
s^3 & 1 & 2000 \\
s^2 & 120 & 1600K \\
s^1 & 200 - 13.33K & \Rightarrow K < 150 \\
s^0 & 1600K & \Rightarrow K > 0
\end{array}
\qquad b_1 = -\frac{1}{120}\begin{vmatrix} 1 & 200 \\ 120 & 1600K \end{vmatrix} = \frac{1}{120}(240{,}000 - 1600K)
$$

$\therefore stable: \; 0 < k < 150$

Marginal stability : $\underline{K = 0}, \quad \underline{K = 150}$

(b) $K = 0$ – no closed-loop system.

$K = 150$, auxiliary equation: $120s^2 + 1600(150) = 0$

$$\therefore \; s^2 + 2000 = 0 \Rightarrow s = \pm j44.72$$

$$\therefore \; w_0 = 44.72\,rad/\sec; \quad T = \frac{2\pi}{w_0} = \underline{0.1405s}$$

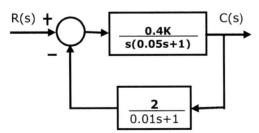

Figure P6.8 System for problem.

Problem 6.9

Given the closed-loop system of Figure P6.9

(a) Find the steady-state error as a function of K for a constant input of unity, assuming that the system is stable.
(b) Find the range of K for which the result of (a) applies.
(c) Determine the value of K (K ≠ 0) that will result in the system being marginally stable.
(d) Find the location of all roots of the system characteristic equation for that value of K found in (c). Do not use a calculator or computer for calculations.
(e) Verify the value of K found in (c) and the jω-axis roots in (d) by running a simulation of the system. Excite the system with only initial condition.

Solution 6.9

(a) System is type 1, $\therefore$ $e_{ss} = 0$

(b) $1 + \frac{K}{s^3 + 5s^2 + 6s} = 0 \Rightarrow s^3 + 5s^2 + 6s + K = 0$

$$
\begin{array}{c|cc}
s^3 & 1 & 6 \\
s^2 & 5 & K \\
s^1 & \frac{30-K}{3} & \\
s^0 & K &
\end{array}
\qquad
b_1 = -\frac{1}{5}\begin{vmatrix} 1 & 6 \\ 5 & K \end{vmatrix} = \frac{30-k}{5}
$$

$\Rightarrow K < 30$
$\Rightarrow K > 0$ $\therefore$ *stable for* $0 < k < 30$

(c) K = 30

(d) Auxiliary equation: $5s^2 + 30 = 5(s^2 + 6)$

$$\therefore \quad s = \pm jw = \pm j\sqrt{6} = \underline{\pm j2.45}$$

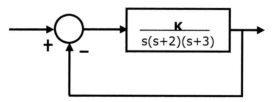

Figure P6.9   System for problem.

$$
\begin{array}{r}
s + 5 \\
s^2 + 6 \overline{\smash{\big)}\, s^3 + 5s^2 + 6s + 30} \\
\underline{s^3 \qquad\quad + 6s} \\
5s^2 \qquad + 30 \\
\underline{\qquad\qquad + 30}
\end{array}
$$

$$\Rightarrow \text{roots}: \ s = -5, \pm j\,2.45$$

(e) $T = \frac{2\pi}{w} = \frac{2\pi}{2.45} = \underline{2.565s}$

Problem 6.10

Consider the closed-loop system of Figure P6.10.

(a) Determine the range of K for stability. Consider both positive and negative values of K.
(b) Just for the purposes of this question, suppose that the system accurately models a ship-steering system, with the system input the wheel position controlled by the person at the helm and the output the heading of the ship. The disturbance D(s) models the effects of ocean currents on the ship's position. In testing the physical system, the value of gain K is set somewhat lower than the lower limit found in (a) for a short period of time. Give a sketch of a typical path that the ship would travel.
(c) Repeat (b) for the case that the value of the gain K is set somewhat higher than the upper limit found in (a). Even though the system is unstable in each case, note the difference in the nature of the instability.

Solution 6.10

(a) $1 + G(s) = 1 + \frac{K(s-2)}{s^3 + 3s^2 + 4s + 2} = 0$

$\therefore \ Q(s) = s^3 + 3s^2 + (K+4)s + (2 - 2\,K) = 0$

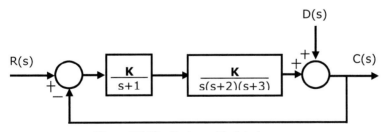

Figure P6.10 System with disturbances.

$$\begin{array}{c|c}
s^3 & 1 \quad K+4 \\
s^2 & 3 \quad 2-2K \\
s^1 & (5k+10)/3 \\
s^0 & 2-2K
\end{array} \quad \therefore stable\ for\ -2 < k < 1$$

(b) $K = -2, \quad Qa(s) = 3s^2 + 6 \quad \therefore sustianed\ oscillation$

$\therefore \quad K < -2$, response of form $e^{at} \cos(wt + \theta), \ a > 0$.

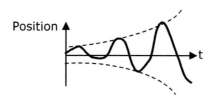

(c) For K>1, one root in the RHP. This root must be real $\therefore$ response is of form beat, a > 0.

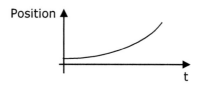

Problem 6.11
Consider the system of Figure P6.11, but with the disturbance D(s) = 0 and the open-loop function given by

$$KG_p(s) = \frac{2K}{(s+4)(s+2)(s-1)} = \frac{2K}{s^3 + 5s^2 + 2s - 8}$$

(a) Find the steady-state error for a constant input of unity, assuming that the system is stable.
(b) Find the range of K for which the results of (a) apply.
(c) Determine the values of K (K ≠ 0) that will result in the system being marginally stable.
(d) Find the locations of all roots of the system characteristic equation for the values of K found in (C). Do not use a calculator or computer for calculations.

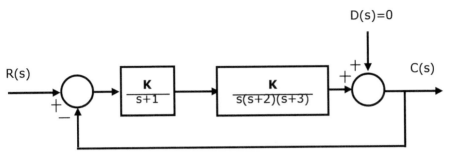

Figure P6.11   Satellite control system.

Solution 6.11

(a) $K_p = \lim_{s\to 0} K\, G_p(s) = -\frac{k}{4}$

$$e_{ss} = \frac{1}{1+k_p} = \frac{1}{1 - k/4} = \frac{4}{4-k}$$

(b) $1 + KG_p(s) = 1 + \frac{2K}{s^3+5s^2+2s-8} \Rightarrow s^3 + 5s^2 + 2s + (2k-8) = 0$

$\therefore$

$$
\begin{array}{c|cc}
s^3 & 1 & 2 \\
s^2 & 5 & 2K-8 \\
s^1 & \frac{18-2K}{5} & \Rightarrow K < 9 \\
s^0 & 2K-8 & \Rightarrow K > 4
\end{array}
\qquad
b_1 = -\frac{1}{5}\begin{vmatrix} 1 & 2 \\ 5 & 2K-8 \end{vmatrix} = \frac{1}{5}(18-2k)
$$

$\therefore stable\ for\ 4 < K < 9$

(c) $K = 4$ and $K = 9$

(d) $K = 4$: Auxiliary equation: $s^3 + 5s^2 + 2s = s(s+4.562)(s+0.4385)$

$$\therefore s = 0, -0.4385, -4.562$$

$K = 9$: Auxiliary equation: $5s^2 + 10 = 0 \Rightarrow s = \pm j\sqrt{2}$

$$
\begin{array}{r}
s+5 \\
s^2+2\overline{)\, s^3 + 5s^2 + 2s + 10} \\
\underline{s^3 \qquad\quad +2s} \\
5s^2 \qquad +10
\end{array}
\qquad \Rightarrow (s+5)(s^2+2) = 0
$$

$\therefore\ s = \pm j2, -5$

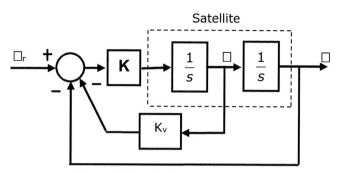

Figure P6.12   Satellite control system.

Problem 6.12

Consider the rigid satellite of Figure P6.12. The thrustors develop a torque τ, and the resultant rotation is modelled by the equation $\wp[\tau(t)] = T(s) = [j/s^2]\Theta(s)$. We consider the design of the altitude-control system for the satellite that utilizes rate feedback as shown in Figure P6.12. As a first step, we determine the range of the gains K and K_v for which the system is stable.

(a) Find the ranges of the gains K and K_v for which the system is stable.
(b) Suppose that the rate sensor has dynamics, with the transfer function $K_v/(as+1)$; that is, let K_v be replaced with $K_v/(as+1)$. Find the ranges of the gains K and K_v and the parameter a for which the system is stable.
(c) Suppose that the power amplifier has dynamics, with the transfer function $K/(bs+1)$. For this part, the rate sensor has no dynamics. Find the ranges of the gains K and K_v as a function of b for which the system is stable.

Solution 6.12

(b) $\Delta = 1 + \frac{KK_v}{s} + \frac{K}{s^2} = 0, \quad \therefore Q(s) = s^2 + KK_vs + K = 0$

$\therefore K > 0, \quad KK_v > 0 \Rightarrow \underline{K_v > 0}$

(b) $\Delta = 1 + \frac{KK_v}{s(as+1)} + \frac{K}{s^2} = 0, \quad \therefore Q(s) = as^3 + s^2 + (KK_v + Ka)s + k$

$$
\begin{array}{c|cc}
s^3 & a & KK_v + K_a \\
s^2 & 1 & K \\
s^1 & KK_v & \\
s^0 & K &
\end{array}
\qquad
\therefore
\boxed{
\begin{array}{l}
a > 0 \\
K > 0 \\
K_\nu > 0
\end{array}
}
$$

(c) $\Delta = 1 + \frac{KK_v}{s(bs+1)} + \frac{K}{s^2(bs+1)} = 0$, $\therefore Q(s) = bs^3 + s^2 + KK_v s + K = 0$

$$
\begin{array}{c|l}
s^3 & b \quad KK_v \\
s^2 & 1 \quad K \\
s^1 & K(K_v - b) \\
s^0 & K
\end{array}
\qquad
\therefore
\boxed{\begin{array}{l} b > 0 \\ K > 0 \\ K_v > 0 \end{array}}
$$

Problem 6.13

Shown in Figure P6.13 is the control system for one joint of a robot arm. The controller is a PD compensator, with $G_c(s) = (K_p + K_D s)$. K_p is the proportional gain and K_D is the derivative gain.

(a) Shown that the plant transfer function is given by

$$\frac{\Theta_L(s)}{E_a(s)} = \frac{0.475}{s^3 + 11s^2 + 10s}$$

(b) As a first step in the system design, find the ranges of the compensator gains K_p and K_D, with these gains positive, such that the closed-loop system is stable.

Solution 6.13

(a) $\frac{\theta_L}{E_a} = \frac{\frac{38/20}{(2s+21)(2s+1)s}}{1 + \frac{38/2}{(2s+21)(2s+1)}} = \frac{0.475}{s^3 + 11s^2 + 10s}$

(b) $\Delta(s) = 1 + \frac{20(K_p + K_0 s)(0.475)}{s^3 + 11s^2 + 10s} = 0$

$\therefore Q(s) = s^3 + 11s^2 + (10 + 9.5K_0) s + 9.5K_p = 0$

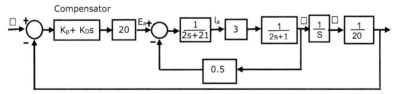

Figure P6.13   Joint control system for robot arm.

$$\begin{array}{c|ll}
s^3 & 1 & 10 + 9.5k_p \\
s^2 & 11 & 9.5k_p \\
s^1 & a & \qquad a = \frac{1}{11}(110 + 104.5k_D - 9.5k_p) \\
s^0 & 9.5k_p &
\end{array}$$

$\therefore\ k_p > 0, \quad 110 + 104.5k_D > 9.5\,k_p$

$\therefore\ 0 < k_p < \dfrac{110 + 104.5k_D}{9.5}, \quad \therefore\ k_D > -\dfrac{110}{104.5} = -1.05$

Problem 6.14

Consider the closed-loop system of Figure P6.14.

(a) Find the value of K that will result in the system being marginally stable.
(b) With K equal to the value found in (a), find the frequency at which the system will oscillate.

Solution 6.14

Char. eq.$\cdot\,1 + GH(s) = 1 + \dfrac{K}{s(10s+1)(s+1)} = 0$

$$\therefore\ Q(s) = 10s^3 + 11s^2 + s + K = 0$$

(a)

$$\begin{array}{c|ll}
s^3 & 10 & 1 \\
s^2 & 11 & K \qquad \therefore stable\ for\ 0 < k < {}^{11}/_{10} \\
s^1 & (11 - 10k)/11 & \\
s^0 & K &
\end{array}$$

marginally stable for $K = \frac{11}{10}$

(b) $K = {}^{11}/_{10}, \quad Q_a(s) = 11s^2 + {}^{11}/_{10} = 11(s^2 + 0.1)$

$\therefore\ $ roots at s $= \pm j\sqrt{0.1} = \pm0.316, \quad w = 0.316\,\text{rad/s}$

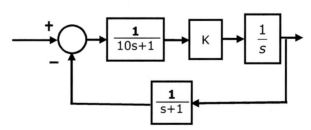

Figure P6.14   System for problem.

Problem 6.15

(a) A control system with the parameter K as the characteristic equation

$$s^3 + (14 - K)s^2 + (6 - K)s + 79 - 18K = 0$$

Find the range of K for which this system is stable.

(b) A system with the parameter K has the characteristic equation

$$s^3 + (14 - K)s^2 + (K - 6)s - 79 + 18K = 0$$

Find the range of K for which this system is stable.

Solution 6.15

(a)

$$
\begin{array}{c|cc}
s^3 & 1 & 6-k \\
s^2 & 14-k & 79-18K \\
s^1 & a \\
s^0 & 79-18K
\end{array}
\qquad a = \frac{K^2-2k+5}{14-k}
$$

$$\therefore K < 14$$

$$K < {}^{79}\!/_{18} = 4.39$$

$$K^2 - 2K + 5 > 0; \text{ satisfied for all real } K$$

$$\therefore \underline{-\infty < K < 4.39}$$

(b)

$$
\begin{array}{c|cc}
s^3 & 1 & K-6 \\
s^2 & 14-K & 18K-79 \\
s^1 & a \\
s^0 & 18K-79
\end{array}
\qquad a = \frac{-[K^2-2k+5]}{14-k}
$$

(i) 14 − K negative for all K > 14
(ii) a negative for all K < 14
    $\therefore$ unstable for all K

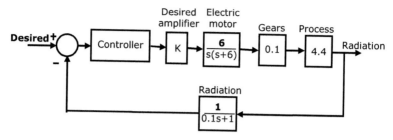

Figure P6.16   Control-rod positioning system.

Problem 6.16

Shown in Figure P6.16 is a closed-loop control rod positioning system for a nuclear reactor. The system positions control rods in a nuclear reactor to obtain a desired radiation level. The gain of 4.4 is the conversion factor from rod position to radiation level. The radiation sensor has a dc gain of unity and a time constant of 0.1 s. Let the controller transfer function be unity; that is, let $G_c(s) = 1$.

(a) Find the range of K of the power amplifier for which the system is stable.
(b) Suppose that the gain K is set to a value of 20. In addition, suppose that the sensor has a general time constant of τ rather than the 0.1 s value given. The transfer function of the sensor is then $H(s) = 1/(\tau s + 1)$. Find the allowable range for the time constant τ such that the system is stable.
(c) Suppose that, in (b), the gain K is also variable. Find the gain K as a function of the sensor time constant τ such that the system is stable.
(d) Consider that K is plotted along the s-axis and τ along the y-axis of a plane (called the parameter plane). Show the regions of this plane in which the system is stable.
(e) Show that $\tau = 1$ and $K = 15.91$ is on the stability boundary in the parameter plane.

Solution 6.16

(a) $\Delta(s) = 1 + \dfrac{K[0.6 \times 4.4 \times 10]}{s(s+6)(s+10)} = 0,$

$$\therefore \quad s^3 + 16s^2 + 60s + 26.4K = 0$$

$$\begin{array}{c|ll}
s^3 & 1 & 60 \\
s^2 & 16 & 26.4K \\
s^1 & (960 - 26.4k)/16 \\
s^0 & 26.4K
\end{array}
\qquad \therefore\ 0 < K < \tfrac{960}{26.4} = 36.36$$

(b) $\Delta(s) = 1 + \dfrac{K[0.6 \times 4.4]}{(s^2 + 6s)(Ts + 1)} = 0,$

$$\therefore\ Q(s) = \underline{Ts^3 + (1 + 6T)s^2 + 6s + 2.64K}$$

$$\begin{array}{c|ll}
s^3 & T & 6 \\
s^2 & 1 + 6T & 2.64K \\
s^1 & (6 + 36T - 2.64kT)/(1 + 6T) \\
s^0 & 2.64K
\end{array}$$

$\therefore\ T > 0$

$(2.64k - 36)T < 6, \quad T < \dfrac{6}{52.8 - 36} = 0.357$

$\therefore\ \underline{0 < T < 0.357}$

(c) From (b), $2.64k\,T < 6 + 36\,T$

$$\therefore\ \underline{0 < T < \dfrac{6 + 36T}{2.64T}} \quad \text{with} \quad \underline{T > 0}$$

(d)

T	K
0.1	36.4
0.3	21.2
0.5	18.2
1	15.9
5	14.1
∞	13.6

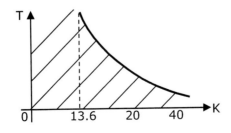

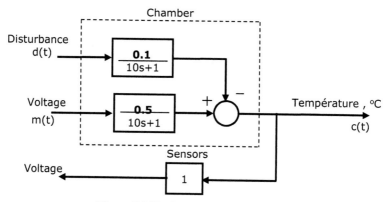

Figure P6.17   Thermal stress chamber.

$$K = \frac{6 + 36T}{2.64T} = \frac{6 + 36}{2.64} = \underline{15.91}$$

Problem 6.17

Shown in Figure P6.17 is the temperature-control system of a large test chamber. The controller is PI, which is commonly employed in temperature-control systems. The transfer function of the controller is given by $G_c(s) = (K_p + K_1/s)$, where K_p is the proportional gain and K_1 is the integral gain. The disturbance $D(s)$ models the opening of the door of the chamber and is the step function $d(t) = 60u(t)$ when the door is opened at $t = 0$. If the door is closed, $d(t) = 0$.

(a) Assume that $m(t)$ is constant; hence, the system is essentially an open loop. If the door is opened and remains open, find the steady-state change in the chamber's temperature in degrees Celsius.

(b) Repeat (a) for the closed-loop system, with $K_1 = 0$ and $K_p > 0$. The result is a function of K_p.

(c) Repeat (a) for the closed-loop system, with $K_1 > 0$ and $K_p > 0$.

(d) The results of (b) and (c) are based on the assumption that the system is stable. Find the ranges of K_p and K_1, with both gains positive, such that the system is stable.

Solution 6.17

(a) $G_d(s) = \frac{-0.1}{10s+1};$ dc gain $= G_d(s) = \underline{-0.1}$

$\therefore C_{dss} = (60)(-0.1) = \underline{-6^0 C}$

(b) $T_d(s) = \dfrac{\frac{-0.1}{10s+1}}{1+k_p(\frac{0.5}{10s+1})} = \dfrac{-0.1}{10s+(1+0.5K_p)}$

$T_d(0) = \dfrac{-0.1}{1+0.5k}; C_{dss} = (60)\left(\dfrac{-0.1}{1+0.5K}\right) = \dfrac{-6}{1+0.5K}$

(c) $G_c(s) = k_p + \frac{k_I}{s} = \frac{k_p s + K_I}{s}$

$T_d(s) = \dfrac{\frac{-0.1}{10s+1}}{1+\left(\frac{0.5}{10s+1}\right)\left(\frac{K_p s + K_I}{s}\right)} = \dfrac{-0.1s}{10s^2 + s + 0.5K_p s + 0.5K_I}$

$= \dfrac{-0.1s}{10s^2 + (1+0.5K_p)s + 0.5K_I}$

$\therefore$ dc gain $= T_d(0) = 0 \Rightarrow \underline{c_{dss} = 0}$

(d)

s^2	10 $0.5K_I$
s^1	$1 + 0.5k_p$
s^0	$0.5K_I$

$\therefore \underline{K_p > 0}$ (problem constraint)

$\underline{K_I \geq 0}$ $K_I = 0$ is stable, from (b).

7

Fourier Series

This chapter provides comprehensive details regarding the introduction to periodic function Fourier synthesis, constructing a waveform, trigonometric form of the Fourier series, use of symmetry, complex form of the Fourier series, Gibbs phenomenon and problems and solutions.

7.1 Periodic Function and Fourier Synthesis

Jean Baptiste Joseph Baron de Fourier (1768–1830), a French physicist, discovered that any periodic waveform can be broken down into a combination of sinusoidal waves. All the sinusoidal waves that can be combined together to produce any periodic waveform are collectively called a basis and each individual wave is an element of the basis.

The amplitudes of the sinusoids are called the Fourier coefficients. The relationship between the frequencies of sinusoids is such that they are all harmonics (integer multiples) of a fundamental frequency. The amplitude of the periodic signal as a function of time is its time-domain representation, whereas the Fourier coefficients correspond to its frequency-domain representation.

7.2 Constructing a Waveform with Sine Waves

In Figures 7.1(a) to (c) given hereunder shows three basic sine waves 2, 4 and 6 Hz with amplitude of 7, 2 and 4, respectively. They add together to form the wave in Figure 7.1(d).

We can write the equation for the wave in Figure 7.1(d) as

$$s(t) = 7\sin(2\pi 2t) + 2\sin(2\pi 4t) + 4\sin(2\pi 6t) \qquad (7.1)$$

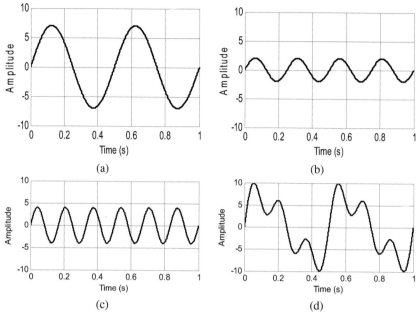

Figure 7.1 (a) Sine wave 1 with f = 2 Hz, amplitude = 7   (b) Sine wave 2 with f = 4 Hz, amplitude = 2   (c) Sine wave 3 with f = 6 Hz, amplitude = 4   (d) Addition of sine waves 1,2,3.

In this case the fundamental frequency is 2 Hz, the frequency of the second harmonic is 4 Hz and the frequency of the third harmonic is 6 Hz.

If the fundamental frequency is denoted by f_1, then (7.1) can be written as

$$s(t) = 7\sin(2\pi[f_1]t) + 2\sin(2\pi[2f_1]t) + 4\sin(2\pi[3f_1]t) \qquad (7.2)$$

Therefore, s(t) is a combination of a fundamental and the second and the third harmonics. Observe that the frequency of the more complicated periodic wave s(t) is 2 Hz, which is the same as the frequency of the lowest frequency sine wave (the fundamental).

7.3 Constructing a Waveform with Cosine Waves

In fact, the waveform in Figure 7.2(d) has its maximum value at t = 0, so it can be created using a combination of cosine wave. Once again the frequency of the more complicated periodic wave, which is 2 Hz, which is

the same as frequency as the frequency of the lowest frequency cosine wave
(the fundamental)

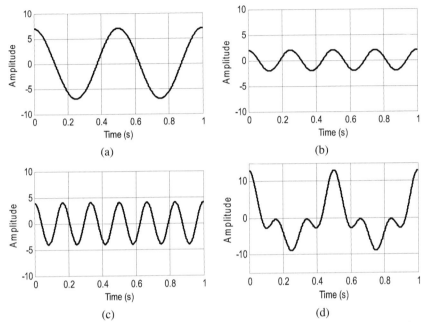

Figure 7.2   (a) Cosine wave 1 with f = 2 Hz, amplitude = 7   (b) Cosine wave 2 with f = 4
Hz, amplitude = 2   (c) Cosine wave 3 with f = 6 Hz, amplitude = 4   (d) Addition of cosine
waves in (a), (b), and (c). The maximum value occurs at t = 0.

Now look at Figure 7.2(d). The waveform does not start at zero; further,
the value is not equal to zero at t = 0. Since sine waves always have a value of
zero at t = 0, we cannot use combination of sine wave to produce it. In fact,
this waveform has its maximum value at t = 0, so it can be created using
combination of cosine waves. The equation for the waveform of Figure 7.2 is

$$c(t) = 7\cos(2\pi 2t) + 2\cos(2\pi 4t) + 4\cos(2\pi 6t) \qquad (7.3)$$

Therefore, s(t) is a combination of a fundamental, second and third
harmonics. In this case the fundamental frequency is 2 Hz, the frequency of
the second harmonic is 4 Hz and the frequency of the third harmonic is 6 Hz.
If the fundamental frequency is denoted by f_1, then (7.3) can be written
as

$$c(t) = 7\cos(2\pi[f_1]t) + 2\cos(2\pi[2f_1]t) + 4\cos(2\pi[3f_1]t) \qquad (7.4)$$

Observe that the frequency of the more complicated periodic wave c(t) is 2 Hz, which is the same as the frequency of the lowest frequency cosine wave (the fundamental).

7.4 Constructing a Waveform with Cosine and Sine Waves

What would a waveform look like that was the combination of three sine waves in Figures 7.1 (a) to (c) and three cosine waves in Figures 7.2 (a) to (c)? This is shown in Figure 7.2(d). Observe that the value at $t = 0$ is neither zero nor maximum value of the wave. This brings us a very important point: by adding sines and cosines of appropriate amplitude we construct a periodic waveform starting with at any value at $t = 0$. The equation for the waveform for Figure 7.3 is

f(t) = cosine wave 1+ cosine wave 2+ cosine wave 3+ sine wave 1+ sine wave 2+sine wave 3

$$f(t) = 7\cos(2\pi[f_1]t) + 2\cos(2\pi[2f_1]t) + 4\cos(2\pi[3f_1]t)$$
$$+ 7\sin(2\pi[f_1]t) + 2\sin(2\pi[2f_1]t) + 4\sin(2\pi[3f_1]t)$$

(7.5)

We have kept these examples simple and are only using the fundamental and the second and the third harmonics to produce periodic waves. Periodic signals actually encountered in practice can be combinations of extremely large number of harmonics. Therefore equation for such harmonics waves

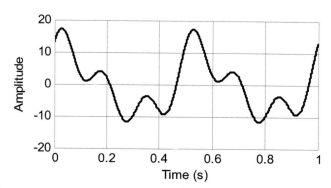

Figure 7.3   Combination of sine waves from 7.1 (a) to (d) and the cosine waves from 7.2 (a) to (d). The value at $t = 0$ is neither 0 nor maximum, but somewhere in between.

can be written as

$$f(t) = a_1 \cos(2\pi[f_1]t) + a_2 \cos(2\pi[2f_1]t) + \cdots a_n \cos(2\pi[nf_1]t)$$
$$+ b_1 \sin(2\pi[f_1]t) + b_2 \sin(2\pi[2f_1]t) + \cdots b_n \sin(2\pi[nf_1]t) \tag{7.6}$$

where

$a_1, a_2, a_3, \ldots$ are the amplitudes of cosine wave
$b_1, b_2, b_3, \ldots$ are the amplitudes of sine wave
$f_1, f_2, f_3, \ldots$ are the frequencies of the fundamental and harmonics

7.5 Constructing a Waveform with Both Sine and Cosine Waves and a DC Component

Finally, this waveform will be like the same as of Figure 7.3, except it would be shifted upwards due to addition of a DC value. This leads finally to (7.7):

$$f(t) = a_0 + a_1 \cos(2\pi[f_1]t) + a_2 \cos(2\pi[2f_1]t) + \cdots a_n \cos(2\pi[nf_1]t)$$
$$+ b_1 \sin(2\pi[f_1]t) + b_2 \sin(2\pi[2f_1]t) + \cdots b_n \sin(2\pi[nf_1]t) \tag{7.7}$$

(7.7) is known as Fourier series. It consists of a series of frequencies used to form any periodic function. The amplitudes of sines and cosines are given a special name – they are called the Fourier coefficients. Thus a_1, a_2, , $\cdots$ a_n, $b_1, b_2, \cdots$ b_{rmn} are the Fourier coefficients.

The sines and cosines, those have a fundamental frequency f_1, and all those corresponding harmonics, are said to form the basis for all periodic waveforms. Each of sine and cosine waveform is an element of basis. This means that not only can sines and cosines be combined to construct any complicated periodic waveform, but also any complicated periodic waveform can be broken down into sum of sines and cosines.

7.6 Trigonometric From of the Fourier Series

Consider a periodic function f(t), defined by the functional relationship

$$f(t) = f(t + T)$$

where T is the period. We further assume that the function f(t) satisfies the following properties:

1 f(t) is single-valued everywhere; that is, f(t) satisfies the mathematical definition of a function.
2 The integral $\int_{t_0}^{t_0+T} |f(t)|\,dt$ exists (i.e., is not infinite) for any choice of t_0.
3 f(t) has a finite number of discontinuities in any one period.
4 f(t) has a finite number of maxima and minima in any one period.

We shall consider f(t) to represent a voltage or current waveform, and any voltage or current waveform which we can actually produce must satisfy these conditions. Certain mathematical functions which we might hypothesize may not satisfy these conditions, but we shall assume that these four conditions are always satisfied.

Given such a periodic function f(t), and f(t) may be represented by the infinite series

$$f(t) = a_0 + a_1 \cos \omega_0 t + a_2 \cos 2\omega_0 t + \ldots$$
$$+ b_1 \sin \omega_0 t + b_2 \sin 2\omega_0 t + \ldots$$
$$= a_0 + \sum_{n=1}^{\infty}(a_n \cos n\omega_0 t + b_n \sin n\omega_0 t) \qquad (7.8)$$

where the fundamental frequency ω_0 is related to the period T a by

$$\omega_0 = \frac{2\pi}{T}$$

and where a_0, a_n and b_n are constants which depend upon n and f(t). Equation (7.8) is the trigonometric form of the Fourier series for f(t), and the process of determining the values of the constants a_0, a_n and b_n is called Fourier analysis. Our objective is not the proof of this theorem, but only a simply development of the procedures of Fourier analysis and a feeling that the theorem is plausible.

Before we evaluate the constants appearing in the Fourier series, let us collect a set of useful trigonometric integrals. We let both n and k represent any element of the set of integers 1, 2, 3, ... In the following integrals, 0 and T are used as the integration limits, but it is understood that any interval of one period is equally correct. Since the average value of a sinusoid over one period is zero

$$\int_0^T \sin n\omega_0 t\, dt = 0 \qquad (7.9)$$

and

$$\int_0^T \cos n\omega_0 t \, dt = 0 \tag{7.10}$$

It is also a simple matter to show that the following three definite integrals are zero:

$$\int_0^T \sin k\omega_0 t \cos n\omega_0 t \, dt = 0 \tag{7.11}$$

$$\int_0^T \sin k\omega_0 t \sin n\omega_0 t \, dt = 0 \quad (k \neq n) \tag{7.12}$$

$$\int_0^T \cos k\omega_0 t \cos n\omega_0 t \, dt = 0 \quad (k \neq n) \tag{7.13}$$

Those cases which are accepted in equations (7.12) and (7.13) are also easily evaluated; we now obtain

$$\int_0^T \sin^2 n\omega_0 t \, dt = \frac{T}{2} \tag{7.14}$$

$$\int_0^T \cos^2 n\omega_0 t \, dt = \frac{T}{2} \tag{7.15}$$

The evaluation of the unknown constants in the Fourier series may now be accomplished readily. We first attack a_0. If we integrate each side of Equation (7.8), over a full period, we obtain

$$\int_0^T f(t)dt = \int_0^T a_0 dt + \int_0^T \sum_{n=1}^{\infty}(a_n \cos n\omega_0 t + b_n \sin n\omega_0 t)dt$$

But every term in the summation is of the form of Equation (7.9) or (7.10), and thus

$$\int_0^T f(t)dt = a_0 T$$

or

$$a_0 = \frac{1}{T}\int_0^T f(t)dt \tag{7.16}$$

This constant a_0 is simply the average value of f(t) over a period, and we therefore describe it as the dc component of f(t).

To evaluate one of the cosine coefficients – say a_k, the coefficient of $\cos k\omega_0 t$ – we first multiply each side of Equation (7.8) by $\cos k\omega_0 t$ and then integrate both sides of the equation over a full period:

$$\int_0^T f(t) \cos k\omega_0 t \, dt = \int_0^T a_0 \cos k\omega_0 t \, dt + \int_0^T \sum_{n=1}^\infty a_n \cos k\omega_0 t \cos n\omega_0 t \, dt$$

$$+ \int_0^T \sum_{n=1}^\infty b_n \cos k\omega_0 t \sin n\omega_0 t \, dt$$

From Equations (7.10), (7.11) and (7.12) we note that every term on the right-hand side of this equation is zero except for the single a_n term where k $= n$. We evaluate that term using Equation (7.15), and in so doing we find a_k or a_n:

$$a_n = \frac{2}{T} \int_0^T f(t) \cos(n\omega_0 t) dt \tag{7.17}$$

This result is twice the average value of the product f(t) $\cos(n\omega_0 t)$ over a period. In a similar way, we obtain b_k by multiplying by $\sin n\omega_0 t$, integrating over a period, noting that all but one of the terms on the right-hand side is zero, and performing that single integration by Equation (7.15). The result is

$$b_n = \frac{2}{T} \int_0^T f(t) \sin(n\omega_0 t) dt \tag{7.18}$$

which is twice the average value of f(t) $\sin(n\omega_0 t)$ over a period.

Equations (7.16) to (7.18) now enable us to determine values for a_0 and all the a_n and b_n in the Fourier series, Equation (7.8):

$$f(t) = a_0 + \sum_{n=1}^\infty (a_n \cos n\omega_0 t + b_n \sin n\omega_0 t)$$

$$\omega_0 = \frac{2\pi}{T} = 2\pi f_0 \tag{7.19}$$

$$a_0 = \frac{1}{T} \int_0^T f(t) dt$$

$$a_n = \frac{2}{T} \int_0^T f(t) \cos n\omega_0 t \, dt$$

$$b_n = \frac{2}{T} \int_0^T f(t) \sin n\omega_0 t \, dt$$

Example 7.1

The 'half-sinusoidal' waveform shown in Figure 7.4 represents the voltage response obtained at the output of a half-wave rectifier circuit, a non-linear circuit whose purpose is to convert a sinusoidal input voltage to a (pulsating) dc output voltage.

Find the Fourier series representation of this waveform.

Solution 7.1

In order to represent this voltage as a Fourier series, we must first determine the period and then express the graphical voltage as an analytical function of time. From the graph, the period is seen to be

$$T = 0.4s \quad \text{and thus} \quad f_0 = 2.5 Hz \quad \text{and} \quad w_0 = 5\pi \, \text{rad/s}$$

With these three quantities determined, we now seek an appropriate expression for f(t) or v(t) which is valid throughout the period. Obtaining this equation or set of equations proves to be the most difficult part of Fourier analysis for many students. The source of the difficulty is apparently either the inability to recognize the given curve, carelessness in determining multiplying constants within the functional expression or a negligence which results in not writing the complete expression. In this example, the statement of the problem implies that the functional form is a sinusoid whose amplitude is V_m. The radian frequency has already been determined as 5π, and only the positive portion of the cosine wave is present. The functional expression over

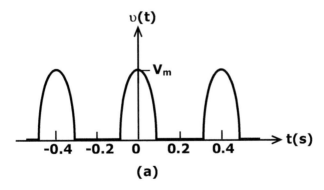

(a)

Figure 7.4(a)   The output of a half-wave rectifier to which a sinusoidal input is applied.

the period $t = 0$ to $t = 0.4$ is therefore

$$v(t) = \begin{cases} V_m \cos 5\pi t & 0 \le t \le 0.1 \\ 0 & 0.1 \le t \le 0.3 \\ V_m \cos 5\pi t & 0.3 \le t \le 0.4 \end{cases}$$

It is evident that the choice of the period extending from $t = -0.1$ to $t = 0.3$ will result in fewer equations and, hence, fewer integrals:

$$v(t) = \begin{cases} V_m \cos 5\pi t & -0.1 \le t \le 0.1 \\ 0 & 0.1 \le t \le 0.3 \end{cases} \tag{1}$$

This form is preferable, although either description will yield the correct results. The zero-frequency component is easily obtained:

$$a_0 = \frac{1}{0.4} \int_{-0.1}^{0.3} v(t)dt = \frac{1}{0.4} \left[\int_{-0.1}^{0.1} V_m \cos 5\pi t \, dt + \int_{0.1}^{0.3} (0)dt \right] a_0 = \frac{V_m}{\pi} \tag{2}$$

Notice that integration over an entire period must be broken up into subintervals of the period, in each of which the functional form of $v(t)$ is known. The amplitude of a general cosine term is

$$a_n = \frac{2}{T} \int_0^T f(t) \cos n\omega_0 t \, dt \quad a_n = \frac{2}{0.4} \int_{-0.1}^{0.1} V_m \cos 5\pi t \cos 5\pi n t \, dt$$

The form of the function we obtain upon integrating is different when n is unity than it is for any other choice of n. if $n = 1$, we have

$$a_1 = 5V_m \int_{-0.1}^{0.1} \cos^2 5\pi t \, dt = \frac{V_m}{2} \tag{3}$$

whereas if n is not equal to unity, we find

$$a_n = 5V_m \int_{-0.1}^{0.1} \cos 5\pi t \cos 5\pi n t \, dt$$

Using trigonometric relationship

$$\cos(a + b) = \cos a \cos b + \sin a \sin b$$
$$\cos(a - b) = \cos a \cos b - \sin a \sin b$$
$$2 \cos a \cos b = \cos(a + b) + \cos(a - b)$$

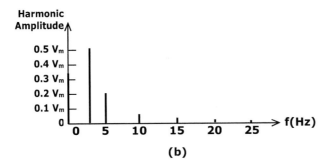

Figure 7.4(b) The discrete line spectrum of the waveform in part a.

$$a_n = 5V_m \int_{-0.1}^{0.1} \frac{1}{2}[\cos 5\pi(1+n)t + \cos 5\pi(1-n)t]dt$$

$$a_n = \frac{5V_m}{2} \cdot \left[-\frac{\sin 5\pi(1+n)t}{5\pi(1+n)} - \frac{\sin 5\pi(1-n)t}{5\pi(1-n)} \right]_{-0.1}^{0.1}$$

or

$$a_n = \frac{2V_m}{\pi} \frac{\cos(\pi n/2)}{1-n^2} \quad (n \neq 1) \tag{4}$$

The expression for a_n when $n \neq 1$ will yield the correct result for $n = 1$ in the limit as $n \to 1$. A similar integration shows that $b_n = 0$ for any value of n, and the Fourier series thus contains no sine terms. The Fourier series is therefore obtained from Equations (2), (3) and (4):

$$v(t) = \frac{V_m}{\pi} + \frac{V_m}{2}\cos 5\pi t + \frac{2V_m}{3\pi}\cos 10\pi t - \frac{2V_m}{15\pi}\cos 20\pi t$$
$$+ \frac{2V_m}{35\pi}\cos 30\pi t - \dots \tag{5}$$

In Figure 7.4(a), $v(t)$ is shown graphically as a function of time; in Equation (1), $v(t)$ is expressed as an analytical function of time. Either of these representations is a time-domain representation. Equation (5), the Fourier series representation of $v(t)$, is also a time-domain expression, but it may be transformed easily into a frequency-domain representation.

Example 7.2

A periodic waveform f(t) is described as follows: $f(t) = -4, 0 < t < 0.3$; $f(t) = 6, 0.3 < t < 0.4$; $f(t) = 0, 0.4 < t < 0.5$; $T = 0.5$. Evaluate: (a) a_0; (b) a_3 and (c) b_1.

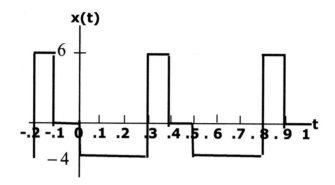

Solution 7.2

T = 0.5 s for the above example from the given example.

$$\omega_0 = 2\pi f \Rightarrow \omega_0 = \frac{2\pi}{0.5} = \frac{2\pi}{0.5} = 4\pi.$$

Values such as a_0 and a_n. b_n would come out to be zero, which has been verified in this example:

$$\omega_0 = \pi \quad \Rightarrow \quad a_0 = \frac{1}{T}\int_0^T f(t)dt$$

$$= \frac{1}{0.5}\left[\int_0^{0.3}(-4)dt + \int_{0.3}^{0.4}6dt + \int_{0.4}^{0.5}0dt\right] = -1.2$$

Similarly

$$a_n = \frac{2}{T}\int_0^T f(t)\cos n\omega_0 t\, dt$$

$$a_3 = \frac{2}{0.5}\left[\int_0^{0.3}(-4)\cos 12\pi t\, dt + \int_{0.3}^{0.4}(6)\cos 12\pi t\, dt\right] = 1.383$$

$$b_n = \frac{2}{T}\int_0^T f(t)\sin n\omega_0 t\, dt$$

$$\therefore b_1 = \frac{2}{0.5}\left[\int_0^{0.3}(-4)\sin 4\pi t\, dt + \int_{0.3}^{0.4}(6)\sin 4\pi t\, dt\right]$$

$$= 4(-1.109) = -4.44$$

7.6.1 Use of Symmetry

It is possible to anticipate the absence of certain terms in a Fourier series, before any integration is performed, by an inspection of the symmetry of the given function.

The two types of symmetry which are most readily recognized are even-function symmetry and odd-function symmetry, or simply even symmetry and odd symmetry. We say that f(t) possesses the property of even symmetry if

$$f(t) = f(-t) \qquad (7.20)$$

Such functions as t^2, cos 3t, in (cos t), $\sin^2 7t$, and a constant C all possess even symmetry; the replacement of t by (-t) does not change the value of any of these functions. This type of symmetry may also be recognized graphically, for if f(t) = f(-t) then mirror symmetry exists about the f(t) axis. The function shown in Figure 7.8(a) possesses even symmetry; if the figure were to be

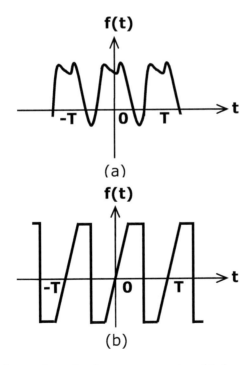

(a)

(b)

Figure 7.8 (a) A waveform showing even symmetry. (b) A waveform showing odd symmetry.

folded along the f(t) axis, then the portions of the graph for positive and negative time would fit exactly, one on top of the other.

We define odd symmetry by stating that if odd symmetry is a property of f(t), then

$$f(t) = -f(-t) \tag{7.21}$$

In other words, if t is replaced by (-t), then the negative of the given function is obtained; for example, t, sin t, cos 70t, $t\sqrt{1+t^2}$, and the function sketched in Figure 17-4b are all odd functions and possess odd symmetry. The graphical characteristics of odd symmetry are apparent if the portion of f(t) for t > 0 is rotated about the positive t-axis and the resultant figure is then rotated about the f(t) axis; the two curves will fit exactly, one on top and f on the other. That is, we now have symmetry about the origin, rather than about the f(t) axis as we did for even functions.

Having definitions for even and odd symmetry, we should note that the product of two functions with even symmetry, or of two functions with odd symmetry, yields a function with even symmetry. Furthermore, the product of an even and an odd function gives a function with odd symmetry.

Now let us investigate the effect that even symmetry produces in a Fourier series. If we think of the expression which equates an even function f(t) and the sum of an infinite number of sine and cosine functions, then it is apparent that the sum must also be an even function. A sine wave, however, is an odd function, and no sum of sine waves can produce any even function other than zero (which is both even and odd). It is thus plausible that the Fourier series of any even function is composed of only a constant and cosine functions. Let us now show carefully that $b_n = 0$. We have

$$b_n = \frac{2}{T} \int_{-T/2}^{T/2} f(t) \sin n\omega_0 t \, dt$$

$$= \frac{2}{T} \left[\int_{-T/2}^{0} f(t) \sin n\omega_0 t \, dt + \int_{0}^{T/2} f(t) \sin n\omega_0 t \, dt \right]$$

Now let us replace the variable t in the first integral by $-\tau$, or $\tau = -t$, and make use of the fact that $f(t) = f(-t) = f(\tau)$:

$$b_n = \frac{2}{T} \left[\int_{T/2}^{0} f(-\tau) \sin(-n\omega_0 t)(-d\tau) + \int_{0}^{T/2} f(t) \sin n\omega_0 t \, dt \right]$$

$$= \frac{2}{T} \left[-\int_{0}^{T/2} f(\tau) \sin n\omega_0 \tau \, d\tau + \int_{0}^{T/2} f(t) \sin n\omega_0 t \, dt \right]$$

But the symbol we use to identify the variable of integration cannot affect the value of the integral. Thus

$$\int_0^{T/2} f(\tau) \sin n\omega_0\tau\, d\tau = \int_0^{T/2} f(t) \sin n\omega_0 t\, dt \tag{7.22}$$

and $\qquad\qquad bn = 0 \quad (even\ sym.)$

No sine terms are present. Therefore, if f(t) shows even symmetry, then $b_n = 0$; conversely, if $b_n = 0$, the f(t) must have even symmetry.

A similar examination of the expression for a_n leads to an integral over the half-period extending from $t = 0$ to $t = \frac{1}{2}T$:

$$a_n = \frac{4}{T} \int_0^{T/2} f(t) \cos n\omega_0 t\, dt \quad (even\ sym.) \tag{7.23}$$

The fact that a_n may be obtained for an even function by taking 'twice the integral over half the range' should seem logical.

A function having an odd symmetry can contain no constant term or cosine terms in its Fourier expansion. Let us prove the second part of this statement. We have

$$a_n = \frac{2}{T} \int_{-T/2}^{T/2} f(t) \cos n\omega_0 t\, dt$$

$$= \frac{2}{T} \left[\int_{-T/2}^0 f(t) \cos n\omega_0 t\, dt + \int_0^{T/2} f(t) \cos n\omega_0 t\, dt \right]$$

and we now let $t = -\tau$ in the first integral:

$$a_n = \frac{2}{T} \left[\int_{T/2}^0 f(-\tau) \cos(-n\omega_0\tau)(-d\tau) + \int_0^{T/2} f(t) \cos n\omega_0 t\, dt \right]$$

$$= \frac{2}{T} \left[\int_0^{T/2} f(-\tau) \cos n\omega_0\tau\, d\tau + \int_0^{T/2} f(t) \cos n\omega_0 t\, dt \right]$$

But $f(-\tau) = -f(\tau)$, and therefore

$$a_n = 0 \quad (odd\ sym.) \tag{7.24}$$

A similar, but simpler, proof shows that

$$a_0 = 0 \quad (odd\ sym.)$$

With the odd symmetry, therefore, $a_n = 0$ and $a_0 = 0$; conversely, if $a_n = 0$ and $a_0 = 0$, the odd symmetry is present.

The values of b_n may again be obtained by integrating over half the range:

$$b_n = \frac{4}{T} \int_0^{T/2} f(t) \sin n\omega_0 t \, dt \quad (odd \ sym.) \tag{7.25}$$

Examples of an even and odd symmetry were afforded by Examples 7.2 and 7.3. In both the examples, a square wave of the same amplitude and period is the given function. The time origin, however, is selected to provide odd symmetry in part a and even symmetry in part b, and the resultant series contain, respectively, only sine and cosine terms. It is also noteworthy that the point at which t = 0 could be selected to provide neither an even nor an odd symmetry; the determination of the coefficients of the terms in the Fourier series would then take at least twice as long.

The Fourier series for both of these square waves have one other interesting characteristic: neither contains any even harmonics. That is, the only frequency components present in the series have frequencies which are odd multiples of the fundamental frequency; a_n and b_n are zero for even values of n. This result is caused by another type of symmetry, called half-wave symmetry. We shall say that f(t) possesses the half-wave symmetry if

$$f(t) = -f(t - \tfrac{1}{2}T)$$

or the equivalent expression, $f(t) = -f(t + \tfrac{1}{2}T)$

Except for a change of sign, each half-cycle is like the adjacent half-cycles. Half-wave symmetry, unlike the even and odd symmetry, is not a function of the choice of the point t = 0. Thus, we can state that the square wave (Figure 7.6) shows the half-wave symmetry. Neither waveform shown in Figure 7.9 has the half-wave symmetry, but the two somewhat similar functions plotted in Figure 7.9 do possess the half-wave symmetry.

It may be shown that the Fourier series of any function which has half-wave symmetry contains only odd harmonics. Let us consider the coefficients, a_n. We have again

$$a_n = \frac{2}{T} \int_{-T/2}^{T/2} f(t) \cos n\omega_0 t \, dt$$

$$= \frac{2}{T} \left[\int_{-T/2}^{0} f(t) \cos n\omega_0 t \, dt + \int_{0}^{T/2} f(t) \cos n\omega_0 t \, dt \right]$$

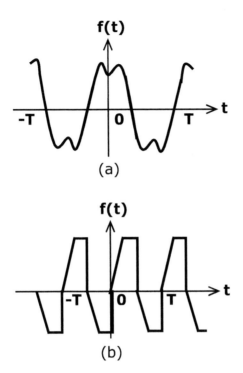

Figure 7.9 (a) A waveform somewhat similar to the one shown in Figure 7.9(a), but possessing half-wave symmetry. (b) A waveform somewhat similar to the one shown in Figure 7.9(b), but possessing half-wave symmetry.

Constant vigilance is required to avoid confusion between an even function and an even harmonic, or between an odd function and an odd harmonic.

We may represent the same as

$$a_n = \frac{2}{T}(I_1 + I_2)$$

Now we substitute the new variable $\tau = t + \frac{1}{2}T$ in the integral I_1:

$$I_1 = \int_0^{T/2} f\left(\tau - \frac{1}{2}T\right) \cos n\omega_0 \left(\tau - \frac{1}{2}T\right) \, d\tau$$

$$= \int_0^{T/2} -f(\tau) \left(\cos n\omega_0\tau \cos \frac{n\omega_0 T}{2} + \sin n\omega_0\tau \sin \frac{n\omega_0 T}{2}\right) \, d\tau$$

But $\omega_0 T$ is 2π, and thus

$$\sin \frac{n\omega_0 T}{2} = \sin n\pi = 0$$

Hence

$$I_1 = -\cos n\pi \int_0^{T/2} f(\tau) \cos n\omega_0 \tau d\tau$$

After noting the form of I_2, we therefore may write

$$a_n = \frac{2}{T}(1 - \cos n\pi) \int_0^{T/2} f(t) \cos n\omega_0 t\, dt$$

The factor $(1 - \cos n\pi)$ indicates that a_n is zero if n is even. Thus

$$a_n = \begin{cases} \frac{4}{T}\int_0^{T/2} f(t) \cos n\omega_0 t dt & n\ odd \\ 0 & n\ even \end{cases} \qquad (\text{half} - \text{wave symmetry})$$

$$(7.26)$$

A similar investigation shows that b_n is also zero for all even n, and therefore

$$b_n = \begin{cases} \frac{4}{T}\int_0^{T/2} f(t) \sin n\omega_0 t dt & n\ odd \\ 0 & n\ even \end{cases} \qquad (\text{half} - \text{wave symmetry})$$

$$(7.27)$$

It should be noted that the half-wave symmetry may be present in a waveform which also shows the odd symmetry or the even symmetry. The waveform sketched in Figure 7.7(a), for example, possesses both the even symmetry and the half-wave symmetry. When a waveform possesses the half-wave symmetry and either the even or the odd symmetry, then it is possible to reconstruct the waveform if the function is known over any quarter-period interval. The value of a_n or b_n may also be found by integrating over any quarter period. Thus

$$\left. \begin{array}{ll} a_n = \frac{8}{T}\int_0^{T/4} f(t) \cos n\omega_0 t\, dt & n\ odd \\ a_n = 0 & n\ even \\ b_n = 0 & all\ n \end{array} \right\} (\text{half} - \text{wave and even symmetry})$$

$$(7.28)$$

$$\left. \begin{array}{ll} a_n = 0 & all\ n \\ b_n = \frac{8}{T}\int_0^{T/4} f(t) \sin n\omega_0 t\, dt & n\ odd \\ b_n = 0 & n\ even \end{array} \right\} (\text{half} - \text{wave and odd symmetry})$$

$$(7.29)$$

It is always worthwhile spending a few moments investigating the symmetry of a function for which a Fourier series is to be determined.

The following table for trigonometric form shows that which of the coefficients for Fourier series have to be calculated. *It is always worthwhile spending few minutes investigating the symmetry of a function for which a Fourier series for the wave form is to be determined.*

Even symmetry	$a_0 = \frac{1}{T} \int_o^T f(t)\, dt$ $a_n = \frac{2}{T} \int_o^T f(t) \cos nw_0 t\, dt$	$b_n = 0$
Odd symmetry	$a_0 = 0,\ a_n = 0$	$b_n = \frac{2}{T} \int_o^T f(t) \sin nw_0 t\, dt$
Half-wave symmetry	$a_n = \frac{4}{T} \int_o^{T/2} f(t) \cos nw_0 t\, dt;\ n\ odd$ $= 0\quad n\ even$	$b_n = \frac{4}{T} \int_o^{T/2} f(t) \sin nw_0 t\, dt;\ n\ odd$ $= 0\quad n\ even$
Half-wave and even symmetry	$a_n = \frac{8}{T} \int_o^{T/4} f(t) \cos nw_0 t\, dt;\ n\ odd$ $a_n = 0\quad n\ even$	$b_n = 0\qquad for\ all\ n$
Half-wave and odd symmetry	$a_n = 0\quad for\ all\ n$	$b_n = \frac{8}{T} \int_o^{T/4} f(t) \sin nw_0 t\, dt\quad n\ odd$ $b_n = 0\quad n\ even$

Example 7.3
Write the Fourier series for the voltage waveform shown in Figure 7.5.

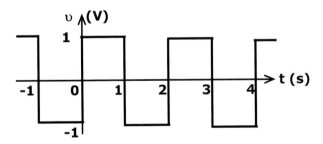

Figure 7.5

Solution 7.3
$T = 2$ s for the above example from the given example.

$$w_0 = 2\,\pi\, f \Rightarrow w_0 = \frac{2\pi}{T} = \frac{2\pi}{2} = \pi.$$

The given waveform is an odd function, and also half-wave symmetry, therefore only b_n *has to be calculated*, other values such as a_0 and a_n would come out to be zero, which has been verified in this example.

$$\omega_0 = \pi \Rightarrow a_0 = \frac{1}{T}\int_0^T f(t)\,dt = \frac{1}{2}\left[\int_0^1 1\,dt + \int_1^2 -1\,dt\right]$$

$$a_0 = \frac{1}{2}\left[\int_0^1 1\,dt + \int_1^2 -1\,dt\right] = \frac{1}{2}\left[\{t\}_0^1 - \{t\}_1^2\right] = \frac{1}{2}[1-1] = 0$$

$$a_n = \frac{2}{T}\int_0^T f(t)\cos n\,\omega_0\,t\,dt = \frac{2}{2}\left[\int_0^1 \cos n\pi t\,dt - \int_1^2 \cos n\pi t\,dt\right]$$

$$= \frac{1}{n\pi}(0) = 0$$

$$= \frac{2}{2}\left[\left\{\frac{\sin n\pi t}{n\pi}\right\}_0^1 - \left\{\frac{\sin n\pi t}{n\pi}\right\}_1^2\right] = \left[\frac{1}{n\pi}\{2\sin n\pi\} - \{\sin 2n\pi\}\right]$$

$$= 0$$

$$b_n = \frac{2}{T}\int_0^T f(t)\sin\,n\omega_0 t\,dt$$

$$\therefore b_n = \frac{2}{T}\left[\int_0^1 \sin n\pi t\,dt - \int_1^2 \sin n\pi t\,dt\right]$$

$$= \left[\left\{\frac{\cos n\pi\,t}{-n\pi}\right\}_0^1 + \left\{\frac{\cos n\pi\,t}{n\pi}\right\}_1^2\right]$$

$$= \left[-\frac{1}{n\pi}\{\cos n\pi - 1\} + \frac{1}{n\pi}\{\cos 2n\pi - \cos n\pi\}\right]$$

$$b_n = -\frac{1}{n\pi}(\cos n\pi - 1) + \frac{1}{n\pi}(1 - \cos n\pi)$$

$$= \frac{4}{n\pi}(n \text{ odd and } 0 \text{ } n \text{ even})$$

$$b_1 = \frac{4}{\pi} \quad b_3 = \frac{4}{3\pi} \quad b_5 = \frac{4}{5\pi}.$$

$$\therefore v(t) = \frac{4}{\pi}\left(\sin \pi\,t + \frac{1}{3}\sin 3\pi\,t + \frac{1}{5}\sin 5\pi\,t + \cdots\right) V.$$

Example 7.4
Write the Fourier series for the voltage waveform shown in the figure provided.

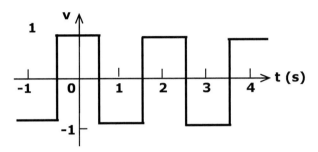

Solution 7.4
$T = 2$ s for the above example from the given example.

$$\omega_0 = 2\,\pi\,f = \frac{2\pi}{T} = \frac{2\pi}{2} = \pi. \quad a_0 = 0$$

The given waveform is an even function and half-wave a_n *has to be calculated for odd values:*

$$a_n = \frac{2}{T}\int_0^T f(t)\cos n\pi\,t\,dt = \frac{2}{2}\left[\int_{-\frac{1}{2}}^{+\frac{1}{2}}\cos n\pi t\,dt - \int_{+\frac{1}{2}}^{\frac{3}{2}}\cos n\pi t\,dt\right]$$

$$a_n = \frac{2}{n\pi}\sin\frac{n\pi}{2} - \frac{1}{n\pi}\left(\sin\frac{3n\pi}{2} - \sin\frac{n\pi}{2}\right) = 0, \ (n\ even)$$

$$a_n = \frac{3}{n\pi}\sin\frac{n\pi}{2} - \frac{1}{n\pi}\sin\frac{3n\pi}{2}$$

$$= \frac{4}{n\pi}\sin\frac{n\pi}{2} = \frac{4}{n\pi}(-1)^{(n-1)/2}(for\ n\ odd).$$

$$a_1 = \frac{4}{\pi} \quad a_3 = -\frac{4}{3\pi} \quad a_5 = \frac{4}{5\pi}.$$

$$b_n = \frac{2}{T}\int_0^T f(t)\sin n\omega_0 t\,dt$$

$$b_n = \int_{-\frac{1}{2}}^{+\frac{1}{2}}\sin n\pi\,t\,dt - \int_{\frac{1}{2}}^{\frac{2}{3}}\sin n\pi\,t = \frac{1}{n\pi}\left[\cos\frac{n\pi}{2} - \cos\left(-\frac{n\pi}{2}\right)\right]$$

$$+ \frac{1}{n\pi}\left[\cos\frac{3n\pi}{2} - \cos\frac{n\pi}{2}\right] = 0$$

$$\therefore v(t) = \frac{4}{\pi}\left(\cos \pi t - \frac{1}{3}\cos 3\pi t + \frac{1}{5}\cos 5\pi t - \cdots\right)$$

Example 7.5

Write the Fourier series for the voltage waveform shown in Figure 7.7.

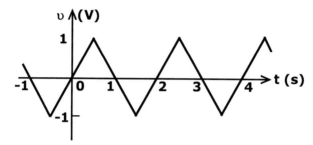

Figure 7.7   Voltage waveform.

Solution 7.5

T = 2 s for the above example from the given example.

$$\omega_0 = 2\pi f \Rightarrow \omega_0 = \frac{2\pi}{T} = \frac{2\pi}{2} = \pi.$$

The given waveform is an odd function and half-wave symmetry also exists, therefore only b_n has to be calculated, other values such as a_0 and a_n would come out to be zero, which has been verified in this example.

Substituting these values in f(t), which has been developed from the graph

$$f(t) = 2t \qquad 0 \le t \le 0.5$$
$$f(t) = 1 \qquad -1 \le t \le 1$$
$$f(t) = 2 - t \qquad 1 \le t \le 2$$

$$a_n = \frac{2}{T}\int_0^T f(t)\cos n\pi t\, dt$$

$$a_n = \frac{2}{2}\left[\int_{-\frac{1}{2}}^{\frac{1}{2}} 2t\cos n\pi t\, dt + \int_{\frac{1}{2}}^{\frac{3}{2}} 2(1-t)\cos n\pi t\, dt\right]$$

$$= 2\int_{-\frac{1}{2}}^{+\frac{1}{2}} t\cos n\pi t\, dt - \int_{\frac{1}{2}}^{\frac{3}{2}} 2t\cos n\pi t\, dt + 2\int_{\frac{1}{2}}^{\frac{3}{2}} \cos n\pi t\, dt$$

$$\text{or} = 2\int_{-\frac{1}{2}}^{\frac{1}{2}} t\cos n\pi t\, dt = \left[\frac{2}{n^2\pi^2}\cos n\pi t + \frac{2t}{n\pi}\sin n\pi t\right]_{-\frac{1}{2}}^{+\frac{1}{2}} = 0$$

$$\text{or} - \int_{\frac{1}{2}}^{\frac{3}{2}} 2t\cos n\pi\, t\, dt = -2\left[\frac{1}{n^2\pi^2}\cos n\pi\, t + \frac{t}{n\pi}\sin n\pi\, t\right]_{\frac{1}{2}}^{\frac{3}{2}}$$

$$= -2\left[\frac{1}{n^2\pi^2}\left(\cos\frac{3n\pi}{2} - \cos\frac{n\pi}{2}\right) + \frac{3}{2n\pi}\sin\frac{3n\pi}{2} - \frac{1}{2n\pi}\sin\frac{n\pi}{2}\right]$$

$$= 0 - \frac{3}{n\pi}\sin\frac{3n\pi}{2} + \frac{1}{n\pi}\sin\frac{n\pi}{2}$$

$$2\int_{\frac{1}{2}}^{\frac{3}{2}}\cos n\pi\, t\, dt = \frac{2}{n\pi}\sin n\pi\, t\int_{\frac{1}{2}}^{\frac{3}{2}} = \frac{2}{n\pi}\sin\frac{3n\pi}{2} - \frac{2}{n\pi}\sin\frac{n\pi}{2}.$$

$$\therefore a_n = \left(\frac{2}{n\pi} - \frac{3}{n\pi}\right)\sin\frac{3n\pi}{2} + \left(\frac{1}{n\pi} - \frac{2}{n\pi}\right)\sin\frac{n\pi}{2}$$

$$= -\frac{1}{n\pi}\left(\sin\frac{3n\pi}{2} - \sin\frac{n\pi}{2}\right) = 0.$$

$$b_n = \frac{2}{T}\int_0^T f(t)\sin n\omega_0 t\, dt \, b_n = \int_{-\frac{1}{2}}^{\frac{1}{2}} 2t\sin n\pi\, t\, dt$$

$$+ \int_{\frac{1}{2}}^{\frac{3}{2}} 2(1-t)\sin n\pi\, t\, dt.$$

$$b_n = 4\int_0^{\frac{1}{2}} t\sin n\pi\, t\, dt + \int_{-\frac{1}{2}}^{\frac{1}{2}} 2\tau\sin n\pi(\tau+1)d\tau \quad when \quad t = \tau+1$$

$$b_n = 4\int_0^{\frac{1}{2}} t\sin n\pi t\, dt - \int_{-\frac{1}{2}}^{\frac{1}{2}} \tau w_s n\pi\sin n\pi\tau d\tau$$

$$b_n = 4\int_0^{\frac{1}{2}} t\sin n\pi t\, dt - 2\int_{-\frac{1}{2}}^{\frac{1}{2}} \tau\cos n\pi t\, d\tau$$

$$= 4\int_0^{\frac{1}{2}} t\sin n\pi t\, dt - 4\int_0^{\frac{1}{2}} \tau(-1)^n\sin n\pi\tau\, d\tau\, 1$$

$$= 4[1-(-1)^n]\int_0^{\frac{1}{2}} t\sin n\pi\, t\, dt:, \; b_n = 0 \; for \; n \; even.$$

For n odd

$$b_n = 8 \left[\frac{1}{n^2\pi^2} \sin n\pi\tau - \frac{1}{n\pi} \cos n\pi t \right]_0^{\frac{1}{2}}$$

$$= 8 \left[\frac{1}{n^2\pi^2} \sin \frac{n\pi}{2} - \frac{1}{2n\pi} \cos \frac{n\pi}{2} \right]$$

$$\therefore b_n = \frac{8}{n^2\pi^2} \sin \frac{n\pi}{2} \ for \ n \ odd \ \ \therefore b_1 = \frac{8}{\pi^2} \ \ b_3 = -\frac{8}{9\pi^2} \ \ b_5 = \frac{8}{25\pi^2}, \ldots$$

$$\therefore v(t) = \frac{8}{\pi^2} \left(\sin \pi t - \frac{1}{9} \sin 3\pi t + \frac{1}{25} \sin 5\pi t - \ldots \right) V.$$

The Fourier series for both of these square wares have one other interesting characteristic it does not contain even values. That is, the only frequency component, present in the series has frequencies which are odd multiple of fundamental frequency, a_n and b_n are zero, for even value of n. This result is caused by another type of symmetry, called the half-wave symmetry.

Example 7.6
Write the Fourier series for the voltage waveforms shown in the figure provided.

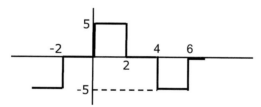

Solution 7.6

$$\begin{array}{ccc} \text{no,} & \text{no,} & \text{yes} \\ / & \backslash & \backslash \\ \text{Even} & \text{odd} & \text{half-wave} \end{array}$$

This is the case of the half-wave symmetry, so following the row three of the above table, both a_n and b_n are to be calculated for n odd values only.

$$a_n = \begin{cases} \frac{4}{T} \int_0^{\frac{T}{2}} f(t) \cos n\omega_0 t \, dt & n \ odd \\ 0 & n \ even \end{cases}$$

$$b_n = \begin{cases} \frac{4}{T} \int_0^{\frac{T}{2}} f(t) \sin n\omega_0 t \, dt & n \text{ odd} \\ 0 & n \text{ even} \end{cases}$$

$$a_n = \frac{4}{8} \int_0^2 5 \cos \frac{n\pi t}{4} \, dt = 2.5 \times \frac{4}{n\pi} \sin \frac{n\pi t}{4} \Bigg|_0^2 = \frac{10}{n\pi} \sin \frac{n\pi}{2} (n \text{ odd})$$

$$a_n = \frac{10}{n\pi} \sin \frac{n\pi}{2} (n \text{ odd})$$

$$b_n = \frac{4}{8} \int_0^2 5 \sin \frac{n\pi t}{4} \, dt = \left[2.5 \times \frac{4}{n\pi} \cos \frac{n\pi t}{4} \right]_0^2$$

$$= \frac{10}{n\pi} \left[\cos \frac{n\pi}{2} - 1 \right] (n \text{ odd})$$

$$b_n = \frac{10}{n\pi} \left(\cos \frac{n\pi}{2} - 1 \right) (n \text{ odd})$$

$$f(t) = \frac{10}{\pi} \sum_{n=odd}^{\infty} \left(\frac{1}{n} \sin \frac{n\pi}{2} \cos \frac{n\pi t}{4} + \frac{1}{n} \sin \frac{n\pi t}{4} + \cdots \right)$$

Example 7.7

Write the Fourier series for the voltage waveforms shown in Figure 7.11.

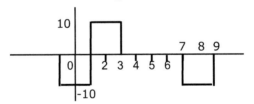

Solution 7.7

$$\begin{array}{ccc} \text{no,} & \text{no,} & \text{yes} \\ \diagup & \diagdown & \diagdown \\ \text{Even} & \text{odd} & \text{half-wave} \end{array}$$

This is the case of the half-wave symmetry, so following the row three of the above table, both a_n and b_n are to be calculated for n odd values only.

$$a_n = \begin{cases} \frac{4}{T} \int_0^{\frac{T}{2}} f(t) \cos n\omega_0 t \, dt & n \text{ odd} \\ 0 & n \text{ even} \end{cases}$$

$$b_n = \begin{cases} \frac{4}{T} \int_0^{\frac{T}{2}} f(t) \sin n\omega_0 t \, dt & n \text{ odd} \\ 0 & n \text{ even} \end{cases}$$

$$a_n = a_n = \frac{4}{8} \left[\int_{-1}^{1} -10 \cos \frac{n\pi t}{4} \, dt + \int_{1}^{3} 10 \cos \frac{n\pi t}{4} \, dt \right]$$

$$= \frac{1}{2} \left[-\frac{40}{n\pi} \sin \frac{n\pi t}{4} \Big|_{-1}^{1} + \frac{40}{n\pi} \sin \frac{n\pi t}{4} \Big|_{1}^{3} \right]$$

$$= \frac{20}{n\pi} \left[-\sin \frac{n\pi}{4} - \sin \frac{n\pi t}{4} + \sin \frac{3n\pi}{4} - \sin \frac{n\pi}{4} \right]$$

$$= \frac{20}{n\pi} \left(\sin \frac{3n\pi}{4} - 3\sin \frac{n\pi}{4} \right)$$

$$b_n = \frac{4}{8} \left[\int_{-1}^{1} -10 \sin \frac{n\pi t}{4} \, dt + \int_{1}^{3} 10 \sin \frac{n\pi t}{4} \, dt \right]$$

$$= \frac{20}{n\pi} \left(\cos \frac{n\pi}{4} - \cos \frac{n\pi}{4} - \cos \frac{3n\pi}{4} + \cos \frac{n\pi}{4} \right)$$

$$f(t) = \sum_{n=1}^{\infty} \frac{10}{n\pi} \left[\left(\sin \frac{3n\pi}{4} - 3\sin \frac{n\pi}{4} \right) \cos \frac{n\pi t}{4} \right.$$

$$\left. \times 4 \left(\cos \frac{n\pi}{4} - \cos \frac{3n\pi}{4} \right) \sin \frac{n\pi t}{4} \right] + \cdots$$

7.6.2 Complex Form of the Fourier Series

Let us take the trigonometric forms for the sines and cosines.

$$f(t) = a_0 + \sum_{n=1}^{\infty} (a_n \cos n\omega_0 t + b_n \sin n\omega_0 t) \tag{7.30}$$

and then substitute the exponential forms for the sines and cosines. After rearranging

$$f(t) = a_0 + \sum_{n=1}^{\infty} (e^{jn\omega_0} \frac{a_n - jb_n}{2} + e^{-jn\omega_0} \frac{a_n + jb_n}{2}) \tag{7.31}$$

Now, we define the complex constant c_n

$$c_n = \frac{a_n - jb_n}{2} \quad n = 1, 2, 3 \ldots \tag{7.32}$$

$$a_n = \frac{2}{T} \int_0^T f(t) \cos \omega_0 n\, t\, dt \quad \Rightarrow a_{-n} = a_n \tag{7.33}$$

$$b_n = \frac{2}{T} \int_0^T f(t) \sin n\omega_0 t\, dt \quad \Rightarrow b_{-n} = -b_n. \tag{7.34}$$

$$c_{-n} = \tfrac{1}{2}(a_n + jb_n) \quad n = 1, 2, 3, .. \tag{7.35}$$
$$c_n = c_{-n}^*$$

Thus, $c_0 = a_0$.
We may therefore express f(t) as

$$f(t) = c_0 + \sum_{n=1}^{\infty} c_n e^{jn\omega_0 t} + \sum_{n=1}^{\infty} c_{-n} e^{-jn\omega_0 t} \tag{7.36}$$

$$f(t) = \sum_{n=0}^{\infty} c_n e^{jn\omega_0 t} + \sum_{n=1}^{\infty} c_{-n} e^{-jn\omega_0 t}. \tag{7.37}$$

Finally, instead of summing the second series over the integers from 1 to ∞, let us sum over the negative integers from -1 to $-\infty$.

$$f(t) = \sum_{n=0}^{\infty} c_n e^{jn\omega_0 t} + \sum_{n=-1}^{\infty} c_n e^{+jn\omega_0 t} \tag{7.38}$$

$$f(t) = \sum_{n=-\infty}^{\infty} c_n e^{jn\omega_0 t}. \tag{7.39}$$

By agreement, a summation from $-\infty$ to ∞ is understood to include from n = 0.

Further, (7.39) is the complex form of the Fourier series of f(t); its conciseness is one of the most important reason for its use. In order to obtain the expression by which a particular complex coefficients c_n may be evaluated

$$c_n = \frac{1}{T} \int_{-\frac{T}{2}}^{+\frac{T}{2}} f(t) \cos n\omega_0 t\, dt - j\frac{1}{T} \int_{-\frac{T}{2}}^{+\frac{T}{2}} f(t) \sin n\omega_0 t\, dt. \tag{7.40}$$

and then we use the exponential equivalent of the sine and cosine and simplify.

$$c_n = \frac{1}{T} \int_{-\frac{T}{2}}^{+\frac{T}{2}} f(t) e^{-jn\omega_o t} \, dt. \tag{7.41}$$

Collecting the two basic relationships for the exponential form of the Fourier series, we have

$$f(t) = \sum_{n=-\infty}^{\infty} c_n e^{jn\omega_o t}. \quad c_n = \frac{1}{T} \int_{-\frac{T}{2}}^{+\frac{T}{2}} f(t) e^{-jn\omega_o t} \, dt. \tag{7.42}$$

The following table for the exponential form shows that which of the coefficients for Fourier series have to be calculated. *It is always worthwhile spending few minutes investigating the symmetry of a function for which a Fourier series for the waveform is to be determined:*

$$f(t) = \sum_{n=-\infty}^{\infty} c_n e^{jn\omega_o t}. \quad .f(t) = \sum_{n=0}^{\infty} c_n e^{jn\omega_o t} + \sum_{n=-1}^{\infty} c_n e^{+jn\omega_o t}$$

Even symmetry	$c_n = \frac{2}{T} \int_0^{\frac{T}{2}} f(t) \cos n\omega_0 t \, dt$
Odd symmetry	$c_n = -\frac{j2}{T} \int_0^{\frac{T}{2}} f(t) \sin n\omega_0 t \, dt$
Half-wave symmetry	$c_n = \begin{cases} \frac{2}{T} \int_0^{\frac{T}{2}} f(t) e^{-jn\omega_0 t} dt & n \text{ odd} \\ 0 & n \text{ even} \end{cases}$
Half-wave and even symmetry	$c_n = \begin{cases} \frac{4}{T} \int_0^{\frac{T}{4}} f(t) \cos n\omega_0 t \, dt & n \text{ odd} \\ 0 & n \text{ even} \end{cases}$
Half-wave and odd symmetry	$c_n = \begin{cases} \frac{-j4}{T} \int_0^{\frac{T}{4}} f(t) \sin n\omega_0 t \, dt & n \text{ odd} \\ 0 & n \text{ even} \end{cases}$

Example 7.8
Determine the general coefficients c_n in the complex Fourier series for the waveform shown in the figure provided.

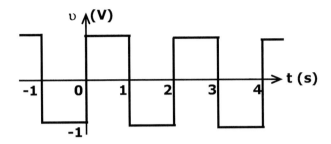

Solution 7.8

$T = 2$ s

$$\omega_0 = 2\pi f \Rightarrow \omega_0 = \frac{2\pi}{T} = \frac{2\pi}{2} = \pi.$$

Odd and half-wave symmetry

$$c_n = \begin{cases} \frac{-j4}{T} \int_0^{\frac{T}{4}} f(t) \sin n\omega_0 t \, dt & n \text{ odd}, \quad half-wave \ and \ odd \\ 0 & n \text{ even}, \quad half-wave \ and \ odd \end{cases}$$

$$c_n = -j2 \int_0^{0.5} 1 \sin n\pi = j2\frac{1}{n\pi} \cos n\pi t \ \Big|_0^{0.5}$$

$$c_n = j\frac{2}{n\pi}\left(\cos \frac{n\pi}{2} - 1\right) = -j\frac{2}{n\pi} \ for \ n \ odd,$$

$$= 0 \quad for \ n \ even$$

Example 7.9

Determine the general coefficients c_n in the complex Fourier series for the waveform shown in the figure provided.

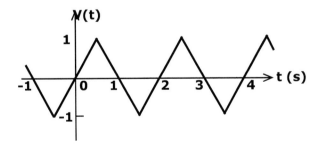

Solution 7.9
This example is of case of the odd and half-wave symmetry:

$$T = 2, \quad \omega_0 = \pi$$

Using the exponential Fourier coefficient formula

$$c_n = -j\frac{4}{T}\int_0^{T/4} f(t)\sin n\omega_0\,dt \quad \text{for } n \text{ odd}$$
$$= 0 \qquad\qquad\qquad\qquad \text{for } n \text{ even}$$

$$c_n = -j2\int_0^{0.5} 2t\sin n\pi t\,dt = -j4\left[\frac{1}{n^2\pi^2}\sin n\pi t - \frac{1}{n\pi}\cos n\pi t\right]_0^{0.5}$$

$$c_n = -j4\left[\frac{1}{n^2\pi^2}\sin\frac{n\pi}{2} - \frac{1}{2n\pi}\cos\frac{n\pi}{2} + 0\right]$$

n *is odd*
$$c_n = -j\frac{4}{n^2\pi^2}\sin\frac{n\pi}{2}$$

Example 7.10
Write the Fourier series for the voltage waveforms shown in the figure provided.

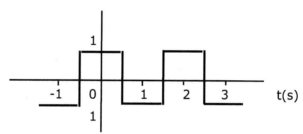

Solution 7.10

$$T = 2, \omega_0 = \frac{2\pi}{2} = \pi$$

$$c_n = \frac{1}{T}\int_{-\frac{T}{2}}^{+\frac{T}{2}} f(t)e^{-jn\omega_0 t}\,dt.$$

$$= \frac{1}{2}\left[\int_{-1}^{-0.5} e^{-jn\omega_0 0 t}\,dt + \int_{-0.5}^{0.5} e^{-jn\omega_0 t}\,dt - \int_{0.5}^{1} e^{-jn\omega_0 t}\,dt\right.$$

$$= \frac{1}{2}\left[\frac{-1}{-jn\pi}\right]\left(e^{-jn\pi t}\right)_{-1}^{0.5} + \frac{1}{-jn\pi}\left(e^{-jn\pi t}\right)_{-0.5}^{+0.5} + \frac{-1}{-jn\pi}\left(e^{-jn\pi t}\right)_{0.5}^{1}$$

$$= \frac{1}{j2n\pi}\left(e^{jn\pi/2} - e^{-jn\pi} - e^{-jn\pi/2} + e^{+jn\pi/2} + e^{-jn\pi} - e^{-jn\pi/2}\right).$$

$$c_n = \frac{1}{j2n\pi}\left(2e^{jn\pi/2} - 2e^{-jn\pi/2}\right) = \frac{2}{n\pi}\sin\frac{n\pi}{2}$$

$$c_0 = 0, \quad c_1 = \frac{2}{\pi} \quad c_2 = 0, \quad c_3 = -\frac{2}{3\pi} \quad c_4 = 0, \quad c_5 = \frac{2}{5\pi} \ldots$$

symmetry (even and half-wave)

$$c_n = \frac{4}{T}\int_0^{\frac{T}{4}} f(t)\cos n\omega_0 t\, dt$$

$$= \frac{4}{2}\int_0^{0.5} \cos n\pi t\, dt = \frac{2}{n\pi}(\sin n\pi)_0^{0.5}$$

$$= \begin{cases} \frac{2}{n\pi}\sin\frac{n\pi}{2} & (n \quad odd) \\ 0 & (n \quad even) \end{cases}$$

7.7 Discrete Time Fourier Series of Periodic Signals

If the signal is periodic with period N and the expansion interval is one period, then the discrete time Fourier series equals the sign for all time. That is

For N odd

$$x[n] = \sum_{k=-(N-1)/2}^{(N-1)/2} a_k\, e^{jk(2\pi n/N)}$$

For N even

$$x[n] = \sum_{k=N/2-1}^{N/2} a_k\, e^{jk(2\pi n/N)}$$

where

$$a_k = \frac{1}{N}\sum_{n=n_1+1}^{n_1+n} x[n]\, e^{-jk(2\pi n/N)}$$

Equality for all time holds since the series expansion equals the signal over the expansion interval and both signal and series repeat outside this interval.

Example 7.11
Determine the Fourier series coefficient a_0 and a_1 for the following signal as x(n) illustrated in the figure provided.

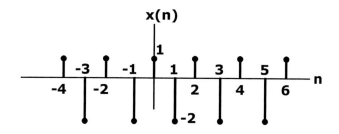

Solution 7.11

An impulse train in discrete time is given as

$$x[n] = \sum_{k=0}^{N-1} a_k e^{jk(2\pi n/N)} \quad x[n] = \sum_{k=0}^{1} a_1 e^{jk(2\pi n/2)} = \sum_{k=0}^{1} a_k e^{jk\pi n}$$

$$a_k = \frac{1}{2} \sum_{n=0}^{N-1} x[n] e^{-jk(2\pi/N)n} = \frac{1}{2} \sum_{n=0}^{1} x[n] e^{-jk\pi n}$$

$$a_k = \frac{1}{2} \left[x(0) e^{-jk\pi 0} + x(1) e^{-jk\pi} \right]$$

$$a_1 = \frac{1}{2} \left[(1)(1) + (-2)e^{-j\pi} \right] = \frac{1}{2} \left[1 - 2e^{-j\pi} \right] = \frac{1}{2} \left[1 - (-2) \right] = \frac{3}{2}$$

$$a_k = \frac{1}{2} \left[x(0) e^{-jk\pi 0} + x(1) e^{-jk\pi} \right]$$

$$a_0 = \frac{1}{2} \left[(1)(1) + (-2)e^{-j(0)\pi} \right] = \frac{1}{2} \left[1 - 2 \right] = -\frac{1}{2}$$

7.8 Gibbs' Phenomenon

Fourier series can be used to construct any complicated periodic waveform. In general, the more the number of elements in the Fourier series, the better is the construction of corresponding waveform.

However, there is limitation in that if the periodic waveform has any discontinuity, that is the vertical edges, then even an infinite number of elements in the Fourier series cannot construct that discontinuity exactly.

Adding more and more harmonics to the sine waves, one gets a better representation of a square wave. Keep on adding the higher harmonics, the resulting wave looks more and more ideal square wave. By including the

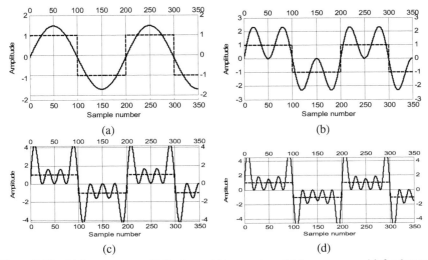

Figure 7.15 (a) Square wave with fundamental harmonic (b) Square wave with fundamental and third harmonic (c) Square wave with fundamental, third, fifth and seventh harmonic (d) Square wave with fundamental, third, fifth, seventh and ninth harmonic.

higher odd harmonics, the following changes in the resulting waveform as the number of harmonics is increased:

1. The number of oscillation increases, but the amplitudes of the oscillations decrease.
2. The vertical edge gets steeper.
3. Overshoot exists at the vertical edges.
4. The approximation is worst at the vertical edges.

It is observed that the overshoots at the vertical edges do not really go away, irrespective how many harmonics are added. Even if an infinite number of harmonics is added, the overshoot remains and its size settles to about 8.95% of the size of the vertical edge. This phenomenon is called the Gibbs phenomenon in honour of Willard Gibbs who described it occurrence in the late nineteenth century.

7.9 Problems and Solutions

Problem 7.1
Determine the Fourier series representations for the following signal x(t) as illustrated in Figure P7.1 ; x(t) periodic with period 2.

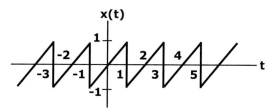

Figure P7.1   Signal x(t).

Solution 7.1

$$\omega_0 = 2\pi f = 2\pi/T = 2\pi/2 = \pi$$

This is the case of an odd and half-wave symmetry, b_n for odd values has to be calculated:

$$b_n = \tfrac{8}{T} \int_o^{T/4} f(t) \sin n w_0\, t\, dt \quad n\ odd$$
$$b_n = 0 \quad n\ even$$

Substituting these values in $f(t) = t$, which has been developed from the graph

$$b_n = \tfrac{8}{T} \int_o^{T/4} t \ \sin n\pi t\, dt \quad n\ odd \qquad b_n = \frac{8}{T} \int_o^{T/4} t\ \sin n\pi t\, dt$$
$$b_n = 0 \quad n\ even$$

$$b_n = 4 \left[\left\{ t\left(-\frac{\cos n\pi t}{n\pi}\right) \right\}_0^{1/2} + \frac{1}{n\pi} \left\{ \int_0^{1/2} 1\cdot(\cos n\pi t\, dt) \right\} \right]$$

$$b_n = -\frac{4}{n\pi} \left[\{t \cos n\pi t\}_0^{1/2} + \frac{1}{n\pi} \left\{ \frac{\sin n\pi t}{n\pi} \right\}_0^{1/2} \right]$$

$$b_n = -\frac{4}{n\pi} \left[\left\{ \frac{1}{2}\left(\cos \frac{n\pi}{2}\right) - 0 \right\} + \frac{1}{n^2\pi^2} \left\{ \sin n\frac{\pi}{2} \right\} \right]$$

Values of $\cos n\pi/2$ will be zero

$$b_n = -\frac{4}{n\pi} \left[\frac{1}{n^2\pi^2} \left\{ \sin n\frac{\pi}{2} \right\} \right] = -\frac{4}{\pi^3} \left[\frac{1}{n^3} \left\{ \sin n\frac{\pi}{2} \right\} \right]$$

For odd values:
For $n = 1, 3, 5, \ldots$

$$b_n = -\frac{4}{\pi^3} \left[\left\{ \sin \frac{\pi}{2} - \frac{1}{27} \sin \frac{3\pi}{2} + \frac{1}{125} \sin \frac{5\pi}{2} \right\} - \cdots \right]$$

Problem 7.2

Determine the Fourier series representations for the following signal as x(t) illustrated in Figure P.7.2.

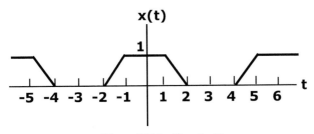

Figure P.7.2 Signal x(t).

Solution 7.2

$$\omega_0 = 2\pi f = 2\pi/T = 2\pi/6 = \pi/3$$

This is the case of an odd and half-wave symmetry, b_n for odd values has to be calculated:

$$b_n = \frac{8}{T} \int_0^{T/4} f(t) \sin n\omega_0 t\, dt \quad n\, odd$$
$$b_n = 0 \quad n\, even$$

Substituting these values in f(t), which has been developed from the graph

$$f(t) = t + 2 \quad -2 \le t \le -1$$
$$f(t) = 1 \quad -1 \le t \le 1$$
$$f(t) = 2 - t \quad 1 \le t \le 2$$

The given waveform is an even function, therefore only a_n *has to be calculated, other values such as* b_0 *would come out to be zero:*

$$a_0 = \frac{1}{T} \int_0^T f(t)\, dt = \frac{1}{6} \left[\int_{-1}^2 (t+2)dt + \int_{-1}^1 1 dt + \int_1^2 (2-t)dt \right] = \frac{1}{2}$$

$$a_n = \frac{2}{T} \int_0^T f(t) \cos n\,\omega_0\, t\, dt$$

$$a_n = \frac{2}{6} \left[\int_{-2}^{-1} (t+2) \cos n\frac{\pi}{3}t\, dt - \int_{-1}^1 (1) \cos n\frac{\pi}{3}t\, dt \right.$$

$$\left. + \int_1^2 (2-t) \cos n\frac{\pi}{3}t\, dt \right]$$

$$f(t) = \frac{1}{2} + \left[\frac{4}{\pi} \sin \frac{n\pi}{3} \cos \frac{\pi}{3}t + \frac{4}{3\pi} \sin \pi \cos \pi t \right.$$

$$\left. + \frac{4}{5\pi} \sin \frac{5\pi}{3} \cos \frac{5\pi}{3}t + \cdots \right]$$

Problem 7.3

Determine the Fourier series representations for the following signal as x(t) illustrated in Figure P.7.3.

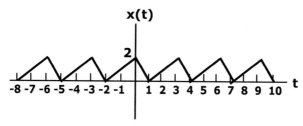

Figure P.7.3 Signal x(t).

Solution 7.3

$$\omega_0 = 2\pi f = 2\pi/T = 2\pi/6 = 2\pi/3$$

This is the case of neither an odd nor an even nor half-wave symmetry. Substituting these values in f(t), which has been developed from the graph

$$f(t) = t - 1 \qquad 1 \le t \le 3$$
$$f(t) = -2t + 8 \qquad 3 \le t \le 4$$

$$f(t) = a_0 + \sum_{n=1}^{\infty} (a_n \cos n\omega_0 t + b_n \sin n\omega_0 t)$$

$$\omega_0 = \frac{2\pi}{T} = 2\pi f_0 \qquad (7.19)$$

$$a_0 = \frac{1}{T} \int_0^T f(t)\, dt \quad a_n = \frac{2}{T} \int_0^T f(t) \cos n\omega_0 t\, dt$$

$$b_n = \frac{2}{T} \int_0^T f(t) \sin n\omega_0 t\, dt$$

Substituting these values in f(t)

$$a_0 = \frac{1}{T} \int_0^T f(t)\, dt = \frac{1}{3}\left[\int_1^3 (t-1)dt + \int_3^4 (-2t+8)dt \right] = 1$$

$$a_n = \frac{2}{T} \int_0^T f(t) \cos n\, w_0\, t\, dt$$

$$a_n = \frac{2}{3}\left[\int_1^3 (t-1)\cos n\frac{2\pi}{3} t\, dt + \int_3^4 (-2t+8)\cos n\frac{2\pi}{3} t\, dt \right]$$

$$a_n = \begin{bmatrix} \frac{2}{n\pi}\sin 3n(\frac{2\pi}{3}) + \frac{3}{2n^2\pi^2}\cos 3n(\frac{2\pi}{3}) - \frac{3}{2n^2\pi^2}\cos n(\frac{2\pi}{3}) + \frac{-14}{n\pi}\sin 3n(\frac{2\pi}{3}) \\ -\frac{6}{2n^2\pi^2}\cos 4n(\frac{2\pi}{3}) + \frac{6}{2n^2\pi^2}\cos 3n(\frac{2\pi}{3}) \end{bmatrix}$$

its $b_n = 0$

Problem 7.4
Determine the Fourier series representations for the following signal as x(t) illustrated in Figure P.7.4.

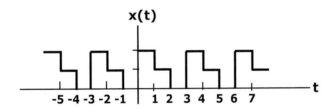

Figure P.7.4 Signal x(t).

Solution 7.4
$T = 2$ s for the above example from the given example.

$$w_0 = 2\,\pi\, f \Rightarrow w_0 = \frac{2\pi}{3}.$$

The given waveform is neither odd, neither even nor half-wave symmetry.
Substituting these values in f(t), which has been developed from the graph

$$f(t) = 2 \quad 0 \le t \le 1$$
$$f(t) = 1 \quad 1 \le t \le 2$$

$$a_0 = \frac{1}{T} \int_0^T f(t)\, dt = \frac{1}{3} \left[\int_0^1 (2) dt + \int_1^2 (1) dt \right] = 1$$

$$a_n = \frac{2}{T} \int_0^T f(t) \cos n\, \omega_0\, t\, dt$$

$$= \frac{2}{3} \left[\int_0^1 \cos n\pi t\, dt - \int_1^2 \cos n\pi t\, dt \right] = \frac{1}{n\pi}(0) = 0$$

$$= \frac{2}{2} \left[\left\{ \frac{\sin n\pi t}{n\pi} \right\}_0^1 - \left\{ \frac{\sin n\pi t}{n\pi} \right\}_1^2 \right]$$

$$= \left[\frac{1}{n\pi} \{ 2\sin n\pi \} - \{ \sin 2n\pi \} \right] = 0$$

$$b_n = \frac{2}{T} \int_0^T f(t) \sin\ n\omega_0 t\, dt$$

$$\therefore b_n = \frac{2}{T} \left[\int_0^1 \sin n\pi t\, dt - \int_1^2 \sin n\pi t\, dt \right]$$

$$= \left[\left\{ \frac{\cos n\pi\, t}{-n\pi} \right\}_0^1 + \left\{ \frac{\cos n\pi\, t}{n\pi} \right\}_1^2 \right]$$

$$= \left[-\frac{1}{n\pi} \{ \cos n\pi - 1 \} + \frac{1}{n\pi} \{ \cos 2n\pi - \cos n\pi \} \right]$$

$$b_n = -\frac{1}{n\pi}(\cos n\pi - 1) + \frac{1}{n\pi}(1 - \cos\ n\pi) = \frac{4}{n\pi}(n\ odd\ and\ 0\ n\ even)$$

$$b_1 = \frac{4}{\pi} \quad b_3 = \frac{4}{3\pi} \quad b_5 = \frac{4}{5\pi}.$$

$$\therefore v(t) = \frac{4}{\pi} \left(\sin \pi\, t + \frac{1}{3} \sin 3\pi\, t + \frac{1}{5} \sin 5\pi\, t + \cdots \right)\, V.$$

Problem 7.5
Let

$$x(t) = \begin{cases} t, & 0 \le t \le 1 \\ 2 - t, & 1 \le t \le 2 \end{cases}$$

be a periodic signal with fundamental period $T = 2$ and Fourier coefficients a_k.

(a) Determine the value of a_0.

(b) Determine the Fourier series representation of dx(t)/dt.

(c) Use the result of part (b) and the differentiation property of the continuous-time Fourier series to help determine the Fourier series coefficients of x(t).

Solution 7.5

(a) We know that $a_0 = \frac{1}{2}\int_0^1 t\, dt + \frac{1}{2}\int_1^2 (2-t)dt = \frac{1}{2}$

(b) The signal

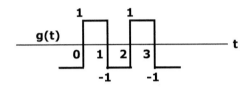

Figure P.7.5   Signal x(t).

$g(t) = \frac{dx(t)}{dt}$ as shown in the figure provided hereunder:
The Fourier series coefficients b_k of g(t) may be found as follows

$$b_0 = \frac{1}{2}\int_0^1 dt - \frac{1}{2}\int_1^2 dt = 0$$

and

$$b_k = \frac{1}{2}\int_0^1 e^{-j\pi kt}\, + \frac{1}{2}\int_1^2 e^{-jk\pi t}.dt = \frac{1}{j\pi k}\left[1 - e^{-j\pi k}\right]$$

(c) We know that

$$g(t) = \frac{dx(t)}{dt} \xleftrightarrow{Fs} b_k = jk\pi\, a_k \quad a_k = \frac{1}{jk\pi}b_k = \frac{1}{\pi^2 k^2}\left\{1 - e^{-j\pi k}\right\}$$

Problem 7.6
In each of the following, we specify the Fourier series coefficients of a signal that is periodic with period N = 8. Determine the signal x[n] in each case.

(a) $a_k = \cos\left(\frac{k\pi}{4}\right) + \sin\left(\frac{3k\pi}{4}\right)$

(b) $a_k = \begin{cases} \sin\left(\frac{k\pi}{3}\right), & 0 \le k \le 6 \\ 0, & k = 7 \end{cases}$

(c) a_k as in Figure P. 7.8 (a)

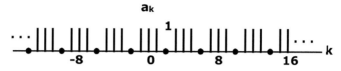

Figure P.7.6 Signal a[k].

Solution 7.6
For N odd

$$x[n] = \sum_{k=-(N-1)/2}^{(N-1)/2} a_k \, e^{jk(2\pi n/N)}$$

For N even

$$x[n] = \sum_{k=N/2-1}^{N/2} a_k \, e^{jk(2\pi n/N)}$$

(a) $N = 8$, over one period ($0 \le n \le 7$)

$$x[n] = \frac{1}{2j} \left[\frac{-e^{-j\frac{3\pi n}{4}} \sin\left\{ \frac{7}{2} \left(\pi \frac{\pi}{4} + \frac{\pi}{3} \right) \right\}}{\sin\left\{ \frac{1}{2} \left(\pi \frac{\pi}{4} + \frac{\pi}{2} \right) \right\}} + e^{j\frac{3\pi n}{4}} \frac{\sin\left\{ \frac{7}{2} \left(\frac{\pi n}{4} + \frac{\pi}{3} \right) \right\}}{\sin\left\{ \frac{1}{2} \left(\frac{\pi n}{4} + \frac{\pi}{3} \right) \right\}} \right]$$

(b) $N = 8$, over one period ($0 \le n \le 7$)

$$x[n] = 1 + (-1)^n + 2\cos\left(\pi \frac{n}{4} \right) + 2\cos\left(\frac{3\pi n}{4} \right)$$

(c) $N = 8$, over one period ($0 \le n \le 7$)

$$x[n] = 2 + 2\cos\left(\frac{\pi n}{4} \right) + \cos\left(\frac{\pi n}{2} \right) + \frac{1}{2}\cos\left(3\frac{\pi n}{4} \right)$$

Problem 7.7
Let

$$x[n] = \begin{cases} 1, & 0 \le n \le 7 \\ 0, & 8 \le n \le 9 \end{cases}$$

be a periodic signal with fundamental period $N = 10$ and the Fourier series coefficients, a_k. Also, let

$$g[n] = x[n] - x[n-1]$$

(a) Show that g[n] has a fundamental period of 10.
(b) Determine the Fourier series coefficients of g[n].
(c) Using the Fourier series coefficients of g[n] and the first difference property determine a_k for $k \neq 0$.

Solution 7.7

(a) g[n] is as shown in the figure provided.
   Clearly g[n] has a fundamental period of 10

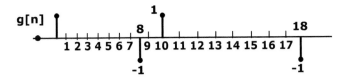

Figure P.7.7 Signal g[n].

(b) The Fourier series coefficients of g[n] are

$$b_k = \left(\frac{1}{10}\right)\left[1 - e^{-j\left(\frac{2\pi}{10}\right)8k}\right]$$

(c) Since $g[n] = x[n] - x[n-1]$
   The Fourier series coefficients a_k and b_k must be related as

$$b_k = a_k - e^{-j\left(\frac{2\pi}{10}\right)k}a_k$$

Therefore

$$a_k = \left(\frac{1}{10}\right)\frac{\left[1 - e^{-j\left(\frac{2\pi}{10}\right)8}k\right]}{1 - e^{-j\left(\frac{2\pi}{10}\right)k}}$$

Problem 7.8
Consider the signal x[n] depicted in Figure P.7.8. The signal is periodic for the period $N = 4$.
   The signal can be expressed in terms of a discrete-time Fourier series as

$$a_k = \frac{1}{N}\sum_{n=n_1+1}^{n_1+n} x[n]\, e^{-jk(2\pi n/N)} \quad \text{and} \quad x[n] = \sum_{k=0}^{3} a_k e^{jk(2\pi n/4)}$$

To determine the Fourier series coefficients as a set of four linear equations (for $n = 0, 1, 2, 3$) in four unknowns (a_0, a_1, a_2, and a_3).

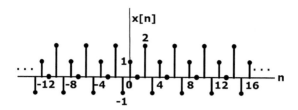

Figure P.7.8 Signal x[n].

(a) Write out these four equations explicitly, and solve them directly using any standard technique for solving four equations in four unknowns.

(b) Check your answer by calculating the a_k directly, using the discrete-time Fourier series analysis equation:

$$a_k = \frac{1}{4} \sum_{n=0}^{3} x[n] e^{-jk(2\pi/4)n}$$

Solution 7.8

If the signal is periodic with period N and the expansion interval is one period, then the discrete time Fourier series equals the sign for all time. The given series is neither even nor odd so taking limit from zero to 3, we have

$$a_k = \frac{1}{N} \sum_{n=n_1+1}^{n_1+n} x[n] \, e^{-jk(2\pi n/N)} \qquad x[n] = \sum_{k=0}^{3} a_k \, e^{jk(2\pi n/4)}$$

Equality for all time holds since the series expansion equals the signal over the expansion interval and both signal and series repeat outside this interval:

$$x(0) = 1, \quad x(1) = 0, \quad x(2) = 2, x(3) = -1$$

(a) The four equations are

$$a_0 + a_1 + a_2 + a_3 = 1 \qquad a_0 - a_1 + a_2 - a_3 = 2$$
$$a_0 + ja_1 - a_2 - ja_3 = 0 \quad a_0 - ja_1 - a_2 + ja_3 = -1$$

Solving, these four equations, we get

$$a_0 = 2, \qquad a_1 = -\frac{1+j}{4} a_2 = -1, \quad \text{and} \quad a_3 = -\frac{1+j}{4}$$

(b) By direct calculations

$$a_k = \frac{1}{4} \left[1 + 0e^{-jk\pi/2} + 2e^{-jk\pi} - e^{-jk3\frac{\pi}{2}} \right] = \frac{1}{4} \left[1 + 2e^{-jk\pi} - e^{-jk3\frac{\pi}{2}} \right]$$

This is the same as the answer we obtained in part (a), for $0 \le k \le 3$.

Problem 7.9

Consider a continuous-time LTI system with impulse response

$$\frac{d}{dt}y(t) + 4y(t) = x(t)$$

Find the Fourier series representation of the output y(t) for each of the following inputs:

(a) $x(t) = \cos 2\pi t$
(b) $x(t) = \sin 4\pi t + \cos(6\pi t + \pi/4)$

Solution 7.9

We will first evaluate the frequency response of the system:

Consider an input x(t) of the form $e^{j\omega t}$.

We know that, the response of this input, will be

$$y(t) = H(j\omega)e^{j\omega t},$$

Therefore, substituting these in the given differential equation we get

$$H(j\omega)[j\omega e^{j\omega t} + 4e^{j\omega t}] = e^{j\omega t}$$

Therefore, $h(j\omega) = \frac{1}{j\omega+4}$

Also

We know that, $y(t) = \sum_{K=-\infty}^{\infty} a_k H(jk\omega_0)e^{jk\omega_0 t}$

When the input is x(t), x(t) has the Fourier series coefficients a_k and fundamental frequency ω_0. Therefore, the Fourier series coefficients of y(t) and H ($jk_{\omega 0}$) (a) here

$\omega_0 = 2\pi$ and the non-zero Fourier series coefficients of x(t) are $a_1 = a_{-1} = \frac{1}{2}$

Therefore, the non-zero Fourier series coefficients of y(t) are

$$b_1 = a_1 H(j2\pi) = \frac{1}{2(4+j2\pi)} \qquad b_{-1} = a_{-1}H(-j2\pi) = \frac{1}{2(4-j2\pi)}$$

(b) Here, $\omega_0 = 2\pi$

and the non-zero Fourier series coefficients of x(t) are

$$a_2 = a_{-2} = \frac{1}{2j} \quad \text{and} \quad a_3 = a_{-3} = \frac{e^{j\frac{\pi}{4}}}{2}$$

Therefore, the non-zero Fourier series coefficients of y(t) are

$$b_2 = a_2 H(j4\pi) = \frac{1}{2j(4+j4\pi)} \qquad b_{-2} = a_{-2}H(-j\pi) = -\frac{1}{2j(4-j4\pi)}$$

$$b_3 = a_3 H(j6\pi) = \frac{e^{j\frac{\pi}{4}}}{2(4+j6\pi)} \qquad b_{-3} = a_{-3}H(-j6\pi) = \frac{e^{-j\frac{\pi}{4}}}{2(4-j6\pi)}$$

Problem 7.10

Consider a causal discrete-time LTI system whose input x[n] and output y[n] are related by the following difference equation:

$$y[n] - \frac{1}{4}y[n-1] = x[n]$$

Find the Fourier series representation of the output y[n] for each of the following inputs:

(a) $x[n] = \sin\left(\frac{3\pi}{4}n\right)$
(b) $x[n] = \cos\left(\frac{\pi}{4}n\right) + 2\cos\left(\frac{\pi}{2}n\right)$

Solution 7.10

We will first evaluate the frequency response of the system. Consider an input x[n] of the form $e^{j\omega n}$.

We know that the response of this input will be

$$y[n] = H(e^{j\omega})e^{j\omega n}$$

Therefore, substituting in the given differential equation, we get

$$H(e^{j\omega})e^{j\omega n} - \frac{1}{4}e^{-j\omega}e^{j\omega n}H(e^{j\omega}) = e^{j\omega n}$$

Therefore

$$H(j\omega) = \frac{1}{1 - \frac{1}{4}e^{-j\omega}}$$

We know that

$$y[n] = \sum_{K=(n)} a_k H\left(e^{j2\pi\frac{k}{N}}\right) e^{jk\left(2\frac{\pi}{N}\right)n}$$

When the input is x[n],

x[n] has the Fourier series coefficients a_k and the fundamental frequency, $2\frac{\pi}{N}$.

Therefore, the Fourier series coefficients of y[n] are $a_k H\left(e^{j2\pi\frac{k}{N}}\right)$

(a) Here, N = 4,

and the non-zero Fourier series coefficients of x[n] are

$$a_3 = a_{-3}^* = \frac{1}{2j}$$

Therefore, the non-zero Fourier series coefficients of y[n] are

$$b_3 = a_{-1} \, H\left(e^{j3\frac{\pi}{4}}\right) = \frac{1}{2j\left(1 - \left(\frac{1}{4}\right)e^{-j3\frac{\pi}{4}}\right)}$$

$$b_{-3} = a_{-1} \, H\left(e^{-j\frac{\pi}{4}}\right) = \frac{1}{2j\left[1 - \left(\frac{1}{4}\right)e^{j3\frac{\pi}{4}}\right]}$$

(b) Here, N = 8,
   and the non-zero Fourier series coefficients of x[n] are

$$a_1 = a_{-1} = \tfrac{1}{2} \quad \text{and}$$
$$a_2 = a_{-2} = 1,$$

Therefore, the non-zero Fourier series coefficients of y(t) are

$$b_1 = a_1 \, H\left(e^{j\frac{\pi}{4}}\right) = \frac{1}{2\left(1 - \left(\frac{1}{4}\right)e^{j\frac{\pi}{4}}\right)}$$

$$b_{-1} = a_{-1} \, H\left(e^{-j\frac{\pi}{4}}\right) = \frac{1}{2\left(1 - \left(\frac{1}{4}\right)e^{j\frac{\pi}{4}}\right)}$$

$$b_2 = a_2 \, H\left(e^{j\frac{\pi}{2}}\right) = \frac{1}{\left[1 - \left(\frac{1}{4}\right)e^{-j\frac{\pi}{2}}\right]}$$

$$b_{-2} = a_{-2} \, H\left(e^{-j\frac{\pi}{2}}\right) = \frac{1}{\left[1 - \left(\frac{1}{4}\right)e^{j\frac{\pi}{2}}\right]}$$

Problem 7.11
Determine the Fourier series representations for the following signal as x(t)
illustrated in Figure P.7.9.

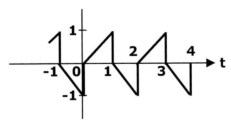

Figure P.7.9 Signal x(t).

Solution 7.11

$$\omega_0 = 2\pi f = 2\pi/T = 2\pi/2 = \pi$$

This is the case of the half-wave symmetry, $a_n \; b_n$ for odd values has to be calculated:

$$a_n = \frac{4}{T} \int_0^{T/2} f(t) \cos n\omega_0 t \, dt \;\; n \; odd \quad b_n = \frac{4}{T} \int_0^{T/2} f(t) \sin n\omega_0 t \, dt \;\; n \; odd$$
$$b_n = 0 \qquad\qquad\qquad\quad n \; even \quad b_n = 0 \qquad\qquad\qquad\qquad n \; even$$

Substituting these values in $f(t) = t$, which has been developed from the graph, for this period

$$a_n = \frac{4}{2} \int_0^{2/2} f(t) \cos n\omega_0 t \, dt \qquad n \; odd$$

$$b_n = 0 \qquad\qquad\qquad\qquad n \; even$$

$$a_n = 2 \int_0^{2/2} f(t) \cos n\omega_0 t \, dt = 2 \int_0^{2/2} t \cos n\omega_0 t \, dt$$

$$a_n = \frac{2}{n^2 \pi^2} [\cos n\pi - 1] \quad for \; n \; odd$$

$$a_1 = \frac{2}{\pi^2} [\cos \pi - 1] = \frac{-4}{\pi^2}; \quad a_3 = \frac{-4}{9\pi^2} = \frac{-4}{\pi^2}; \quad a_5 = \frac{-4}{25\pi^2}$$

$$b_n = \frac{4}{T} \int_0^{T/2} t \sin n\pi t \, dt$$

$$b_n = \frac{4}{2} \int_0^{2/2} t \sin n\pi t \, dt = \frac{-2}{n\pi} \cos n\pi \quad for \;\; n$$

$$b_1 = -\frac{2}{\pi} \cos \pi = \frac{2}{\pi}; \quad b_3 = \frac{2}{3\pi}; \quad b_5 = \frac{2}{5\pi}$$

$$f(t) = a_0 + \sum_{n=1}^{\infty} (a_n \cos n\omega_0 t + b_n \sin n\omega_0 t)$$

$$f(t) = \left[\frac{-4}{\pi^2} \cos \pi t - \frac{-4}{9\pi^2} \cos 3\pi t + \frac{-4}{25\pi^2} \cos 5\pi t + \cdots \frac{2}{\pi} \sin \pi t \right.$$

$$\left. + \frac{2}{3\pi} \sin 3\pi t + \frac{2}{5\pi} \sin 5\pi \; t \cdots \right]$$

8

Fourier Transform

This chapter provides comprehensive details regarding the properties of Fourier transform (FT), some familiar FT pairs, inverse FTs and problems and solutions.

8.1 Introduction

The Fourier transform (FT) and Laplace transform are operations that convert a function of time in to a function of jw (FT or s [Laplace transform]). These transforms are integral transforms which are extremely important in the study of many types of engineering systems.

Let us proceed to define the FT by first recalling the spectrum of the periodic train of rectangular pulses, which was discrete line spectrum, which is the type we must always obtain for periodic function of time. The spectrum was discrete in the sense that it was not a smooth or continuous function of frequency; instead, it has non-zero values only at specific frequencies. There are many important forcing functions; however, that are not periodic function, or rather strange type of function called the impulse function. Frequency spectra may be obtained for such non-periodic function, but they will be continuous spectra in which some energy, in general may be found in any non-zero interval, no matter how small.

This transform pair relationship is most important (see Table 8.1). We should memorize it, draw arrows pointing to it and mentally keep it on the conscious level henceforth and forevermore:

$$X(j\omega) = \int_{-\infty}^{\infty} x(t)\, e^{-j\omega t}.dt \tag{8.1}$$

$$x(t) = \frac{1}{2\pi} \int\limits_{-\infty}^{\infty} X(j\omega)\, e^{j\omega t}.d\omega \tag{8.2}$$

The exponential terms in these two equations carry opposite sign for exponents.

8.2 Some Properties of the FT

Using Euler's identity to replace $e^{-j\mathrm{wt}}$

$$X(j\omega) = \int\limits_{-\infty}^{\infty} x(t)\cos\omega t.dt - j \int\limits_{-\infty}^{\infty} x(t)\sin\omega t.dt \tag{8.3}$$

since f(t), cos wt, and sin wt are all real functions of time, both the integral in (8.3) are real functions of w thus by letting

$$X(jw) = A(w) + jB(w) = \mid X(jw) \mid e^{j\Phi w} \tag{8.4}$$

$$A(\omega) = \int\limits_{-\infty}^{\infty} x(t)\cos\omega t.dt \qquad B(\omega) = \int\limits_{-\infty}^{\infty} x(t)\sin\omega t.dt \tag{8.5}$$

$$X(j\omega) = \sqrt{A^2(\omega) + B^2(\omega)} \qquad \phi(\omega) = \tan^{-1}\frac{B(\omega)}{A(\omega)} \tag{8.6}$$

A(w) and $\mid X(jw)\mid$ are both even functions of w while B(w) and $\Phi(w)$ are both odd functions of w.

Example 8.1
Find out the FT of (a) $\delta(t)$ and (b) $\delta(t - t_0)$.

Solution 8.1

(a) $\Im\{\delta(t)\} = \int\limits_{-\infty}^{\infty} \delta(t)e^{-j\omega t}.dt$

Using integration equalities,
the FT $\delta(t - 0)$ means t = 0, substituting in integral formula

$$\Im\{\delta(t)\} = \int\limits_{-\infty}^{\infty} \delta(t)\, e^{-j\omega t}.dt = 0$$

(b) $\Im\{\delta(t - t_0)\} = \int\limits_{-\infty}^{\infty} \delta(t - t_0)\, e^{-j\omega t}.dt$

Using integration equalities,
the FT $\delta(t - t_0)$ means $t = t_0$, substituting in integral formula

$$\Im\{\delta(t - t_0)\} = \int\limits_{-\infty}^{\infty} \delta(t - t_0)e^{-j\omega t}.dt = e^{-j\omega_0 t}$$

It means that the FT of $\Im\{\delta(t - t_0)\} = e^{-j\omega t_0}$.

Example 8.2

(a) Find out the inverse FT of $e^{-j\omega t_0}$.

(b) Find out the inverse FT of $\{\delta(\omega - \omega_0)\}$

Solution 8.2

(a) $\Im^{-1}\{X(j\omega) = \frac{1}{2\pi} \int\limits_{-\infty}^{=\infty} e^{-j\omega t_0}e^{j\omega t}d\omega = \frac{1}{2\pi} \int\limits_{-\infty}^{=\infty} e^{j\omega(t-t_0)}d\omega$

$$= \delta(t - t_0)$$

(b) $\Im^{-1}\{\delta(\omega - \omega_0) = \frac{1}{2\pi} \int\limits_{-\infty}^{=\infty} e^{j\omega t}\delta(\omega - \omega_0)d\omega = \frac{1}{2\pi}e^{j\omega_0 t}$

$$\frac{1}{2\pi}e^{j\omega_0 t} = \delta(\omega - \omega_0) \quad or \quad e^{j\omega_0 t} = 2\pi\delta(\omega - \omega_0)$$

$$\frac{1}{2\pi}e^{-j\omega_0 t} = \delta(\omega + \omega_0) \quad or \quad e^{j\omega_0 t} = 2\pi\delta(\omega + \omega_0)$$

Example 8.3
Calculate the FTs of:

(a) $e^{-2(t-1)}u(t - 1)$

(b) $e^{-2|t-1|}$

Solution 8.3

(a) Let $x(t) = e^{-2(t-1)}u(t - 1)$.
Then the FT $X(j\omega)$ of x(t) as

$$X(j\omega) = \int\limits_{-\infty}^{\infty} e^{-2(t-1)}u(t-1)e^{-j\omega t}.dt$$

$$= \int\limits_{1}^{\infty} e^{-2(t-1)}e^{-j\omega t}.dt = \frac{e^{-j\omega}}{2+j\omega}$$

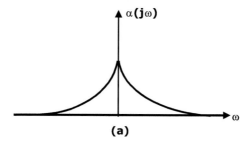

(a)

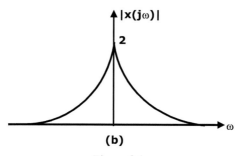

(b)

Figure 8.1

(b) Let $x(t) = e^{-2|t-1|}$
 Then the FT $X(j\omega)$ of x(t) is

$$X(j\omega) = \int\limits_{-\infty}^{\infty} e^{-2|t-1|}e^{-j\omega t}.dt$$

$$= \int\limits_{1}^{\infty} e^{-2|t-1|}e^{-j\omega t}.dt + \int\limits_{-\infty}^{1} e^{2|t-1|}e^{-j\omega t}.dt$$

$$= \frac{e^{-jw}}{(2+jw)} - \frac{e^{-jw}}{(2-jw)} = \frac{4e^{-jw}}{(4+w^2)}$$

$|X(jw)|$ is as shown in the figure.

8.2.1 Linearity

The linearity property of the FT states that if we are given the transform pairs

If $\Im\{x_1(t)\} \leftrightarrow X_1(jw)$ and $\Im\{x_2(t)\} \leftrightarrow X_2(jw)$

Then $\Im[ax_1(t) + bx_2(t)] \leftrightarrow a\, X_1(jw) + b\, X_2(jw)$

$$\alpha_1(jw) = \int_{-\infty}^{\infty} [\delta(t+1) + \delta(t-1)]e^{-jwt}\,.dt = 2\cos w$$

Example 8.4

Use the FT analysis equation to calculate the FTs of:

$$\delta(t+1) + \delta(t-1)$$

Solution 8.4

Let $x(t) = \delta(t+1) + \delta(t-1)$

The FT x(t) is $X(jw)$.

$$X(jw) = \int_{-\infty}^{\infty} [\delta(t-1) + \delta(t+1)]e^{-jwt}dt = 2\cos w$$

Example 8.5

Find the FT $f(t) = B\cos w_0 t$.

Solution 8.5

Using Euler's identity

$$\cos\alpha = \frac{e^{j\alpha} + e^{-j\alpha}}{2}$$

The above expression can be written as

$$f(t) = \frac{B}{2}\left[e^{jw_0 t} + e^{-jw_0 t}\right] = \frac{Be^{jw_0 t}}{2} + \frac{Be^{-jw_0 t}}{2}$$

$$\Im\{B\cos w_0 t\} = \frac{B}{2}\Im\left[e^{jw_0 t}\right] + \frac{B}{2}\Im\left[e^{-jw_0 t}\right]$$

$$\Im\{B\cos w_0 t\} = \pi B\left[\delta(w - w_0)\right] + \pi B\left[\delta(w + w_0)\right]$$

Table 8.1　Some familiar FT pairs

f(t)	f(t)	$\mathcal{F}\{f(t)\}=F(j\omega)$	$\lvert F(j\omega)\rvert$
	$\delta(t - t_0)$	$e^{-j\omega t_0}$	
Complex	$e^{-j\omega_0 t}$	$2\pi\,\delta(\omega - \omega_0)$	
	$\cos\omega_0 t$	$\pi[\,\delta(\omega+\omega_0)+\ \delta(\omega-\omega_0)\,]$	
	1	$2\pi\,\delta(\omega)$	
	$\mathrm{sgn}(t)$	$\dfrac{2}{j\omega}$	
	$u(t)$	$\pi\delta(\omega)\,\dfrac{1}{j\omega}$	
	$e^{-\alpha t}u(t)$	$\dfrac{1}{\alpha + j\omega}$	
	$e^{-\alpha t}\cos\omega_d t \cdot u(t)$	$\dfrac{\alpha + j\omega}{(\alpha + j\omega)^2 + \omega_d^{\,2}}$	
	$u(t + \tfrac{1}{2}T) - u(t - \tfrac{1}{2}T)$	$T\,\dfrac{\sin\frac{\omega T}{2}}{\frac{\omega T}{2}}$	

8.2.2 Time Reversal

The time reversal property of the FT states that if we are given the transform pairs

If $\Im\{x(t)\} \leftrightarrow X(j\omega))$
Then $\Im[x(-t)] \leftrightarrow X(-j\omega)$

Example 8.6

Given that x(t) has the FT $X(j\omega)$, express the FTs of the signals listed hereunder in terms of $X(j\omega)$.

$$x_1(t) = x(1 - t) + x(-1 - t)$$

Solution 8.6

Throughout this problem we assume that

$$x(t) \xleftrightarrow{FT} X(j\omega)$$

Using the time reversal property, we have

$$x(-t) \xleftrightarrow{FT} X(-j\omega)$$

Using the time shifting property on this, we have

$$x(-t + 1) \xleftrightarrow{FT} e^{-j\omega t} . X(-j\omega)$$

and

$$x(-t - 1) \xleftrightarrow{FT} e^{J\omega t} X(-j\omega)$$

Therefore

$$x_1(t) = x(-t + 1) + x(-t - 1) \xleftrightarrow{FT} e^{-j\omega} X(-j\omega)$$

$$+ e^{j\omega} X(-j\omega) \xleftrightarrow{FT} 2X(-j\omega) \cos \omega.$$

8.2.3 Time Scaling

The time scaling property provides that if

If $\Im\{x(t)\} \leftrightarrow X(\omega)$
Then for a constant scale factor, α

$$\Im\{x(\alpha t)\} = \frac{1}{|\alpha|} X(\frac{\omega}{\alpha})$$

$$\Im\{x(\alpha t)\} = \int_{-\infty}^{\infty} x(at) e^{-j\omega_0 t} dt$$

If we substitute $\tau = $ at

$d\tau = \alpha\, dt$ and equation can be written

$$\Im\{x(\tau)\} = \int_{-\infty}^{\infty} x(\tau)e^{-j(\frac{\omega}{\alpha})\tau}d\tau \qquad \Im\{x(\alpha t)\} = \frac{1}{\alpha}X(\frac{\omega}{\alpha})$$

8.2.4 Time Transformation

The property of time scaling and time shifting can be combined into a more general property of time transformation.

$$\text{let} \quad \tau = \alpha t + t_0$$

where α is a scale factor and t_0 is time shift.

$$\Im\{x(\alpha t)\} = \frac{1}{|\alpha|}X(\frac{\omega}{\alpha})$$

Application of time shifting property to the time scaled factor gives the time transformation property:

If $\Im\{x(t)\} \leftrightarrow X(j\omega)$

Then for a constant scale factor, α

$$\Im\{x(\alpha t + t_0)\} = \frac{1}{|\alpha|}X(\frac{j\omega}{\alpha})e^{-jt_0(\omega/\alpha)}$$

Example 8.7

Given that $x(t)$ has the FT $X(j\omega)$, express the FTs of the signals listed hereunder in terms of $X(j\omega)$.

$$x_2(t) = x(3t - 6)$$

Solution 8.7

Throughout this problem we assume that

$$x(t) \overset{FT}{\longleftrightarrow} X(j\omega)$$

Using the time scaling property, we have

$$x(3t) \overset{FT}{\longleftrightarrow} \frac{1}{3}X\left(j\frac{\omega}{3}\right)$$

Using the time shifting property on this we have

$$x_2(t) = x[3(t - 2)] \overset{FT}{\longleftrightarrow} e^{-2j\omega}\frac{1}{3}X\left(j\frac{\omega}{3}\right)$$

8.2.5 Duality

The symmetry of the FT and its inverse in variable t and ω can be compared:

$$\Im\{x(t)\} = \int_{-\infty}^{\infty} x(t)e^{-j\omega_0 t}dt$$

$$\Im^{-1}X(j\omega) = x(t) = \frac{1}{2\pi}\int_{-\infty}^{\infty} X(j\omega)\,e^{j\omega t}.d\omega$$

8.2.6 Frequency Shifting

This property is stated mathematically as

$$\Im\{x(t)e^{-j\omega_0 t}\} = X(\omega - \omega_0)$$

8.2.7 Time Differentiation

This property is stated mathematically as

If $\Im\{x(t)\} \leftrightarrow X(\omega)$

Then $\frac{dx(t)}{dt} = j\omega X(\omega)$

Mathematically it can be stated as

$$\frac{d^n x(t)}{dt} = (j\omega)^n X(\omega)$$

Example 8.8

Given that y(t) has the FT Y(jω), express the FTs of the signals listed hereunder in terms of Y(jω).

$$y(t) = \frac{d^2}{dt^2}x(t-1)$$

Solution 8.8

Throughout this problem we assume that

$$x(t) \xleftrightarrow{FT} X(j\omega)$$

Using the differentiation in time property we have

$$\frac{dx(t)}{dt} \xleftrightarrow{FT} j\omega X(j\omega)$$

Applying the property again, we have

$$\frac{d^2x(t)}{dt^2} \xleftrightarrow{FT} -\omega^2 X(j\omega)$$

Using the tie shifting property, we have

$$x_3(t) = \frac{d^2x(t-1)}{dt^2} \xleftrightarrow{FT} -\omega^2 X(j\omega)\, e^{-j\omega}$$

8.2.8 Frequency Differentiation

The time differentiation property has a dual for the case of differentiation in frequency domain. This property is stated mathematically as

If $\Im\{x(t)\} \leftrightarrow X(\omega)$

Then, mathematically it can be stated as

$$(-j)^n f(t) = \frac{d^n X(\omega)}{d^n \omega}$$

$$\Im\{B\cos\omega_0 t\} = \pi B\left[\delta(\omega - \omega_0)\right] + \pi B\left[\delta(\omega + \omega_0)\right]$$

8.2.9 Convolution Property

The convolution property states that convolution in time domain is equal to multiplication in frequency domain, mathematically defined as

$$\text{If x(t)} \xleftrightarrow{FT} X(\omega) \quad \text{and} \quad h(t) \xleftrightarrow{FT} H(\omega)$$

$$y(t) = x(t) \otimes h(t) \xleftrightarrow{FT} Y(\omega) = X(\omega)\, H(\omega)$$

Example 8.9

Determine the FT of each of the following periodic signals:

(a) $\sin\left(2\pi t + \frac{\pi}{4}\right)$

(b) $1 + \cos\left(6\pi t + \frac{\pi}{8}\right)$

Solution 8.9

(a) The signal $x_1(t) = \sin(2\pi t + \frac{\pi}{4})$ is periodic with a fundamental period of T = 1, this translates to a fundamental frequency of $\omega_0 = \pi$.

The non-zero Fourier series coefficients of this signal may be found by writing it in the form

$$x_1(t) = \frac{1}{2j}\left(e^{j\left(2\pi t + \frac{\pi}{4}\right)} - e^{-j\left(2\pi t + \frac{\pi}{4}\right)}\right)$$

$$= \frac{1}{2j}e^{j\frac{\pi}{4}}e^{j(2\pi t)} - \frac{1}{2j}e^{-j\frac{\pi}{4}}e^{-j2\pi t}$$

Therefore, the non-zero Fourier series coefficients of $x_1(t)$ are

$$a_1 = \frac{1}{2j}e^{j\frac{\pi}{4}}e^{j2\pi t} \qquad a_{-1} = -\frac{1}{2j}e^{-j\frac{\pi}{4}}e^{-j2\pi t}$$

We know that for periodic signals the FT consists of a train of impulse occurring at $k\omega_0$. Furthermore the area under each impulse is 2π times the Fourier series coefficients, a_k.

Therefore for $x_1(t)$ the corresponding FT $X_1(j\omega)$ is given by

$$X_1(j\omega) = 2\pi a_1 \delta(\omega - \omega_0) - 2\pi a_{-1}\delta(\omega - \omega_0)$$

$$= \left(\frac{\pi}{j}\right)e^{j\frac{\pi}{4}}\delta(\omega - 2\pi) - \left(\frac{\pi}{j}\right)e^{-j\frac{\pi}{4}}\delta(\omega - 2\pi)$$

(b) The signal
$x_2(t) = 1 - \cos\left(6\pi t + \frac{\pi}{8}\right)$ is periodic with a fundamental period of $T = \frac{1}{3}$. This translates to a fundamental frequency of $\omega_0 = 6\pi$.
The non-zero Fourier series coefficients of this signal may be found in writing it in the form

$$x_2(t) = 1 + \frac{1}{2}\left(e^{j\left(6\pi t + \frac{\pi}{8}\right)} + e^{-j\left(6\pi t + \frac{\pi}{8}\right)}\right)$$

$$= 1 + \frac{1}{2}e^{j\frac{\pi}{8}}e^{j6\pi t} + \frac{1}{2}e^{-j\frac{\pi}{8}}e^{-j6\pi t}$$

Therefore, the non-zero Fourier series coefficients of $x_2(t)$ are

$$a_0 = 1 \qquad a_1 = \frac{1}{2}e^{j\frac{\pi}{8}}e^{j6\pi t} \qquad a_2 = \frac{1}{2}e^{-j\frac{\pi}{8}}e^{-j6\pi t}$$

We know that for a periodic signal, the FT consists of a train of impulses occurring at $k\omega_0$.

Furthermore the area under each impulse is 2π times the Fourier series coefficients, a_k.

Therefore for $x_2(t)$, the corresponding FT $X_2(j\omega)$ is given by

$$X_2(j\omega) = 2\pi a_0 \delta(\omega) + 2\pi a_1 \delta(\omega - \omega_0) + 2\pi a_{-1} \delta(\omega + \omega_0)$$
$$= 2\pi \delta(\omega) + \pi e^{j\frac{\pi}{8}} \delta(\omega - 6\pi) + \pi e^{j\frac{\pi}{8}} \delta(\omega - 6\pi)$$

Example 8.10

Obtain the impulse response of the system described by

$$H(\omega) = 1 \quad for \quad |\omega| \le \omega_C$$
$$H(\omega) = 0 \quad for \quad \omega_C \le |\omega| \le \pi$$

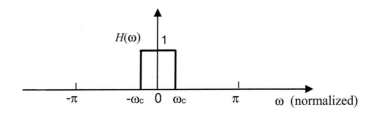

Solution 8.10

Inverse Discrete Time Fourier Transform (IDTFT) is expressed as

$$\text{or} \quad h(n) = \frac{1}{2\pi} \int_{-\pi}^{\pi} H(\omega)\, e^{j\omega n}\, d\omega = \frac{1}{2\pi} \int_{-\omega_C}^{\omega_C} 1 \cdot e^{j\omega n}\, d\omega$$

$$\text{or} \quad h(n) = \frac{1}{2\pi j n}\, e^{j\omega n}\big]_{-\omega_C}^{\omega_C} = \frac{e^{j\omega_C n} - e^{-j\omega_C n}}{2\pi j n} = \frac{\sin \omega_C n}{\pi n}$$

It means that the inverse FT of a rectangular pulse is a sine function.

Example 8.11

Use the FT synthesis equation to determine the inverse FTs of:

(a) $X_1(j\omega) = 2\pi\delta(\omega) + \pi\delta(\omega - 4\pi) + \pi\delta(\omega + 4\pi)$

(b) $X_2(j\omega) = \begin{cases} 2, & 0 \le \omega \le 2 \\ -2, & -2 \le \omega < 0 \\ 0, & |\omega| > 2 \end{cases}$

Solution 8.11

(a) The inverse FT is

$$x_1(t) = \left(\frac{1}{2\pi}\right) \int_{-\infty}^{\infty} [2\pi\delta(\omega) + \pi\delta(\omega - 4\pi) + \pi\delta(\omega + 4\pi)] \, e^{-j\omega t} \, .d\omega$$

$$= \left(\frac{1}{2\pi}\right) [2\pi e^{j0t} + \pi e^{j4\pi t} + \pi e^{-j4\pi t}]$$

$$= 1 + \left(\frac{1}{2}\right) e^{j4\pi t} + \left(\frac{1}{2}\right) e^{-j4\pi t} = 1 + \cos(4\pi t)$$

(b) The inverse FT is

$$x_2(t) = \left(\frac{1}{2\pi}\right) \int_{-\infty}^{\infty} X_2(j\omega) \, e^{j\omega t} \, .d\omega$$

$$= \left(\frac{1}{2\pi}\right) \int_{0}^{2} 2 \, e^{j\omega t} \, d\omega + \left(\frac{1}{2\pi}\right) \int_{-2}^{0} (-2) \, e^{j\omega t} \, .d\omega$$

$$= \frac{(e^{j2t} - 1)}{\pi jt} - \frac{(1 - e^{-j2t})}{\pi jt} = -\frac{(4j \sin^2 t)}{\pi t}$$

Example 8.12
Consider the signal

$$x(t) = \begin{cases} 0, & t < -\frac{1}{2} \\ t + \frac{1}{2}, & -\frac{1}{2} \le t \le \frac{1}{2}. \\ 1, & t > \frac{1}{2} \end{cases}$$

(a) Use the differentiation and integration properties and the FT pair for the rectangular pulse to find a closed-form expression for X(jω).
(b) What is the FT of $g(t) = x(t) - \frac{1}{2}$?

Solution 8.12

(a) The signal x(t) is as shown in the figure:

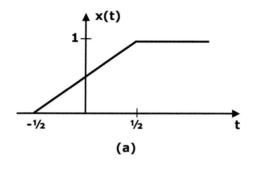

(a)

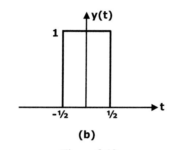

(b)

Figure 8.12

We may express this signal as

$$x(t) = \int_{-\infty}^{t} y(t) \, dt$$

where y(t) is the rectangular pulse shown in the figure; using the integration property of the FT, we have

$$x(t) \xleftrightarrow{FT} X\,(j\omega) = \frac{1}{j\omega}y(j\omega)+\pi y(j0)\delta(\omega) \quad \text{or} \quad y(j\omega) = \frac{2\sin\left(\frac{\omega}{2}\right)}{\omega}$$

Therefore

$$X(j\omega) = \frac{2\sin\left(\frac{\omega}{2}\right)}{j\omega^2} - \pi\delta(\omega)$$

(b) If $g(t) = x(t) + \frac{1}{2}$ the FT G(jω) of g(t) is given by

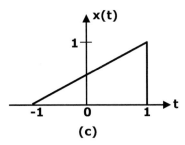

(c)

Figure 8.13

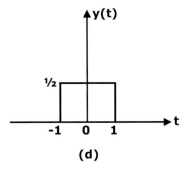

(d)

Figure 8.14

$$G(j\omega) = X\ (j\omega) - \left(\frac{1}{2}\right)2\pi\delta(\omega) = \frac{2\sin\left(\frac{\omega}{2}\right)}{j\omega^2}$$

8.3 Problems and Solutions

Problem 8.1
(a) Determine the FT of the following signal:

$$x(t) = t\left(\frac{\sin t}{\pi t}\right)^2$$

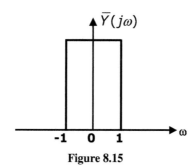

Figure 8.15

Solution 8.1

$$Y(j\omega) = \int\limits_{-\infty}^{\infty} x(t)e^{-j\omega t}.dt$$

$$y(t) = \frac{1}{2\pi} \int\limits_{-\infty}^{\infty} Y(j\omega)e^{j\omega n}.d\omega$$

or $\qquad y(t) = \frac{1}{2\pi} \int_{-1}^{1} Y(\omega)e^{j\omega t}d\omega = \frac{1}{2\pi} \int_{-1}^{1c} 1.e^{j\omega t}d\omega$

or $\qquad y(t) = \frac{1}{2\pi jt}e^{j\omega t}]_1^{-1} = \frac{e^{jt} - e^{-jt}}{2\pi jt} = \frac{\sin t}{\pi t}$

It means that the inverse FT of a rectangular pulse is a sine function. We may write

$$t\left(\frac{\sin t}{\pi t}\right)^2 \overset{FT}{\longleftrightarrow} X(j\omega) = j\frac{d}{d\omega}y_1(j\omega)$$

This is shown in the figure, and X(jω) may be expressed mathematically as

$$X(j\omega) = \begin{cases} \frac{j}{2\pi} & -2 \leq \omega \leq 0 \\ -\frac{j}{2\pi} & 0 \leq \omega_2 \\ 0 & otherwise \end{cases}$$

This is a triangular function $y_1(j\omega)$ as shown in the figure.

Problem 8.2

Given the relationships

$$y(t) = x(t) * h(t) \quad \text{and} \quad g(t) = x(3t) * h(3t),$$

and given that x(t) has the FT $X(j\omega)$ and h(t) has the FT $H(j\omega)$, use the FT properties to show that g(t) has the form

$$g(t) = Ay\,(Bt).$$

Determine the values of A and B.

Solution 8.2

We know that

$$x(3t) \xrightarrow{FT} \frac{1}{3}X\left(\frac{j\omega}{3}\right)$$

$$h(3t) \xrightarrow{FT} \frac{1}{3}H\left(\frac{j\omega}{3}\right)$$

Therefore

$$G(j\omega) = \Im\{x(3t) * h(3t)\} = \frac{1}{9}X\left(\frac{j\omega}{3}\right) H\left(\frac{j\omega}{3}\right)$$

Now, note that

$$y(j\omega) = \Im\{x(t) \otimes h(t)\} = X(j\omega)\,H(j\omega)$$

From this
We may write

$$y\left(j\frac{\omega}{3}\right) = X\left(j\frac{\omega}{3}\right).H\left(j\frac{\omega}{3}\right)$$

we have $G(j\omega) = \frac{1}{9}y\left(j\frac{\omega}{3}\right)$ and $g(t) = \frac{1}{3}y(3t)$. Therefore, $A = \frac{1}{3}$ and $B = 3$.

Problem 8.3

Consider the FT pair

$$e^{-|t|} \xrightarrow{FT} \frac{1}{1+\omega^2}$$

(a) Use the appropriate FT properties to find the FT of $te^{-|t|}$.
(b) Use the result from part (a), along with the duality property, to determine the FT of $\frac{4t}{(1+t^2)^2}$.

Solution 8.3
We know that
$$e^{-|t|} = X(j\omega) \xleftrightarrow{FT} \frac{2}{1+\omega^2}$$

Using the differentiation in frequency property we have
$$t\, e^{-|t|} \xleftrightarrow{FT} j\frac{d}{d\omega}\left(\frac{2}{1+\omega^2}\right) = -\frac{4j\omega}{(1+\omega^2)^2}$$

(b) The duality property state that if
$$g(t) \xleftrightarrow{FT} G(j\omega)$$

Then
$$G(t) \xleftrightarrow{FT} 2\pi\, g(j\omega)$$

Now, since
$$t\, e^{-|t|} \xleftrightarrow{FT} -\frac{4j\omega}{(1+\omega^2)^2}$$

We may use duality to write
$$\frac{-4j\omega}{(1+t^2)^2} \xleftrightarrow{FT} 2\,\pi\omega e^{-|\omega|}$$

Multiplying both sides by (j), then we get
$$\frac{-4jt}{(1+t^2)^2} \xleftrightarrow{FT} j2\pi\omega\, e^{-|\omega|}   \quad Ans.$$

Problem 8.4
Let x(t) be a signal whose FT is
$$X(j\omega) = \delta(\omega) + \delta(\omega - \pi) + \delta(\omega - 5),$$

and let
$$h(t) = u(t) - u(t - 2).$$

(a) Is x(t) periodic?
(b) Is x(t)* h(t) periodic?
(c) Can the convolution of two aperiodic signals be periodic?

Solution 8.4

(a) Taking the inverse FT of X(jω), then we get

$$x(t) = \frac{1}{2\pi} + \frac{1}{2\pi}e^{j\pi t} + \frac{1}{2\pi}e^{j5t}$$

The signal x(t) is therefore a constant summed with two complex exponential whose fundamental frequencies are $2\frac{\pi}{5}$ rad/s and 2 rad/s. These two complex exponentials are not harmonically related. That is, the fundamental frequencies of these complex exponentials can never be integral multipliers of common fundamental frequencies. Therefore, the signal is not periodic.

(b) Consider the signal y(t) = x(t)* h(t), and from the convolution property we know that

$$y(j\omega) = X(j\omega)\,H(j\omega)$$

Also, from h(t) we know that

$$H(j\omega) = e^{-j\omega}\frac{2\sin\omega}{\omega}$$

The function H(jω) is zero, when $\omega = k\pi$ where k is a non-zero integer.

$$y(j\omega) = X(j\omega)\,H(j\omega) = \delta(\omega) + \delta(\omega - 5)$$

This gives

$$y(t) = \frac{1}{2\pi} + \frac{1}{2\pi}e^{j5t}$$

Therefore, y(t) is a complex exponential summed with a constant. We know that a complex exponential is periodic. Adding a constant to the complex exponential does not affect its periodicity. Therefore, y(t) will be a signal with a fundamental frequency of 2π/5.

(c) From the result of parts (a) and (b), we see that the answer is yes.

Problem 8.5

Consider a signal x(t) with the FT X(jω). Suppose we are given the following facts:

1. x(t) is real and non-negative.
2. $F^{-1}\{(1 + j\omega)X(j\omega)\} = Ae^{-2t}u(t)$, where A is independent of t.
3. $\int_{-\infty}^{\infty}|X(j\omega)|^2 d\omega = 2\pi$

Determine a closed-form expression for x(t).

Solution 8.5

Taking the FT of both sides of the equation

$$F^{-1}\{(1+j\omega)X(j\omega)\} = A2^{-2t}u(t)$$

We obtain

$$X(j\omega) = \frac{A}{(1+j\omega)(2+j\omega)} = A\left\{\frac{1}{1+j\omega} - \frac{1}{2+j\omega}\right\}$$

Taking the inverse FT of the above equation

$$x(t) = Ae^{-t}u(t) - Ae^{-2t}u(t)$$

We know that

$$\int_{-\infty}^{\infty} |X(j\omega)|^2 d\omega = 2\pi \int_{-\infty}^{\infty} |x(t)|^2 \, dt$$

using the fact that

$$\int_{-\infty}^{\infty} |X(j\omega)|^2 dt = 2\pi,$$

we have

$$\int_{-\infty}^{\infty} |x(t)|^2 dt = 1$$

substituting the previously obtained expression for x(t) in the above equation, we get

$$\int_{-\infty}^{\infty} |A^2 e^{-2t} + A^2 e^{-4t} - 2A^2 e^{-3t} |u(t) \, dt = 1$$

$$\int_{0}^{\infty} |A^2 e^{-2t} + A^2 e^{-4t} - 2A^2 e^{-3t}| \, dt = 1$$

$$A^2 \frac{1}{12} = 1$$

$$\Rightarrow \quad A = \sqrt{12}$$

We choose A to be $\sqrt{12}$ instead of $-\sqrt{12}$ because we know that x(t) is non-negative.

Problem 8.6

Let x(t) be a signal with the FT X(jω). Suppose we are given the following facts:

1. x(t) is real.
2. $x(t) = 0 \; for \; t \leq 0$.
3. $\frac{1}{2\pi} \int_{-\infty}^{\infty} Re\{X(jw)\}e^{jwt}dw = |t|e^{-|t|}$.

Determine a closed-form expression for x(t).

Solution 8.6

$$\Im\{x_e(t)\} = \frac{x(t) + x(-t)}{2} \xrightarrow{FT} Re\{X(jw)\}$$

We are given that

$$\Im\{Re[X(jw)]\} = |t|e^{-|t|}$$

Therefore

$$\{x_e(t)\} = \frac{x(t) + x(-t)}{2} = |t|e^{-|t|}$$

We also know that x(t) = 0, for t ≤ 0, this implies that x(-t) is zero, for t > 0 we may conclude that

$$x(t) = 2(t) \, e^{-|t|} \; for \; t \geq 0$$

Therefore

$$x(t) = 2te^{-t}u(t)$$

Problem 8.7

Determine whether each of the following statements is true or false. Justify your answers.

(a) An odd and imaginary signal always has an odd and imaginary FT.
(b) The convolution of an odd FT with an even FT is always odd.

Solution 8.7

(a) We know that a real odd signal x(t) has a purely imaginary and odd FT X(jω). Let us now consider the purely imaginary and odd signal jx(t). Using linearity, we obtain the FT of this signal to be $|X(jw)|$. The function X(jω) will clearly be real and odd. Therefore, the given statement is false.

(b) An odd FT corresponds to an odd signal, while an even FT corresponds to an even signal. The convolution of an even FT with an odd FT may be viewed in the time domain as a multiplication of an even and an odd signal. Such a multiplication will always result in an odd signal. The FT of this odd signal will always be odd. Therefore, the given statement is true.

Problem 8.8

Consider a causal LTI system with frequency response

$$H(j\omega) = \frac{1}{j\omega + 3}.$$

For a particular input x(t) this system is observed to produce the output

$$y(t) = e^{-3t}u(t) - e^{-4t}u(t).$$

Determine x(t).

Solution 8.8

We know that

$$H(j\omega) = \frac{Y(j\omega)}{X(j\omega)}$$

Since it is given that

$$y(t) = e^{-3t}u(t) - e^{-4t}u(t)$$

We can compute y(jω)

$$Y(j\omega) = \frac{1}{3 + j\omega} - \frac{1}{4 + j\omega} = \frac{1}{(3 + j\omega)(4 + j\omega)}$$

Since

$$H(j\omega) = \frac{1}{(3 + j\omega)}, \quad we\ have$$

$$X(j\omega) = \frac{Y(j\omega)}{H(j\omega)} = \frac{1}{(4 + j\omega)}$$

Taking the inverse FT of X(jω), we have

$$x(t) = e^{-4t}u(t)$$

Problem 8.9

Compute the FT of each of the following signals:

(a) $[e^{-\alpha t} \cos \omega_0 t] \, u(t), \ \alpha > 0$

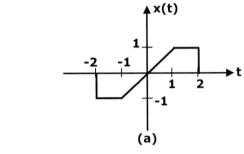

(a)

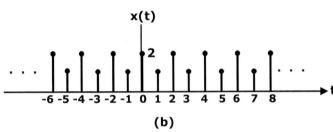

(b)

Figure P8.9

(b) $e^{-3|t|} \sin 2t$

(c) $x(t) = \begin{cases} 1 + \cos \pi t, & |t| \leq 1 \\ 0, & |t| > 1 \end{cases}$

(d) $\sum_{k=0}^{\infty} \alpha^k \delta(t - kT), |\alpha| < 1$

(e) $[te^{-2t} \sin 4t] \, u(t)$

(f) $\left[\frac{\sin \pi t}{\pi t}\right] \left[\frac{\sin 2\pi(t-1)}{\pi(t-1)}\right]$

(g) x(t) as shown in Figure P. 8.10 (a)

(h) x(t) as shown in Figure 8.10 (b)

(i) $x(t) = \begin{cases} 1 - t^2, & 0 < t < 1 \\ 0, & \text{otherwise} \end{cases}$

Solution 8.9

(a) The given signal is

$$e^{-\alpha t}\cos(\omega_0 t)\,u(t) = \frac{1}{2}\,e^{-\alpha t}e^{j\omega_0 t}\,u(t) - \frac{1}{2}\,e^{-\alpha t}e^{j\omega_0 t}\,u(t)$$

$$e^{-\alpha t}\cos(\omega_0 t)\,u(t) = \frac{1}{2}\left[e^{-(\alpha+j\omega_0)t} - e^{-(\alpha-j\omega_0)t}\right]$$

Therefore

$$X(j\omega) = \frac{1}{2(\alpha - j\omega_0 + j\omega)} - \frac{1}{2(\alpha - j\omega_0 - j\omega)}$$

(b) The given signal is

$$x(t) = e^{-3t}\sin(2t)\,u(t) + e^{3t}.\sin(2t)\,u(-t)$$

we have

$$x_1(t)e^{-3t}.\sin(2t).u(t) \overset{FT}{\longleftrightarrow} X_1(j\omega) = \frac{\frac{1}{2j}}{3 - 2j + j\omega} - \frac{\frac{1}{2j}}{3 + j2 - j\omega}$$

Also

$$x_2(t) = e^{3t}\sin(2t)\,u(-t)$$

$$= -x_1(-t) \overset{FT}{\longleftrightarrow} X_2(j\omega) = -X_1(-j\omega)$$

$$= \frac{\frac{1}{2j}}{3 - 2j - j\omega} - \frac{\frac{1}{2j}}{3 + j2 - j\omega}$$

Therefore

$$X(j\omega) = X_1(j\omega) + X_2(j\omega) = \frac{3j}{9 + (\omega + 2)^2} - \frac{3j}{9 - (\omega + 2)^2}$$

(c) Using the FT analysis, we have

$$X(j\omega) = \frac{2\sin\omega}{\omega} + \frac{\sin\omega}{\pi - \omega} - \frac{\sin\omega}{\pi + \omega}$$

(d) Using the FT analysis, we have

$$X(j\omega) = \frac{1}{1 - \alpha\,e^{-j\omega T}}$$

(e) We have

$$x(t) = \left(\frac{1}{2j}\right) t e^{-2t} e^{j4t} . u(t) - \left(\frac{1}{2j}\right) t e^{-2t} . e^{-j4t} . u(t)$$

Therefore

$$X(j\omega) = \frac{\frac{1}{2j}}{(2 - j4 + j\omega)^2} - \frac{\frac{1}{2j}}{(2 + j4 - j\omega)^2}$$

(f) We have

$$x_1 t = \frac{\sin t}{\pi t} \xleftrightarrow{FT} X_1(j\omega) = \begin{cases} 1 & |\omega| < \pi \\ 0 & \text{otherwise} \end{cases}$$

Also

$$x_2 t = \frac{\sin 2\pi(t-1)}{\pi(t-1)} \xleftrightarrow{FT} X_2(j\omega) = \begin{cases} e^{-2\omega} & |\omega| < 2\pi \\ 0 & \text{otherwise} \end{cases}$$

$$x(t) = x_1(t).x_2(t) \xleftrightarrow{FT}$$

$$X(j\omega) = \frac{1}{2\pi}\{X_1(j\omega) * X_2(j\omega)\}$$

Therefore

$$X(j\omega) = \begin{cases} e^{-j\omega} & |\omega| < \pi \\ \left(\frac{1}{2\pi}\right)(3\pi + \omega)e^{-j\omega} & -3\pi < \omega < -\pi \\ \left(\frac{1}{2\pi}\right)(3\pi - \omega)e^{-j\omega} & \pi < \omega < 3\pi \\ 0 & \text{otherwise} \end{cases}$$

(g) Using the FT analysis, then we get

$$X(j\omega) = \frac{2j}{\omega}\left[\cos 2\omega - \frac{\sin \omega}{\omega}\right]$$

(h) If

$$x_1(t) = \sum_{k=-\infty}^{\infty} \delta(t - 2k) \quad \text{Then,} \quad x(t) = 2\,x_1(t) + x_1(t-1)$$

Therefore

$$X(j\omega) = X_1(j\omega)[2 + e^{-\omega}] = \pi \sum_{k=-\infty}^{\infty} \delta(\omega - k\pi)\,[2 + (-1)^k]$$

(i) Using the FT analysis, we get

$$X(j\omega) = \frac{1}{j\omega} + \frac{2e^{-j\omega}}{-\omega^2} - \frac{2e^{-j\omega} - 2}{j\omega^2}$$

where X(jω) is the FT of one period of x(t). That is

$$X(j\omega) = \frac{1}{1 - e^{-2}} \left[\frac{1 - e^{-2(1+j\omega)}}{1 + j\omega} - \frac{e^{-2}[1 - e^{-2(1-j\omega)}]}{1 - j\omega} \right]$$

Problem 8.10
Determine the continuous-time signal corresponding to each of the following transforms.

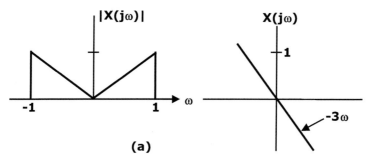

(a)

Figure 8.10

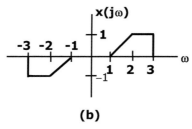

(b)

Figure 8.11

(a) $X(j\omega) = \frac{2\sin[3(\omega - 2\pi)]}{(\omega - 2\pi)}$

(b) $X(j\omega) = \cos(4\omega + \pi/3)$

(c) X(jω) as given by the magnitude and phase plots of Figure P8.11(a)

(d) $X(j\omega) = 2[\delta(\omega - 1) - \delta(\omega - 2\pi)] + 3[\delta(\omega - 2\pi) + \delta(\omega + 2\pi)]$

(e) $X(j\omega)$ as in Figure 8.20 (b).

Solution 8.10

(a) $x(t) = \begin{cases} e^{j2\pi t} & |t| < 3 \\ 0 & \text{otherwise} \end{cases}$

(b) $x(t) = \frac{1}{2}e^{-j\frac{\pi}{3}}\delta(t - 4) + \frac{1}{2}e^{-j\frac{\pi}{3}}\delta(t + 4)$

(c) The FT synthesis may be written as

$$x(t) = \frac{1}{2\pi} \int_{-\infty}^{\infty} |X(j\omega)| e^{-j\pi X(j\omega)} e^{j\omega t}.d\omega$$

From the given figure, we get

$$x(t) = \frac{1}{\pi}\left[\frac{\sin(t - 3)}{t - 3} + \frac{\cos(t - 3) - 1}{(t - 3)^2}\right]$$

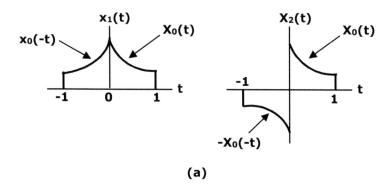

(a)

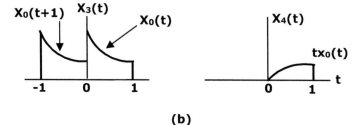

(b)

Figure P8.11

(d) $x(t) = \frac{2j}{\pi} \sin t + \frac{3}{\pi} \cos(2\pi t)$

(e) Using the transform synthesis equation

$$x(t) = \frac{\cos 3t}{j\pi t} + \frac{\sin t - \sin 2t}{j\pi t^2}$$

Problem 8.11

Consider the signal

$$x_0(t) = \begin{cases} e^{-t}, & 0 \le t \le 1 \\ 0, & elsewhere \end{cases}$$

Determine the FT of each of the signals shown in Figure 8.12. You should be able to do this by explicitly evaluating only the transform of $x_0(t)$ and then using properties of the FT.

Solution 8.11

For the given signal $x_0(t)$, we use the FT, to evaluate the corresponding FT:

$$X_0(j\omega) = \frac{1 - e^{-(1+j\omega)}}{1 + j\omega}$$

(i) We know that

$$x_1(t) = x_0(t) - x_0(-t)$$

Using the linearity and time reversal property of the FT, we have

$$X_1(j\omega) = X_0(j\omega) + X_0(-j\omega) = \frac{2 - 2e^{-1} \cos \omega - 2\omega e^{-1} \sin \omega}{1 + \omega^2}$$

(ii) We know that

$$x_2(t) = x_0(t) - x_0(-t)$$

Using the linearity and time reversal properties of the FT we have

$$X_2(j\omega) = X_0(j\omega) - X_0(j\omega) = j\left[\frac{-2\omega + 2e^{-1} \sin \omega + 2\omega e^{-1} \cos \omega}{1 + \omega^2}\right]$$

(iii) We know that

$$x_3(t) = x_0(t) + x_0(t+1)$$

Using the linearity and time shifting properties of the FT, we have

$$X_3(j\omega) = X_0(j\omega) + e^{j\omega} X_0(j\omega) = \frac{1 + e^{j\omega} - e^{-1}(1 + e^{-j\omega})}{1 + j\omega}$$

(iv) We know that
$$x_4(t) = t.x_0(t)$$

Using the differentiation in frequency property
$$X_4(j\omega) = tx_0(t)$$

Using the differentiation in frequency property
$$X_4(j\omega) = j\frac{d}{d\omega}X_0(j\omega)$$

Therefore,
$$X_4(j\omega) = \frac{1 - 2e^{-1}e^{-j\omega} - j\omega e^{-1}e^{-j\omega}}{(1+j\omega)^2}.$$

Problem 8.12

(a) Determine which, if any, of the real signals depicted in the given figure have FTs that satisfy each of the following conditions:

  (1) $Re\{X(j\omega)\} = 0$

  (2) $Im\{X(j\omega)\} = 0$

  (3) There exists a real α such that $e^{j\alpha\omega} X(j\omega)$ is real.

  (4) $X(j\omega) \, d\omega = 0$

  (5) $\omega X(j\omega) \, d\omega = 0$

  (6) $X(j\omega)$ is periodic

(b) Construct a signal that has properties (1), (4) and (5) and does not have the others.

Solution 8.12

(a) (i) For Re $\{X(j\omega)\}$ to be 0, the signal x(t) must be real and odd. Therefore signals in figures (a) and (c) have this property.

   (ii) For Im $\{X(j\omega)\}$ to be 0, the signal x(t) must be real and even. Therefore signals in figures (e) and (f) have this property.

   (iii) For there to exist a real α such that $e^{j}X(j\omega)$ is real. We require that x(t − α) be a real and an even signal. Therefore signals in figures (a), (b), (e) and (f) have this property.

   (iv) For this condition to be true x(0) = 0, therefore signals in figures (a), (b), (c), (d) and (f) have this property.

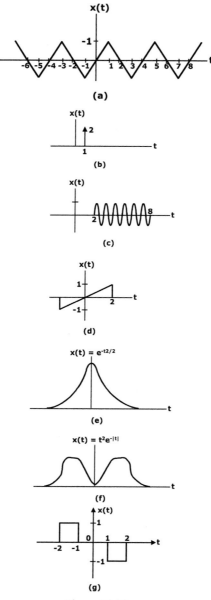

Figure P8.12

(v) For this condition to be true the derivative of x(t) has to be zero at t = 0, therefore signals in figures (b), (c), (e) and (f), have this property.

(vi) For this to be true the signal x(t) has to be periodic only if the signal in figure (a) has this property.

(b) For a signal to satisfy only properties, (1), (4) and (5). It must be real and odd and

$$x(t) = 0, \ x'(0) = 0$$

Example 8.13
Let X(jω) denote the FT of the signal x(t) depicted in Figure P8.13 (a). Now

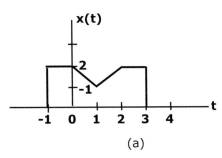

(a)

Figure P8.13

(a) Find X(jω)

(b) Find X(j0)

(c) Find $X \int_{-\infty}^{\infty} X(j\omega) \, d\omega$.

(d) Evaluate $\int_{-\infty}^{\infty} X(j\omega) \frac{2\sin \omega}{\omega} e^{j2\omega} \, d\omega$.

(e) Evaluate $\int_{-\infty}^{\infty} |X(j\omega)|^2 \, d\omega$.

(f) Sketch the inverse FT of (Re $\{X (j\omega)\}$.

Solution 8.13
(a) Note that y(t) = x(t + 1) is a real and an even signal, therefore y(jω) is also real and even. This implies that

$$y(j\omega) = 0$$

Also, since $y(j\omega) = e^{j\omega}X(\omega)$
We know that

$$X(j\omega) = -\omega$$

(b) We have

$$X(j0) = \int_{-\infty}^{\infty} x(t) \, dt = 7$$

(c) We get

$$\int_{-\infty}^{\infty} x(j\omega) \, d\omega = 2\pi \, x(0) = 4\pi$$

(d) Let $Y(j\omega) = \frac{2\sin\omega}{\omega} e^{2j\omega}$

The corresponding signal y(t) is

$$y(t) = \begin{cases} 1 & -3 < t < -1 \\ 0 & otherwise \end{cases}$$

Then, the given integral is

$$\int_{-\infty}^{\infty} |X(j\omega)|^2 \, d\omega = 2\pi \int_{-\infty}^{\infty} |X(t)|^2 \, dt = 26\,\pi$$

(f) The inverse FT at Re $\{X(j\omega)\}$ is the $\{x_e(t)\}$ which is $\frac{[x(t)+x(-t)]}{2}$. This is as shown in the provided figure:

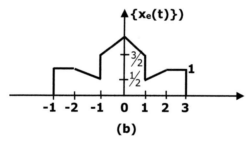

(b)

Figure P8.13

Problem 8.14
Find the FT of $x(t) = e^{-at}u(t)$.

Solution 8.14
The FT does not converge for $a \leq 0$ since x(t) is not absolutely integrable, as shown by

$$\int_{0}^{\infty} e^{-at} dt = \infty, \ a \leq 0$$

For a > 0, we have

$$X(j\omega) = \int_{-\infty}^{\infty} e^{-at}u(t)\, e^{-j\omega t} = \int_{-\infty}^{\infty} e^{-(a+j\omega)t}dt$$

$$= -\frac{1}{a+j\omega}e^{-(a+j\omega)}\bigg|_{0}^{\infty} = \frac{1}{a+j\omega}$$

Converting to polar form, the magnitude and phase of X(jω)

$$|X(j\omega)| = \frac{1}{(a^2+\omega^2)^{1/2}} \quad \arg\{X(j\omega)\} = -\arg\tan\left(\frac{\omega}{a}\right)$$

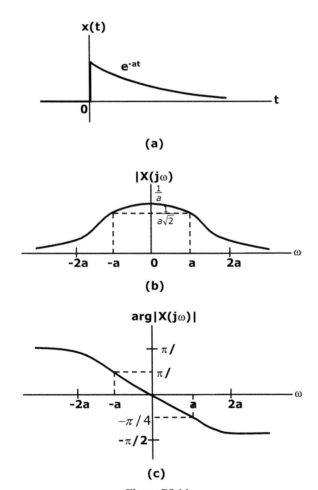

(a)

(b)

(c)

Figure P8.14

Problem 8.15

Consider the rectangular pulse depicted in Figure P8.15(a) and mathematically defined as

$$x(t) = \begin{cases} 1, & -T \le t \le T \\ 0, & |T| > T \end{cases}$$

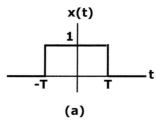

Figure P8.15

Find FT of the pulse.

Solution 8.15

The rectangular pulse x(t) is absolutely integrable provided $T < \infty$. For $\omega \ne 0$, we have

$$X(j\omega) = \int_{-\infty}^{\infty} x(t)\, e^{-j\omega t} dt = \int_{-T}^{T} e^{-j\omega t} dt = \frac{1}{j\omega} e^{-j\omega t} \Big|_{-T}^{T}$$

$$= \frac{2}{\omega} \sin(\omega T), \quad \omega \ne 0$$

For $\omega = 0$, the integral simplifies to 2T. It is straightforward to show using L' Hopital's rule that

$$\lim_{\omega \to 0} \frac{2}{\omega} \sin(\omega T) = 2T$$

Thus we write for all ω, $X(j\omega) = \frac{2}{\omega} \sin(\omega T)$

With the understanding that the value at $\omega = 0$ is obtained by evaluating a limit, $X(j\omega)$ is real. It is depicted in Figure 7.44.

(b) The magnitude spectrum is

$$|X(j\omega)| = 2 \left| \frac{\sin(\omega T)}{\omega} \right|$$

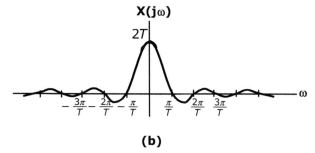

(b)

Figure P8.15

and the phase spectrum is

$$\arg\{X(j\omega)\} = \begin{cases} 0, & \frac{\sin(\omega T)}{\omega} > 0 \\ \pi, & \frac{\sin(\omega T)}{\omega} < 0 \end{cases}$$

We may write $X(j\omega)$ using the sine c function notation as

$$X(j\omega) = 2T \sin c \left(\frac{\omega T}{\pi} \right)$$

9

Solution of Difference Equations

This chapter provides comprehensive details regarding the homogenous and particular solution, solution of difference equations and problems and solutions.

9.1 Constant-Coefficient Difference Equation

Discrete time systems are modelled by a set of difference equations and the transform used in the analysis of linear time-invariant (LTI) discrete systems is the Z-transform.

A discrete-time (DT) system is described by the difference equation. Infinite duration impulse response (IIR) requires an infinite number of memory locations, multiplication and additions. So it is impossible to implement the IIR system by applying convolution. Further, IIR systems are more conveniently described by difference equations.

A recursive system is described with an input–output equation

$$y(n) = ay(n-1) + x(n) \tag{9.1}$$

where 'a' is a constant.

The system described by the first order difference equation in (9.1) is the simplest possible recursive system in the general class of recursive systems described by linear constant coefficient difference equations. The general formula for such an equation is

$$y(n) = -\sum_{k=1}^{N} a_k \, y(n-k) + \sum_{k=0}^{M} b_k x(n-k) \tag{9.2}$$

or equivalently

$$\sum_{k=0}^{N} a_k\, y(n-k) = \sum_{k=0}^{M} b_k\, x(n-k), \quad a_0 = 1 \tag{9.3}$$

the integer N is called the order of the difference equation or the order of the system. The negative sign on the right-hand side of (9.2) is introduced as a matter of convenience to allow us to express the difference equation in (9.3) without any negative sign.

Given a linear constant-coefficient difference equation as the input–output relationship describing the LTI system, the objective is to determine the explicit expression for the output y(n).

Basically, the goal is to determine the output y(n), $n \geq 0$, of the system for the specific input x(n), $n \geq 0$ and a set of initial conditions.

9.2 Solution of Difference Equation

Three techniques are used for solving LTI difference equations. The first two techniques are discussed here, while the third is included in Chapter 10.

   (i) Sequential procedure: This technique used in the digital computer solution of difference equation.
   (ii) Classical technique: It consists of finding the complementary and the particular solution as in case of solution of differential equation.
(iii) Using Z-transform

   In this chapter, sequential and classical methods are discussed.

9.2.1 Using Sequential Procedure

Example 9.1
We solve for m(k) for the equation $m(k) = e(k) - e(k-1) - m(k-1)$, $k \geq 0$.

$$\text{Where} = e(k) = \begin{cases} 1, & k \text{ even} \\ 0, & k \text{ odd.} \end{cases}$$

and both $e(-1)$ and $m(-1)$ are zero

Solution 9.1

$$k = 0, \quad m(0) = e(0) - e(-1) - m(-1) = 1$$
$$k = 1, \quad m(1) = e(1) - e(0) - m(0) = 0 - 1 - 1 = -2$$
$$k = 2, \quad m(2) = e(2) - e(1) - m(1) = 1 - 0 - (-2) = 3$$
$$k = 3, \quad m(3) = e(3) - e(2) - m(2) = 0 - 1 - 3 = -4$$
$$k = 4, \quad m(4) = e(4) - e(3) - m(3) = 1 - 0 - (-4) = 5$$

Example 9.2
Solve the difference equation $x(k) - 4x(k - 1) + 3x(k - 2) = e(k)$, using the sequential technique. Find $x(k)$ for $k \leq 4$.

$$\text{where} \quad e(k) = \begin{cases} 2, & k = 0, 1 \\ 0, & k \geq 2 \end{cases} \qquad x(-2) = x(-1) = 0.$$

Solution 9.2

$$x(k) - 4x(k - 1) - 3x(k - 2) = e(k)$$
$$k = 0 \qquad x(0) = 0 - 0 + 2 = 2$$
$$k = 1 \qquad x(1) = 4 \times 2 - 0 + 2 = 10$$
$$k = 2 \qquad x(2) = 4 \times 10 - 6 + 0 = 34$$
$$k = 3 \qquad x(3) = 4 \times 34 - 30 + 0 = 106$$

Generally, it is written as $x(k) = -2 + 4(3)^k$.

9.2.2 Classical Technique

The classical technique which is also called the direct solution method assumes that the total solution is the sum of two parts:

$$y(n) = y_h(n) + y_p(n)$$

where $y_h(n)$ is the homogeneous or complementary solution, where $y_p(n)$ is called the particular solution.

9.2.2.1 The homogeneous solution of difference equation

The homogeneous solution here is same as the complementary function in differential equation. It is calculated when the forcing function is kept zero, it is also called the zero-input response.

For finding the homogeneous solution, the input x(n) is assumed to be zero:

$$\sum_{k=0}^{n} a_k y(n-k) \tag{9.4}$$

The procedure of solving a linear constant coefficient difference equation directly is very similar to the procedure for solving a linear constant coefficient difference equation. Basically we assume that the solution is in the form of an exponential, that is

$$y_h(n) = \lambda^n \tag{9.5}$$

where the subscript h on y(n) is used to denote the solution to the homogeneous difference equation. Substituting the assumed solution of (9.5), we obtain the polynomial equation:

$$\sum_{k=0}^{N} a_k \lambda^{n-k} = 0 \tag{9.6}$$

This polynomial is called the characteristics polynomial of the system. In general it has N roots, which is denoted as $\lambda_1, \lambda_2, \lambda_3, \lambda_4, \ldots \ldots \ldots \lambda_N$.

The roots may be real or complex valued, if the coefficients $a_1, a_2, a_3, a_4, \ldots a_N$ are real. Some of the N roots may be identical, in which case we have multiple order roots.

For the moment, let us assume that the roots are distinct; that is, there are no multiple roots. The most general solution to the homogeneous difference equation in (9.6) is

$$y_h(n) = C_1 \lambda_1^n + C_2 \lambda_2^n + C_2 \lambda_3^n + \ldots \ldots + C_N \lambda_N^n \tag{9.7}$$

where $C_1, C_2, C_3, \ldots, C_N$ are weighting coefficients. These coefficients are determined from the initial conditions specified for the system. Since the input $x(n) = 0$, (9.6) may be used to obtain the zero-input response of the system. The following example illustrates the procedure.

Example 9.3
Determine the homogeneous solution of the system described by the first-order difference equation:

$$y(n) + a_1 y(n-1) = x(n) \tag{1}$$

Solution 9.3

The assumed solution obtained by setting $x(n) = 0$ is

$$y_h(n) = \lambda^n$$

When we substitute this solution in (1), we obtain [with $x(n) = 0$]

$$\lambda^n + a_1 \lambda^{n-1} = 0$$

$$\lambda^{n-1}(\lambda + a_1) = 0, \quad \lambda = -a_1$$

Therefore, to the homogeneous difference equation is

$$y_h(n) = C\lambda^n = C(-a_1)^n \tag{2}$$

The zero-input response of the system can be determined from (1) and (2). With $x(n) = 0$, (2) yields

$$y(0) = -a_1 y(-1)$$

On the other hand, from (2) we have

$$y_h(0) = C$$

Equating y(0) and $y_h(0)$ $C = (-a_1)y(-1)$
and finally the zero-input response of the system is

$$y_{zi}(n) = (-a_1)^{n+1} y(-1), \quad n \geq 0 \tag{3}$$

Example 9.4

Determine the zero-input response of the system described by the homogeneous second-order difference equation, with initial condition $y(-1) = 5$ and $y(-2) = 0$:

$$y(n) - 3y(n-1) - 4y(n-2) = 0 \tag{1}$$

Solution 9.4

First determine the solution to the homogeneous equation. We assume the solution to be exponential. The assumed solution obtained by setting $x(n) = 0$ is

$$y_h(n) = \lambda^n$$

When we substitute this solution in (1), we obtain [with $x(n) = 0$]

$$\lambda^n - 3\lambda^{n-1} - 4\lambda^{n-2} = 0$$

$$\lambda^{n-2}(\lambda^2 - 3\lambda - 4) = 0$$

Therefore, the roots are $\lambda = -1, 4$ and the general form of the solution to the homogeneous equation is

$$y_h(n) = C_1\lambda_1^n + C_2\lambda_2^n$$
$$y_h(n) = C_1(-1)^n + C_2(4)^n \tag{2}$$

The zero-input response of the system can be determined from the homogeneous solution by evaluating the constants in (2), given the initial conditions $y(-1)$ and $y(-2)$. From the difference Equation (1), we have

$$y(0) = 3y(-1) + 4y(-2), y(1) = 3y(0) + 4y(-1)$$

$$y(1) = 3[3y(-1) + 4y(-2)] + 4y(-1) = 13y(-1) + 12y(-2)$$

On the other hand, from (2) we obtain $y(0) = C_1 + C_2, y(1) = -C_1 + 4C_2$.

By equating these two sets of relation, we have

$$C_1 + C_2 = 3y(-1) + 4y(-2)$$
$$-C_1 + 4C_2 = 13y(-1) + 12y(-2)$$

The solution of these two equations is

$$C_1 = -1/5 \ y(-1) + 4/5 \ y(-2)$$
$$C_2 = 16/5 \ y(-1) + 16/5 \ y(-2)$$

Therefore, the zero-input response of the system is

$$y_{zi}(n) = [-\frac{1}{5}y(-1) + \frac{4}{5}y(-2)](-1)^n + [\frac{16}{5}y(-1) + \frac{16}{5}y(-2)](4)^n, \quad n \geq 0 \tag{3}$$

If we substitute here $y(-1) = 5$ and $y(-2)$ $C_1 = -1, C_2 = 16$ and hence

$$y_{zi}(n) = (-1)^{n+1} + (4)^{n+2}, \quad n \geq 0$$

9.2.2.2 The particular solution of difference equation

The particular solution $y_p(n)$ is required to satisfy the difference equation (9.4) for the specific input signal $x(n) \geq 0$. In other words, $y_p(n)$ is any solution satisfying

$$\sum_{k=0}^{N} a_k y_p(n-k) = \sum_{k=0}^{M} b_k x(n-k), \quad a_0 = 1 \tag{9.8}$$

To solve (9.8) we assume for $y_p(n)$ a form that depends on the form of the input $x(n)$. The following examples illustrate the procedure and show also the rule for choosing the particular solution.

9.2.2.3 Rules for choosing particular solutions

As is the case with the solution of differential equations, there are a set of rules one must follow to form appropriate particular solutions while solving difference equations, as summarized in Table 9.1.

Example 9.5

To find out the general solution of the second-order difference equation

$$y(n) - \frac{5}{6} y(n-1) + \frac{1}{6} y(n-2) = 5^{-n}, \quad n \geq 0 \tag{1}$$

with initial conditions $y(-2) = 25$ and $y(-1) = 6$.

Solution 9.5

In (1), y(n) may be interpreted as the response (output) of a DT system to input (forcing) function 5^{-n} for $(n \geq 0)$, where n is a time index. It is apparent that (1) is a second-order difference equation since it expresses the output y(n) at time n as a linear combination of two previous outputs $y(n-1)$ and $y(n-2)$.

The general (or closed-form) solution y(n) of (1) is obtained in three steps that are similar to those used for solving second-order differential equations. They are as follows:

Obtain the complementary solution $y_c(n)$ in terms of two arbitrary constants c_1 and c_2.

Obtain the particular solution $y_p(n)$, and write

$$y(n) = y_c(n) + y_p(n) = f(c_1, c_2) + y_p(n) \tag{2}$$

Table 9.1   Rules for choosing particular solutions

Terms in Forcing Function	Choice of Particular Solution
1. A constant	c; c is a constant
2. $b_1 n^k$; b_1 is a constant	$c_0 + c_1 n + c_2 n^2 + \ldots + c_k n^k$; c_i are constants
3. $b_2 d^{\pm n}$; b_2 and d are constants	Proportional to $d^{\pm n}$
4. $\begin{matrix} b_3 & \cos(n\omega n) \\ b_4 & \sin(n\omega n) \end{matrix} \Big\}$ b_3 *and* b_4 *are* *constants*	$c_1 \sin(n\omega) + c_2 \cos(n\omega)$
If a term in any of the particular solutions in this column is a part of the complementary solution, it is necessary to modify the corresponding choice by multiplying it by n before using it. If such a term appears r times in the complementary solution, the corresponding choice must be multiplied by n^r.	

where $y(n) = f(c_1, c_2); y_c(n)$ a function of c_1 and c_2.

Solve for c_1 and c_2 in (2) using two given initial conditions. The steps are elaborated as follows:

We assume that the complementary solution $y_c(n)$ has the form

$$y_c(n) = c_1 a_1^n + c_2 a_2^n \qquad (3)$$

where a_1 and a_2 are the real constants.

Next substitute $y(n) = a^n$ in the homogeneous equation to get

$$a^n - \frac{5}{6}a^{n-1} + \frac{1}{6}a^{n-2} = 0 \qquad (4)$$

Dividing both sides of (4) by a^{n-2}, we obtain

or

$$\left(a - \frac{1}{2}\right)\left(a - \frac{1}{3}\right) = 0$$

which yields the characteristic roots

$$a_1 = \frac{1}{2} \quad \text{and} \quad a_2 = \frac{1}{3}$$

Thus, the complementary solution is

$$y_c(n) = c_1 2^{-n} + c_2 3^{-n}$$

where c_1 and c_2 are arbitrary constants.

The particular solution $y_p(n)$ is assumed to be

$$y_p(n) = c_3 5^{-n}$$

since the forcing function is 5^{-n}; substitution of $y(n) = y_p(n) = c_3 5^{-n}$ in (1) leads to

$$c_3 [5^{-n} - \left(\frac{5}{6}\right) 5^{-(n-1)} + \left(\frac{1}{6}\right) 5^{-(n-2)}] = 5^{-n}$$

Dividing both sides of this equation by 5^{-n}, we obtain

$$c_3 [1 - \left(\frac{5}{6}\right) 5 + \left(\frac{1}{6}\right) 5^2] = 1$$

which implies that $c_3 = 1$. Thus

$$y(n) = y_c(n) + y_p(n)$$
$$= c_1 2^{-n} + c_2 3^{-n} + 5^{-n} \qquad (5)$$

Since the initial conditions are

$$y(-2) = 25 \quad \text{and} \quad y(-1) = 6$$

Using the above given initial conditions (5) yields the simultaneous equations:

$$4c_1 + 9c_2 = 0$$

and
$$2c_1 + 3c_2 = 1 \tag{6}$$

Solving (6) for c_1 and c_2, we obtain

$$c_1 = \frac{3}{2} \quad \text{and} \quad c_2 = -\frac{2}{3} \tag{7}$$

Thus, the desired general solution is given by

$$y(n) = \frac{3}{2}(2^{-n}) - \frac{2}{3}(3^{-n}) + 5^{-n}, \quad n \geq 0 \tag{8}$$

$y(n)$ can be interpreted as the output of a DT system when it is subjected to the exponential input (forcing function) 5^{-n}, which is the right-hand side of the given difference Equation in (1).

Example 9.6
Solve the second-order difference equation

$$y(n) - \frac{3}{2}y(n-1) + \frac{1}{2}y(n-2) = 1 + 3^{-n}, \quad n \geq 0 \tag{1}$$

with the initial conditions $y(-2) = 0$ and $y(-1) = 2$.

Solution 9.6
The solution consists of three steps.

Assume the complementary solution as $y_c(n) = c_1a_1^n + c_2a_2^n$. Substituting $y(n) = a^n$ in the homogeneous counterpart of (1) we obtain the characteristic equation

$$a^2 - \frac{3}{2}a + \frac{1}{2} = 0$$

the roots of which are $a_1 = \frac{1}{2}$ and $a_2 = 1$.
Thus

$$y_c(n) = c_1 2^{-n} + c_2 1^n = c_1 2^{-n} + c_2 \tag{2}$$

To choose an appropriate particular solution, we refer to Table 9.1. From the given forcing function and lines 1 and 3 of Table 9.1, it follows that a choice for the particular solution is $c_3 + c_4 3^{-n}$.

However, we observe that this choice for the particular solution and $y_c(n)$ in (2) have common terms, each of which is a constant; that is, c_3 and c_2, respectively. Thus in accordance with the Footnote of Table 9.1, we modify the choice $c_3 + c_4 3^{-n}$ to obtain

$$y_p(n) = c_3 n + c_4 3^{-n} \tag{3}$$

Next, substitution of $y_p(n)$ in (3) into (1) leads to

$$c_3 n + c_4 3^{-n} - \frac{3}{2} c_3 n + \frac{3}{2} c_3 - \frac{9}{2} c_4 3^{-n} + \frac{1}{2} c_3 n - c_3 + \frac{9}{2} c_4 3^{-n} = 3^{-n} + 1 \tag{4}$$

equating the coefficient of 3^{-n} and constant, from (4) it follows that

$$\frac{1}{2} c_3 = 1$$

and

$$c_4 \left[1 - \frac{9}{2} + \frac{9}{2} \right] 3^{-n} = 3^{-n}$$

which results in

$$c_3 = 2; c_4 = 1$$

Thus, combining (2) and (3), we get

$$y(n) = c_1 2^{-n} + c_2 + 2n + 3^{-n} \tag{5}$$

To evaluate c_1 and c_2 in (5), the given initial conditions are used; that is, $y(-2) = 0$ and $y(-1) = 2$. This leads to the simultaneous equations:

$$4c_1 + c_2 = -5$$

$$2c_1 + c_2 = 1$$

Solving, we obtain $c_1 = -3$ and $c_2 = 7$, which yields the desired solution as

$$y(n) = (-3)2^{-n} + 7 + 2n + 3^{-n}, \quad n \geq 0$$

Example 9.7

Find the general solution of the first-order difference equation.

$$y(n) - 0.9\, y(n-1) = 0.5 + (0.9)^{n-1}, \quad n \geq 0 \tag{1}$$

with $y(-1) = 5$.

Solution 9.7

Substituting $y(n) = a^n$ in the homogeneous equation

$$y(n) - 0.9\, y(n-1) = 0$$

we obtain $\qquad y_c(n) = c_1(0.9)^n \qquad\qquad (2)$

since it is a first-order difference equation.

From the forcing function in (1), the complementary solution in (2), and lines 1 and 3 of Table 9.1, it follows that

$$y_P(n) = c_2 n(0.9)^n + c_3$$

Substitution of $y(n) = y_p(n)$ in (1) results in

$$c_3 + c_2 n(0.9)^n - 0.9 c_2(n-1)(0.9)^{n-1} - 0.9 c_3 = 0.5 + (0.9)^{n-1}$$

equating the coefficient of the constant 0.5 and $(0.9)^{n-1}$ leads to

$$0.1 c_3 = 0.5$$

and $(0.9)^n c_2 = (0.9)^{n-1}$

Thus, we have

$$c_3 = 5 \quad \text{and} \quad c_2 = \frac{10}{9}$$

which implies that

$$y_p(n) = \frac{10}{9} n(0.9)^n + 5 \qquad\qquad (3)$$

Combining (2) and (3), the following equation is obtained

$$y(n) = c_1(0.9) + \frac{10}{9} n(0.9)^n + 5 \qquad\qquad (4)$$

From (4) and the initial condition $y(-1) = 5$, it follows that $c_1 = \frac{10}{9}$. Hence, the desired solution can be written as

$$y(n) = (n+1)(0.9)^{n-1} + 5, \quad n \geq 0$$

Example 9.8

Find the general solution of the second-order difference equation

$$y(n) - 1.8 y(n-1) + 0.81\, y(n-2) = 2^{-n}, \quad n \geq 0 \qquad\qquad (1)$$

   Leave the answer in terms of unknown constants, which one can evaluate if the initial conditions are given.

Solution 9.8
With $y(n) = a^n$ substituted into the homogeneous counterpart, the following is obtained

$$a^2 - 1.8a + 0.81 = 0$$

which results in the repeated roots

$$a_1 = a_2 = 0.9$$

   Thus, as in the case of differential equations, we consider the complementary solution to be

$$y_c(n) = c_1(0.9)^n + c_2 n(0.9)^n \qquad (2)$$

   From the given forcing function in (1), $y_c(n)$ in (2) and line 3 of Table 9.1, it is clear that

$$y_p(n) = c_3 2^{-n} \qquad (3)$$

   Substitution of (3) in (1) leads to

$$c_3[1 - (1.8)(2) + (0.81)(4)]2^{-n} = 2^{-n}$$

which yields $c_3 = \frac{1}{1.72} = 0.5813$.
   Thus the desired solution given by (2) and (3) to be

$$y(n) = c_1(0.9)^n + c_2 n(0.9)^n + (0.5813)\,2^{-n}$$

where c_1 and c_2 can be evaluated if two initial conditions are specified.

Example 9.9
Find the particular solution for the first-order difference equation.

$$y(n) - 0.5y(n-1) = \sin\left(\frac{n\pi}{2}\right), \quad n \ge 0 \qquad (1)$$

Solution 9.9
Since the forcing function is sinusoidal, we refer to 5 of Table 9.1 and choose a particular solution of the form

$$y_p(n) = c_1 \sin\left(\frac{n\pi}{2}\right) + c_2 \cos\left(\frac{n\pi}{2}\right) \qquad (2)$$

Substitution of $y(n) = y_p(n)$ in (1) leads to

$$c_1 \sin\left(\frac{n\pi}{2}\right) + c_2 \cos\left(\frac{n\pi}{2}\right) - 0.5 c_1 \sin\left[\frac{(n-1)\pi}{2}\right] - 0.5 c_2 \cos\left[\frac{(n-1)\pi}{2}\right]$$

$$= \sin\left(\frac{n\pi}{2}\right) \tag{3}$$

using the following identities:

$$\sin\left[\frac{(n-1)\pi}{2}\right] = \sin\left(\frac{n\pi}{2} - \frac{\pi}{2}\right) = -\cos\left(\frac{n\pi}{2}\right) \tag{4}$$

$$\cos\left[\frac{(n-1)\pi}{2}\right] = \cos\left(\frac{n\pi}{2} - \frac{\pi}{2}\right) = \sin\left(\frac{n\pi}{2}\right)$$

Substituting (2) in (1), we obtain

$$(c_1 - 0.5 c_2) \sin\left(\frac{n\pi}{2}\right) + (0.5 c_1 + c_2) \cos\left(\frac{n\pi}{2}\right) = \sin\left(\frac{n\pi}{2}\right)$$

which yields the simultaneous equations.

$$c_1 - 0.5 c_2 = 1$$
$$0.5 c_1 + c_2 = 0 \tag{5}$$

The solution of (5) yields $c_1 = \frac{4}{5}$ and $c_2 = -\frac{2}{5}$ Hence, the desired result given by (2) to be

$$y_p(n) = \frac{4}{5} \sin\left(\frac{n\pi}{2}\right) - \frac{2}{5} \cos\left(\frac{n\pi}{2}\right) \quad n \geq 0$$

Example 9.10
Find the general solution of the second-order difference equation

$$y(n) - 1.8\, y(n-1) + 0.81\, y(n-2) = (0.9)^n, n \geq 0 \tag{1}$$

Leave the answer in terms of unknown constants, which one can evaluate if the initial conditions are given.

Solution 9.10
The complementary solution is given by

$$y_c(n) = c_1 (0.9)^n, + c_2 n (0.9)^n \tag{2}$$

Since the forcing function is $(0.9)^n$, line 3 of Table 9.1 implies that a choice for the particular solution is $c_3(0.9)^n$.

However, since this choice and the preceding $y_c(n)$ have a term in common, we must modify our choice according to the footnote of Table 9.1 to obtain $c_3n(0.9)^n$.

But this choice again has a term in common with $y_c(n)$.

Thus we apply to the footnote of Table 9.1 once again to obtain

$$y_p(n) = c_3n^2(0.9)^n \tag{3}$$

which has no more terms in common with $y_c(n)$.

Hence $y_p(n)$ in (3) is the appropriate choice for the particular solution for the difference equation in (1) when the forcing function is $(0.9)^n$.

9.3 Problems and Solutions

Problem 9.1
Solve the following second-order linear difference equations with constant coefficients.

(a) $2y(n-2) - 3y(n-1) + y(n) = 3, n \geq 0,$ *with* $y(-2) = -3$ *and* $y(-1) = -2.$

(b) $y(n-2) - 2y(n-1) + y(n) = 1, n \geq 0,$ *with* $y(-1) = -0.5$ *and* $y(-2) = 0.$

Hint $: y_c(n) = c_1 + c_2n,$ *and* $y_p(n) = c_3n^2(why?).$

Solution 9.1(a)
Let the assumed $y_c = a^n$ by substitution into original equation, we get

$$2a^{n-2} - 3a^{n-1} + a^n = 0$$

$$a^n - 3a^{n-1} + 2a^{n-2} = 0$$

$$a^2 - 3a + 2 = 0$$

$$a = 2, 1$$

$$y_c(n) = c_1(2)^n + c_2(1)^n \tag{1}$$

$$y_c(n) = c_1 2^n + c_2$$

The assumed $y_p = nc_3$

Substituting (1) in the original equation

$$[c_3(n-2)] - 3[c_3(n-1)] + nc_3 = 3$$

$$c_3 = -3$$

The complete solution

$$y(n) = y_p(n) + y_C(n)$$

$$y(n) = c_1 2^n + c_2 - 3n \qquad (2)$$

Now for finding c_1 and c_2, using the conditions $f(-2) = -3$ and $f(-1) = -2$

$$c_1 + 4c_2 = -44/3,$$

$$c_1 + 2c_2 = -14/3,$$

$$c_1 = 16/3, c_2 = -5 \quad \text{Substituting in (2)}$$

$$y(n) = -13 + 16 \, 2^n - 3n$$

Solution 9.1(b)

Let $y_c(n) = a^n$

By putting into the original equation

$$a^{n-2} - 2a^{n-1} + a^n = 0$$

$$a^n - 2a^{n-1} + a^{n-2} = 0$$

$$a^2 - 2a + 1 = 0$$

$$(a-1)^2 = 0$$

$$a = 1, 1$$

'Since roots turned out to be same' (and also constants of the same type). So

$$y_c(n) = c_1 1^n + nc_2 1^n$$

$$y_c(n) = c_1 + nc_2$$

The assumed $\qquad\qquad y_p = c_3 n^2 \qquad\qquad (3)$

Since in the right-hand side of the original equation we had a constant value too, we multiply it twice to distinguish from c_2.

Put (3) in the original equation

$$[c_3(n-2)^2] - 2[c_3(n-1)^2] + c_3n^2 = 1, c_3 = 1/2$$

The complete solution

$$y(n) = c_1 + nc_2 + \frac{n^2}{2}$$

using condition $f(-1) = -0.5$

$$c_1 - c_2 = -1, \quad \text{and} \quad f(-2) = 0$$

$$c_1 - 2c_2 = -2$$

$$c_1 = 0, c_2 = 1$$

$$y(n) = n + 0.5(n^2)$$

Problem 9.2

Solve the following second-order linear difference equations with constant coefficients:

(a) $y(n) - 0.8y(n-1) = (0.8)^n, n \geq 0, \quad with \quad y(0) = 6.$

(b) $y(n) - y(n-1) = 1 + (0.5)^n, n \geq 0, \quad with \quad y(0) = 1.$

(c) $-5y(n-2) + 5y(n-1) = 1, n \geq 0, \quad with \quad y(-2) = 2$
   $and \ y(-1) = 1.$

The answer may be left in terms of two arbitrary constants:

$$y_p(n) = c_3n^2 + c_4n + c_5.$$

Solution 9.2(a)

$$y(n) - 0.8y(n-1) = 0.8^n, \quad n \geq 0 \quad y(0) = 6$$

the y_c part can be assumed as $y = a^n$
by substitution $a^n - 0.8a^{n-1} = 0$

$$a = 0.8$$

$$y_c = c_1 \, 0.8^n$$

for the y_p part (0.8^n) is the same; so, the assumed solution is $y_p = nc_2 0.8^n$

$$c_2 = 8$$

$$y(n) = y_c + y_p$$

$$\text{but} \ y(n) = c_1 0.8^n + (n0.8^n \times 8)$$

$$= c_1 0.8^n + 8n0.8^n$$

$$\text{Using} \ y(0) = 6c_1 = 6$$

$$y(n) = 6(0.8)^n + 8n(0.8)^n$$

Solution 9.2(b)

$$y(n) - y(n-1) = 1 + 0.5^n$$

for the y_c part, we have equation

$$a^n - a^{n-1} = 0$$

$$a - 1 = 0$$

$$a = 1$$

$$y_c = c_1 (1)^n = c_1$$

for $y_p = nc_2 + c_3 (0.5)^n$
(to distinguish from c_1) by substitution into the original equation

$$(nc_2 + c_3 (0.5)^n) - (c_2 (n-1) + c_3 (0.5)^{n-1}) = 1 + 0.5^n$$

$$c_3 = -1 \ (\text{by comparing} \ 0.5^n)$$

$$\text{and} \ \ c_2 = 1 \ (\text{by comparing coefficients})$$

Total solution

$$y(n) = c_1 + nc_2 + c_3 0.5^n$$

$$y = c_1 + n - 0.5^n$$

using $y(0) = 1$

$$c_1 = 2 \ \text{and} \ y(n) = 2 + n - 0.5^n$$

Solution 9.2(c)

$$y(n-2) - 6y(n-2) + 5y(n) = 1n \geq 0y(-2) = 2y(-1) = 1$$

for $y_c = -5y(n-2) + 5y(n) = 0$ put $y(n) = a^n$

$$-5a^{n-2} + 5a^n = 0$$

$$5a^2 - 5 = 0$$

$$a = \pm 1$$

$$y_c(n) = c_1(1)^n + c_2(-1)^n$$
$$= c_1 + c_2(-1)^n$$

let $y_p = nc_3$ (to distinguish from c_1) by substitution

$$-5((n-2)c_3) + 5(nc_3) = 1$$
$$c_3 = 1/10 = 0.1$$
$$y(n) = y_p + y_c$$
$$= c_1 + c_2(-1)^n + 0.1n$$

using $y(-2) = 2$
$\quad y(-1) = 1$
we get

$$c_1 + c_2 = 2.2$$
$$c_1 - c_2 = 1.1$$
$$c_1 = 1.65, c_2 = 0.55$$
$$y(n) = 1.65 + 0.55(-1)^n + n(0.1)$$

Problem 9.3

Find the particular solution of the second-order difference equation (try it yourself):

$$8y(n) - 6y(n-1) + y(n-2) = 5\sin\left(\frac{n\pi}{2}\right), \quad n \geq 0$$

Problem 9.4

Solve the second-order difference equation (try it yourself)

$$y(n) + ky(n-2) = 0, \quad n \geq 0$$

for the two cases when $k = 1$ and $k = \frac{1}{4}$ with the initial conditions $y(0) = 1$ and $y(-1) = 0$

Problem 9.5

Determine the zero-input response of the system described by the second-order difference equation

$$x(n) - 3y(n-1) - 4y(n-2) = 0$$

Solution 9.5

With $x(n) = 0$, we have $y(n-1) + \frac{4}{3}y(n-2) = 0$

$$y(-1) = -\frac{4}{3}y(-2)$$
$$y(0) = \left(-\frac{4}{3}\right)^2 y(-2)$$
$$y(1) = \left(-\frac{4}{3}\right)^3 y(-2)$$
$$y(k) = \left(-\frac{4}{3}\right)^{k+2} y(-2) \leftarrow \text{Zero} - \text{input response}$$

Problem 9.6

Determine the particular solution of the difference equation

$$y(n) = \frac{5}{6}y(n--1) - \frac{1}{6}y(n-2) + x(n)$$

when the forcing function is $x(n) = 2^n u(n)$.

Solution 9.6

Consider the homogeneous equation:

$$y(n) - \frac{5}{6}y(n-1) + \frac{1}{6}y(n-2) = 0$$

The characteristic equation is $\lambda^2 - \frac{5}{6}\lambda + \frac{1}{6} = 0$, $\lambda = 1/2, 1/3$.
Hence, $y_h(n) = C_1\left(\frac{1}{2}\right)^n + C_2\left(\frac{1}{3}\right)^n$
The particular solution to $x(n) = 2^n u(n)$ is $y_p(n) = K(2^n)u(n)$. Substitute this solution into the difference equation.
Then, we obtain

$$K(2^n)u(n) - K\frac{5}{6}(2^{n-1})u(n-1) + k\frac{1}{6}2^{n-2}u(n-2) = 2^n u(n)$$

For $n = 2, 4k - \frac{5}{3}k + \frac{k}{6} = 4 \Rightarrow k = \frac{8}{5}$
Therefore, the total solution is

$$y(n) = y_p(n) + y_h(n) = \frac{8}{5}(2)^n u(n) + c_1\left(\frac{1}{2}\right)^n u(n) + c_2\left(\frac{1}{3}\right)^n u(n)$$

To determine c_1 and c_2, assume that $y(-2) = y(-1) = 0$.
Then, $y(0) = 1$ and $y(1) = \frac{5}{6}y(0) + 2 = 17/6$
Then, $\frac{8}{5} + c_1 + c_2 = 1 \Rightarrow c_1 + c_2 = -\frac{3}{5}$

$$\frac{16}{5} + \frac{1}{2}c_1 + \frac{1}{3}c_2 = \frac{17}{6} \Rightarrow 3c_1 + 2c_2 = -\frac{11}{5}$$

and, therefore, $c_1 = -1, c_2 = \frac{2}{5}$

The total solution is

$$y(n) = \left[\tfrac{8}{5}(2)^n - (\tfrac{1}{2})^n + \tfrac{2}{5}(\tfrac{1}{3})^n\right] u(n)$$

Problem 9.7

Determine the response y(n), n $\geq$ 0, of the system described by the second-order difference equation

$$y(n) - 3y(n-1) - 4y(n-2) = x(n) + 2x(n-1) \text{ to the input}$$
$$x(n) = 4^n u(n).$$

$$y(n) - 3y(n-1) - 4y(n-2) = x(n) + 2x(n-1)$$

Solution 9.7

The characteristic equation is $\lambda^2 - 3\lambda - 4 = 0$

Hence, $\lambda = 4, -1$, and

$$y_h(n) = c_1 4^n + c_2(-1)^n$$

we assume a particular solution of the form $y_p(n) = kn4^n u(n)$. Then

$$kn\, 4^n u(n) - 3k(n-1)4^{n-1}u(n-1) - 4k(n-2)4^{n-2}u(n-2)$$
$$= 4^n u(n) + 2(4)^{n-1}u(n-1)$$

For n $= 2, k(32 - 12) = 42 + 8 = 24 \Rightarrow k = \tfrac{6}{5}$ The total solution is

$$y(n) = y_p(n) + y_h(n) = [\tfrac{6}{5}n4^n + c_1 4^n + c_2(-1)^n]u(n)$$

To solve for c_1 and c_2, we assume that $y(-1) = y(-2) = 0$.

Then, $y(0) = 1$ and $y(1) = 3y(0) + 4 + 2 = 9$

Hence, $c_1 + c_2 = 1$ and $\tfrac{24}{5} + 4c_1 - c_2 = 9$

$$4c_1 - c_2 = \tfrac{21}{5}$$

Therefore, $c_1 = \tfrac{26}{25}$ and $c_2 = -\tfrac{1}{25}$

The total solution is

$$y(n) = [\tfrac{6}{5}n\, 4^n + \tfrac{26}{25}4^n - \tfrac{1}{25}(-1)^n]u(n)$$

Problem 9.8

Determine the impulse response of the following causal system:

$$y(n) - 3y(n-1) - 4y(n-2) = x(n) + 2x(n-1)$$

Solution 9.8

From Problem 4.7, the characteristic values are $\lambda = 4, -1$. Hence

$$y(n) = c_1 4^n + c_2 (-1)^n$$

When $x(n) = \delta(n)$, we find that $y(0) = 1$ and

$$y(1) - 3y(0) = 2 \quad \text{or} \quad y(1) = 5.$$

Hence $c_1 + c_2 = 1$ and $4c_1 - c_2 = 5$
This yields $c1 = 6/5$ and $c_2 = -1/5$
Therefore, $h(n) = [\frac{6}{5} 4^n - \frac{1}{5}(-1)^n] u(n)$

Problem 9.9

Determine the impulse response and the unit step response of the systems described by the difference equation

(a) $y(n) = 0.6y(n-1) - 0.08y(n-2) + x(n)$

(b) $y(n) = 0.7y(n-1) - 0.1y(n-2) + 2x(n) - x(n-2)$

Solution 9.9

(a) $y(n) - 0.6y(n-1) + 0.08y(n-2) = x(n)$
The characteristic equation is $\lambda^2 - -0.6\lambda + 0.08 = 0$;

$$\lambda = 0.2, 0.4.$$

Hence, $y(n) = c_1 (\frac{1}{5})^n + c_2 (\frac{2}{5})^n$ with $x(n) = \delta(n)$,
the initial conditions are $y(0) = 1$,

$$y(1) - 0.6y(0) = 0 \Rightarrow y(1) = 0.6$$

Hence, $c_1 + c_2 = 1$ and $\frac{1}{5}c_1 + \frac{2}{5}c_2 = 0.6 \Rightarrow c_1 = -1, c_2 = 2$
Therefore, $h(n) = [-(\frac{1}{5})^n + 2(\frac{2}{5})^n] u(n)$
The step response is

$$s(n) = \sum_{k=0}^{n} h(n-k), \quad n \geq 0$$

$$= \sum_{k=0}^{n} [2(\frac{2}{5})^{n-k} - (\frac{1}{5})^{n-k}]$$

$$= 2(\frac{2}{5})^n [(\frac{5}{2})^{n+1} - 1] u(n) - (\frac{1}{5})^n [(5)^{n+1} - 1] u(n)$$

(b) $y(n) - 0.7y(n-1) + 0.1y(n-2) = 2x(n) - x(n-2)$

The characteristic equation is $\lambda^2 - 0.7\lambda + 0.1 = 0 \lambda = \frac{1}{2}, \frac{1}{5}$.

Hence, $y(n) = c_1(\frac{1}{2})^n + c_2(\frac{1}{5})^n$ with $x(n) = \delta(n)$, we have $y(0) = 2$,

$$y(1) - 0.7y(0) = 0 \Rightarrow y(1) = 1.4$$

Hence, $c_1 + c_2 = 2$ and $\frac{1}{2}c_1 + \frac{1}{5}c_2 = 1.4 = \frac{7}{5} \Rightarrow c_1 + \frac{2}{5}c_2 = \frac{14}{5}$

These equations yield $c_1 = 10/3, c_2 = -4/3$

$$s(n) = \sum_{k=0}^{n} h(n-k) = \frac{10}{3}\sum_{k=0}^{n}(\tfrac{1}{2})^{n-k} - \frac{4}{3}\sum_{h=0}^{n}(\tfrac{1}{5})^{n-h}$$

$$= \frac{10}{3}(\tfrac{1}{2})^n \sum_{k=0}^{n} 2^h - \frac{4}{3}(\tfrac{1}{5})^n \sum_{k=0}^{n} 5^k$$

$$= \frac{10}{3}(\tfrac{1}{2})^n (2^{n+1} - 1)u(n) - \frac{1}{3}(\tfrac{1}{5})^n (5^{n+1} - 1)u(n)$$

10

Z-Transform

This chapter provides comprehensive details regarding the Z-transforms, properties of Z-transform, inverse Z-transforms and problems and solutions.

10.1 Introduction

Discrete transformation is a representation of discrete-time signals in the frequency domain or the conversion between time and frequency domain. The spectrum of a signal is obtained by decomposing it into its constituent frequency components using a discrete transform. Conversion between time and frequency domains is necessary in many digital signal processing (DSP) applications. For example, it allows for a more efficient implementation of the DSP algorithms, such as those for digital filtering, convolution and correlation.

10.2 Z-Transform

The Z-transform of a number sequence x(nT) or x(n) is defined as the power series in z^{-n} with coefficient equal to the values x(n), where z is a complex variable.

The Z-transform of a discrete time signal x(n) may be expressed as

$$Z[\{x(n)\}] = X(z) = \sum_{n=-\infty}^{\infty} x(n)z^{-n} \qquad (10.1)$$

The above expression is generally known as two sided Z-transform. The sequence $\{x(n)\}$ is generated from a time function x(t) by sampling every T seconds, x(n) is understood to be x (nT); that is, T is dropped for convenience.

If the discrete time signal is a causal signal $x(n) = 0$ for $n < 0$, then the Z-transform is called as one sided Z-transform and is expressed as

$$Z[\{x(n)\}] = X(z) = \sum_{n=0}^{\infty} x(n)z^{-n} \tag{10.2}$$

In fact, generally we assume that $x(n)$ is a causal discrete time signal unless it is stated. This means that generally we analyse the causal signal.

On the other hand, if $x(n)$ is a non-causal discrete time signal $x(n) = 0$ for $n \geq 0$, then its Z-transform is expressed as

$$Z[\{x(n)\}] = X(z) = \sum_{n=-\infty}^{-1} x(n)z^{-n} \tag{10.3}$$

10.2.1 Region of Convergence

The Z-transform of a sequence $x(n)$ is defined as

$$X(z) = \sum_{n=-\infty}^{\infty} x(n)z^{-n} \tag{10.4}$$

Let us express the complex variable z in polar form as $z = re^{j}$:

$$X(z)|_{z=re^{j\omega}} = \sum_{n=-\infty}^{\infty} [x(n)r^{-n}] e^{-j\omega n} \tag{10.5}$$

From the relationship in (10.5) we note that X(z) can be interpreted as the discrete time Fourier transform of the signal sequence $x(n)r^{-n}$. Now if $r = 1$ then $|z| = 1$, then (10.5) reduces to the discrete transform. Hence the expression in (10.5) will converge if $[x(n)r^{-n}]$ is absolutely summable.

Mathematically

$$\sum_{n=-\infty}^{\infty} |x(n)r^{-n}| \leq \infty \tag{10.6}$$

Hence for $x(n)$ to be finite, the magnitude of Z-transform, X(z) must also be finite.

The region of convergence (RoC) of X(z) is the set of all values of z for which X(z) attains a finite value. Thus any time we cite a Z-transform we should also indicate its RoC. Concepts of RoC are illustrated by some simple examples.

Example 10.1

Determine the Z-transform of the following finite-duration signals.

(a) $x_1(n) = \{1, 2, 5, 7, 0, 1\}$

(b) $x_2(n) = \{1, 2, \underset{\uparrow}{5}, 7, 0, 1\}$

(c) $x_3(n) = \{0, 0, 1, 2, 5, 7, 0, 1\}$

(d) $x_4(n) = \{2, 4, \underset{\uparrow}{5}, 7, 0, 1\}$

(e) $x_5(n) = \delta(n)$

(f) $x_6(n) = \delta(n - k), k > 0$

(g) $x_7(n) = \delta(n + k), K > 0$

Solution 10.1

(a) $X_1(z) = 1 + 2z^{-1} + 5z^{-2} + 7z^{-3} + z^{-5}$, RoC: entire z-plane except $z = 0$

(b) $X_2(z) = z^2 + 2z + 5 + 7z^{-1} + z^{-3}$, RoC: entire z-plane except $z = 0$ and $z = \infty$

(c) $X_3(z) = z^{-2} + 2z^{-3} + 5z^{-4} + 7z^{-5} + z^{-7}$, RoC: entire z-plane except $z = 0$

(d) $X_4(z) = 2z^{-2} + 4z + 5 + 7z^{-1} + z^{-3}$, RoC: entire z-plane except $z = 0$ and $z = \infty$

(e) $X_5(z) = 1[i.e., \delta(n) \overset{z}{\longleftrightarrow} 1], ROC : entire\ z - plane$

(f) $X_6(z) = z^{-k}[i.e., \delta(n-k) \overset{z}{\longleftrightarrow} z^{-k}], k > 0, ROC : entire\ z - plane$ except $z = 0$

(g) $X_7(z) = z^k[i.e., \delta(n + k) \overset{z}{\longleftrightarrow} z^k], k > 0, ROC : entire\ z - plane\ except\ z = \infty$

From this example it is easily seen that the RoC of a *finite-duration signal* is the entire z-plane, except possibly the points $z = 0$ and or $z = \infty$. These points are excluded, because z^k ($k > 0$) becomes unbounded for $z = \infty$ and $z^{-k}(k > 0)$ becomes unbounded for $z = 0$.

10.2.2 Properties of RoC

10.2.2.1 The RoC of X(z) consists of a ring or circle in the z-plane centred about the origin.

10.2.2.2 The RoC does not contain any pole.

10.2.2.3 The Z-transform X(z) converges uniformly if and only if the RoC of the Z-transform X(z) of a given discrete signal includes the unit circle. The RoC of X(z) consists of a ring in the z-plane centred about the origin. This means that the RoC of the Z-transform of x(n) has a value of z for which $[x(n)r^{-n}]$ is absolutely summable

$$\sum_{n=-\infty}^{\infty} \left| x(n)r^{-n} \right| \leq \infty$$

10.2.2.4 If the discrete time signal x(n) is of a finite duration, the RoC will be entire z-plane except $z = 0$ and $z = \infty$.

10.2.2.5 If the discrete time signal x(n) is a right-sided sequence, then the RoC will not include infinity.

10.2.2.6 If the discrete time signal x(n) is a left-sided sequence, then the RoC will not include $z = 0$. But in a case if $x(n) = 0$ for all $n > o$ then the RoC will include $z = 0$.

10.2.2.7 If the discrete time signal x(n) is a two-sided sequence, and if the circle $|z| = r_0$ is in the RoC, then the RoC will consist of a ring in the Z-plane that includes circle $|z| = r_0$. This means that the RoC will include the intersections of the RoC's complements.

10.2.2.8 If the Z-transform X(z) is rational, then the RoC will extend to infinity. It means that the RoC will be bounded by poles.

10.2.2.9 If the discrete time signal is causal, then the RoC will include $z = \infty$.

10.2.2.10 If the discrete time signal is anti-causal, then the RoC will include $z = 0$.

Example 10.2

Find the Z-transform of the discrete time unit impulse $\delta(n)$.

Solution 10.2

We know that the Z-transform is expressed as

$$Z[\{x(n)\}] = X(z) = \sum_{n=0}^{\infty} x(n)z^{-n}$$

We know that the unit impulse sequence $\delta(n)$ is a causal. The signal x(n) consists of a number at n=0, otherwise it is zero. The Z-transform of x(n) is the infinite power series

$$X(z) = 1 + (0)z^{-1} + (0)z^{-2} + \ldots (0)^n z^{-n} + \ldots$$

Therefore, the $Z[\delta(n)] = 1$.

Example 10.3

Find the Z-transform of the discrete time unit step signal u(n).

Solution 10.3

We know that the Z-transform is expressed as

$$Z\left[\{x(n)\}\right] = X(z) = \sum_{n=0}^{\infty} x(n)z^{-n}$$

We know that the unit step sequence u(n) is a causal. The Z-transform of x(n) is the infinite power series

$$X(z) = 1 + (1)z^{-1} + (1)z^{-2} + \dots (1)^n z^{-n} + \dots$$

This is an infinite geometric series. We recall that to sum an infinite series we use the following relation of the geometric series.

$$1 + A + A^2 + A^3 + \dots = \frac{1}{1-A} \quad \text{if } |A| < 1$$

Therefore

$$X(z) = \frac{1}{1 - z^{-1}} \quad ROC \quad |z| > 1$$

Example 10.4

Determine the Z-transform of the signal

$$x(n) = (\tfrac{1}{2})^n u(n)$$

Solution 10.4

The Z-transform of x(n) is the infinite power series

$$X(z) = 1 + \tfrac{1}{2}z^{-1} + (\tfrac{1}{2})^2 z^{-2} + \dots (\tfrac{1}{2})^n z^{-n} + \dots$$

$$= \sum_{n=0}^{\infty} (\tfrac{1}{2})^n z^{-n} = \sum_{n=0}^{\infty} (\tfrac{1}{2}z^{-1})^n$$

Consequently, for $|\tfrac{1}{2}z^{-1}| < 1$, or equivalently, for $|z| > \tfrac{1}{2}$, X(z) converges to

$$X(z) = \frac{1}{1 - \tfrac{1}{2}z^{-1}} \quad ROC \quad |z| > \frac{1}{2}$$

We see that in this case, the Z-transform provides a compact alternative representation of the signal x(n).

Example 10.5

Determine the Z-transform of the signal

$$x(n) = \alpha^n u(n) = \begin{cases} \alpha^n, & n \geq 0 \\ 0, & n < 0 \end{cases}$$

Solution 10.5

From the definition (4.1) we have

$$X(z) = \sum_{n=0}^{\infty} \alpha^n z^{-n} = \sum_{n=0}^{\infty} (\alpha z^{-1})^n$$

If $\alpha z^{-1} < 1$ or equivalently, $|z| > |\alpha|$, this power series converges to $1/(1 - \alpha z^{-1})$. Thus we have the Z-transform pair

$$x(n) = \alpha^n u(n) \overset{z}{\longleftrightarrow} X(z) = \frac{1}{1 - \alpha z^{-1}}, ROC : |z| > |\alpha| \qquad (1)$$

The RoC is the exterior of a circle having radius $|\alpha|$. Figure 10.1 shows a graph of the signal x(n) and its corresponding RoC. If we set $\alpha = 1$ in equation (1), we obtain the Z-transform of the unit step signal.

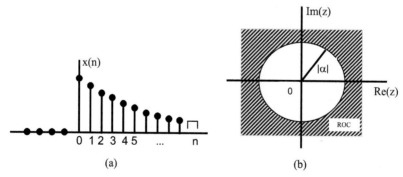

Figure 10.1   (a) The exponential signal $x(n) = \alpha^n u(n)$, (b) and the RoC of its Z-transform.

Example 10.6

An expression of discrete time $x(n) = e^{-anT}$ is given , find X(z).

Solution 10.6

$$X(z) = z[e^{anT}] = \sum_{n=0}^{\infty} x(n) z^{-n}$$

$$= 1 + e^{-aT} z^{-1} + e^{-2aT} z^{-2} + e^{-3aT} z^{-3} + \dots \dots$$

$$= 1 + (e^{-aT} z^{-1}) + (e^{-aT} z^{-1})^2 + \dots \dots$$

$$X(z) = \frac{1}{1 - e^{-aT} z^{-1}} = \frac{z}{z - e^{-aT}}$$

Example 10.7

Find the Z-transform of the function x(n), where $x(nT) = nT$.

Solution 10.7

$$z[\{x(nT)\}] = \sum_{n=0}^{\infty} x(nT) z^{-n} \quad z[\{x(nT)\}] = \sum_{n=0}^{\infty} nT z^{-n}$$

$$X(z) = 0 + Tz^{-1} + 2Tz^{-2} + 3Tz^{-3} + \dots$$

$$X(z) = T[z^{-1} + 2z^{-2} + 3z^{-3} + \dots]$$

$$= T \frac{z^2}{(z-1)^2} \quad - \quad Z - transform \ from \ table$$

10.3 Theorems and Properties of Z-Transform

10.3.1 Multiplication Property

The Z-transform of a number sequence multiplied by constant is equal to the constant multiplied by the Z-transform of the number sequence; that is

$$z[a\{x(n)\}] = az[\{x(n)\}] = aX(z) \tag{10.7}$$

$$Z[a\{x(n)\}] = \sum_{n=0}^{\infty} ax(n) z^{-n} = a \sum_{n=0}^{\infty} x(n) z^{-n} = aX(z)$$

Example 10.8
Determine the Z-transform and the RoC of the signal
$$x(n) = 10(2^n)u(n)$$

Solution 10.8
If we define the signals
$$x_1(n) = 2^n u(n)$$
then x(n) can be written as after its Z-transform
$$X(z) = 10X_1(z)$$

We recall that
$$a^n u(n) \xleftrightarrow{z} \frac{1}{1 - az^{-1}} \quad ROC : |z| > |a| \tag{1}$$

By setting $\alpha = 2$ in (1), we obtain
$$x_1(n) = 2^n u(n) \xleftrightarrow{z} X_1(z) = \frac{1}{1 - 2z^{-1}} \quad ROC : |z| > 2$$

The RoC of X (z) is $|z|| > 2$. Thus the transform X(z) is
$$x(z) = \frac{10}{1 - 2z^{-1}} \quad ROC : |z| > 2$$

Example 10.9
Determine the Z-transform and the RoC of the signal
$$x(n) = 10 \left(\frac{1}{2}\right)^n u(n)$$

Solution 10.9
If we define the signals
$$x_1(n) = \left(\frac{1}{2}\right)^n u(n)$$
then x(n) can be written as after its Z-transform
$$x_1(n) = \left(\frac{1}{2}\right)^n u(n) \xleftrightarrow{z} X_1(z) = \frac{1}{1 - \frac{1}{2}z^{-1}} \quad ROC : |z| > 2$$

Thus, the transform X(z) is
$$x(z) = \frac{10}{1 - \frac{1}{2}z^{-1}} \quad ROC : |z| > \frac{1}{2}$$

10.3.2 Linearity Property

The Z-transform of the sum of number sequences is equal to the sum of the Z-transform of the number sequence; that is

$$Z[\{x_1(n) + x_2(n)\}] = X_1(z) + X_2(z)$$

$$Z[a\{x_1(n) + x_2(n)\}] = a \sum_{n=0}^{\infty} [x_1(n) + x_2(n)] z^{-n}$$

$$X(z) = \sum_{n=0}^{\infty} x_1(n) z^{-n} + \sum_{n=0}^{\infty} x_2(n) z^{-n} = X_1(z) + X_2(z)$$

$$(10.8)$$

Example 10.10

Determine the Z-transform and the RoC of the signal

$$x(n) = [3(2^n) - 4(3^n)]u(n)$$

Solution 10.10

If we define the signals

$$x_1(n) = 2^n u(n), x_2(n) = 3^n u(n)$$

then x(n) can be written as

$$x(n) = 3x_1(n) - 4x_2(n)$$

Its Z-transform is

$$X(z) = 3X_1(z) - 4X_2(z)$$

We recall that

$$a^n u(n) \overset{z}{\longleftrightarrow} \frac{1}{1 - az^{-1}} \quad ROC : |z| > |a|$$

By setting $a = 2$ and $a = 3$, we obtain

$$x_1(n) = 2^n u(n) \overset{z}{\longleftrightarrow} X_1(z) = \frac{1}{1 - 2z^{-1}} \quad ROC : |z| > 2$$

$$x_2(n) = 3^n u(n) \overset{z}{\longleftrightarrow} X_2(z) = \frac{1}{1 - 3z^{-1}} \quad ROC : |z| > 3$$

The intersection of the RoC of $X_1(z)$ and $X_2(z)$ is $|z| > 3$. Thus, the transform $X(z)$ is

$$x(z) = \frac{3}{1 - 2z^{-1}} - \frac{4}{1 - 3z^{-1}} \quad ROC : |z| > 3$$

Example 10.11

A discrete time signal is expressed as x(n)= cos ωn for n ≥ 0. Determine its Z-transform.

Solution 10.11

$$\cos \omega n = \frac{1}{2}[e^{j\omega n} + e^{-j\omega n}]$$

The Z-transform is expressed as

$$Z[\{x(n)\}] = X(z) = \sum_{n=0}^{\infty} x(n)z^{-n}$$

$$Z[e^{j\omega n}] = \frac{1}{1 - e^{j\omega}z^{-1}} \quad for |z| > 1 \quad [e^{j\omega} = 1]$$

In the same way for n ≥ 0

$$Z[e^{-j\omega n}] = \frac{1}{1 - e^{-j\omega}z^{-1}} \quad for \quad |z| > 1 \quad [e^{-j\omega} = 1]$$

$$X(z) = Z[\frac{1}{2}\{e^{j\omega n} + e^{-j\omega n}\}]$$

$$X(z) = \frac{1}{2}Z\{e^{j\omega n}\} + \frac{1}{2}Z\{e^{-j\omega n}\}$$

$$X(z) = \frac{1}{2}[\frac{1}{1 - e^{j\omega}z^{-1}}] + \frac{1}{2}[\frac{1}{1 - e^{-j\omega}z^{-1}}]$$

$$X(z) = \frac{1}{2}[\frac{2 - (e^{j\omega} + e^{-j\omega})z^{-1}}{(1 - e^{j\omega}z^{-1})(1 - e^{-j\omega}z^{-1})}]$$

$$X(z) = \frac{1 - \cos\omega z^{-1}}{1 - 2z^{-1}\cos\omega + z^{-2}} = \frac{z^2 - z\cos\omega}{z^2 - 2z\cos\omega + 1} \quad for \quad |z| > 1$$

Example 10.12

A discrete time signal is expressed as $x(n) = \sin \omega n$ for n ≥ 0. Determine its Z-transform.

Solution 10.12

$$\sin \omega n = \frac{1}{2j}[e^{j\omega n} - e^{-j\omega n}]$$

The Z-transform is expressed as

$$Z[\{x(n)\}] = X(z) = \sum_{n=0}^{\infty} x(n)z^{-n}$$

$$Z[e^{j\omega n}] = \frac{1}{1 - e^{j\omega} z^{-1}} \quad for \quad |z| > e^{j\omega n}$$

or $\qquad Z[e^{j\omega n}] = \frac{1}{1 - e^{j\omega} z^{-1}} \quad for \quad |z| > 1 \quad [e^{j\omega} = 1]$

In the same way for n ≥ 0

$$Z[e^{-j\omega n}] = \frac{1}{1 - e^{-j\omega} z^{-1}} \quad for \quad |z| > e^{-j\omega n}$$

or $\ Z[e^{-j\omega n}] = \frac{1}{1 - e^{-j\omega} z^{-1}} \quad for \quad |z| > 1 \quad [e^{-j\omega} = 1]$

$$X(z) = Z[\frac{1}{2j}\{e^{j\omega n} - e^{-j\omega n}\}] \quad X(z) = \frac{1}{2j}Z\{e^{j\omega n}\} - \frac{1}{2j}Z\{e^{-j\omega n}\}$$

$$X(z) = \frac{1}{2j}[\frac{1}{1 - e^{j\omega} z^{-1}}] - \frac{1}{2j}[\frac{1}{1 - e^{-j\omega} z^{-1}}]$$

$$X(z) = \frac{z^{-1}(e^{j\omega} + e^{-j\omega})/2j}{(1 - e^{j\omega} z^{-1})(1 - e^{-j\omega} z^{-1})}$$

$$X(z) = \frac{z^{-1}\sin\omega}{1 - 2z^{-1}\cos\omega + z^{-2}} = \frac{z\sin\omega}{z^2 - 2z\cos\omega + 1} \quad for \quad |z| > 1$$

10.3.3 Time-Shifting Property

The time-shifting property states that

$$\text{If} \quad x(n) \xrightarrow{Z} X(z) \quad \text{then}$$

$$x(n - n_0) \xrightarrow{Z} z^{-n_0} X(z) \tag{10.9}$$

Example 10.13
Determine the Z-transform of

(a) $x_1(n) = \delta(n)$ (d) $x_4(n) = u(n)$

(b) $x_2(n) = \delta(n-k)$ (e) $x_5(n) = u(n-k)$

(c) $x_3(n) = \delta(n+k)$ (f) $x_6(n) = u(n+k)$

Solution 10.13

(a) $\quad x_1(n) = \delta(n)$ (d) $\quad x_4(n) = u(n)$

$\quad X_1(z) = 1$ $X_4(z) = \frac{1}{1-z^{-1}} = \frac{z}{z-1}$

(b) $\quad x_2(n) = \delta(n-k)$ (e) $\quad x_5(n) = u(n-k)$

$\quad X_2(z) = z^{-k}(1)$ $X_5(z) = z^{-k}\frac{1}{1-z^{-1}}$

$\quad X_2(z) = z^{-k}$ $X_5(z) = z^{-k}\left(\frac{z}{z-1}\right)$

(c) $\quad x_3(n) = \delta(n+k)$ (f) $\quad x_6(n) = u(n+k)$

$\quad X_3(z) = z^{+k}(1)$ $X_6(z) = z^{k}\frac{1}{1-z^{-1}}$

$\quad X_3(z) = z^{k}$ $X_6(z) = z^{k}\left(\frac{z}{z-1}\right)$

Example 10.14
A discrete time signal is expressed as $x(n) = \delta(n+1) + 2\delta(n) + 5\delta(n-3) - 2\delta(n-4)$. Determine its Z-transform.

Solution 10.14
According to linear property

$$X(z) = a_1 X_1(z) + a_2 X_2(z) + a_3 X_3(z) + a_4 X_4(z)$$

or $\quad X(z) = a_1 Z\{x_1(n)] + a_2 Z\{x_2(n)\} + a_3 Z\{x_3(n)\} + a_4 Z\{x_4(n)\}$

$\quad X(z) = Z\{\delta(n+1)\} + 2Z\{\delta(n)\} + 5Z\{\delta(n-3)\} - 2Z\{\delta(n-4)\}.$

$\quad X(z) = z + 2(1) + 5(z^{-3}) - 2(z^{-4})$

$\quad X(z) = z + 2 + 5z^{-3} - 2z^{-4}$

Example 10.15
By applying the time-shifting property, determine the x(n) of the signal.

$$X(z) = \frac{z^{-1}}{1 - 3z^{-1}}$$

Solution 10.15

$$X(z) = \frac{z^{-1}}{1 - 3z^{-1}} = z^{-1}X_1(z)$$

$$X_1(z) = \frac{1}{1 - 3z^{-1}}$$

$$x_1(n) = (3)^n u(n)$$

$$x(n) = (3)^{n-1} u(n-1)$$

Example 10.16

The Z-transform of a particular discrete time signal x(n) is expressed

$$X(z) = \frac{1 + \frac{1}{2}z^{-1}}{1 - \frac{1}{2}z^{-1}}$$

Determine x(n) using the time-shifting property.

Solution 10.16

$$X(z) = \frac{1 + \frac{1}{2}z^{-1}}{1 - \frac{1}{2}z^{-1}} = \frac{1}{1 - \frac{1}{2}z^{-1}} + \frac{\frac{1}{2}z^{-1}}{1 - \frac{1}{2}z^{-1}}$$

$$x(n) = Z^{-1}\left[\frac{1}{1 - \frac{1}{2}z^{-1}} + \frac{\frac{1}{2}z^{-1}}{1 - \frac{1}{2}z^{-1}}\right]$$

$$x(n) = Z^{-1}\{X(z)\} = \left[\frac{1}{1 - \frac{1}{2}z^{-1}} + \frac{\frac{1}{2}z^{-1}}{1 - \frac{1}{2}z^{-1}}\right]$$

$$x(n) = Z^{-1}\{X_1(z) + X_2(z)\}$$

$$X_1(z) = \frac{1}{1 - \frac{1}{2}z^{-1}} \qquad X_2(z) = \frac{\frac{1}{2}z^{-1}}{1 - \frac{1}{2}z^{-1}}$$

$$X_1(z) = \frac{1}{1 - \frac{1}{2}z^{-1}} = Z[(\frac{1}{2})^n u(n)]$$

$$X_2(z) = \frac{\frac{1}{2}z^{-1}}{1 - \frac{1}{2}z^{-1}} = \frac{1}{2}Z^{-1}[X_1(z)]$$

Now using the time-shifting property

$$\text{If} \quad x(n) \xrightarrow{Z} X(z) \quad \text{then}$$

$$x(n - n_0) \xrightarrow{Z} z^{-n_0} X(z)$$

$$X_2(z) = \frac{1}{2} Z^{-1}[X_1(z)]$$

$$X_2(z) = \frac{1}{2} Z[(\frac{1}{2})^{n-1} u(n - 1)]$$

Substituting the values of $X_1(z)$ and $X_2(z)$

$$x(n) = (\frac{1}{2})^n u(n) + \frac{1}{2}(\frac{1}{2})^{n-1} u(n - 1)]$$

$$x(n) = (\frac{1}{2})^n \{u(n) + u(n - 1)\}$$

or
$$x(n) = (\frac{1}{2})^n \{\delta(n) + 2u(n - 1)\}$$

Example 10.17
Given the Z-transform of x(n) as

$$X(z) = \frac{z}{z^2 + 4}$$

Use the Z-transform properties to determine $y(n) = x(n - 4)$.

Solution 10.17
$X(z) = \frac{z}{z^2+4}$, now using the time-shifting property

$$Z[x(n - 4)] = z^{-4}[\frac{z}{z^2 + 4}] = [\frac{z^{-3}}{z^2 + 4}]$$

10.3.4 Scaling Property

$$x(n) \xrightarrow{Z} X(z) \quad a^n x(n) = X(a^{-1}z) \qquad (10.10)$$

Example 10.18
Determine the Z-transform of $x(n) = (-1)^n u(n)$.

Solution 10.18
In the given problem x(n) = $(-1)^n$u(n), here we take x(n) = u(n) and
$X(z) = \frac{1}{1-z^{-1}}$

For scaling the function, we place $z = a^{-1}z$

$$X(z) = \frac{1}{1 - \frac{1}{a^{-1}z}}. \text{ Now, place a} = -1$$

$$X(z) = \frac{1}{1 - \frac{1}{(-1)^{-1}z}}$$

$$Z(-1)^n u(n) = X(z) = \frac{1}{1 + z^{-1}}$$

Example 10.19

A discrete time signal is expressed as $x(n) = 2^n u(n - 2)$. Determine its Z-transform.

Solution 10.19

In the given problem $x(n) = 2^n u(n - 2)$.

Now, $Z\{u(n)\} = \frac{1}{1-z^{-1}}$

Therefore, using the time-shifting property we have

$$Z\{u(n - 2)\} = z^{-2} \frac{1}{1 - z^{-1}}$$

Now to find $Z\{2^n u(n-2)\}$ we shall use the scaling property, which states that

$$x(n) \xrightarrow{Z} X(z) \quad a^n x(n) = X(a^{-1}z)$$

$$Z\{u(n - 2)\} = z^{-2} \frac{1}{1 - z^{-1}} \quad \text{then}$$

$$Z\{2^n u(n - 2)\} = \frac{(2^{-1}z)^{-2}}{1 - (2^{-1}z)^{-1}} = \frac{(2\,z^{-1})^2}{1 - 2\,z^{-1}} = \frac{4\,z^{-2}}{1 - 2\,z^{-1}}$$

Example 10.20

Using the scaling property, determine the Z-transform of $\cos \omega n$ for $n \geq 0$.

Solution 10.20

$$Z\{\cos \omega n\} = \frac{1 - \cos \omega z^{-1}}{1 - 2z^{-1} \cos \omega + z^{-2}} = \frac{z^2 - z \cos \omega}{z^2 - 2z \cos \omega + 1}$$

Using the scaling property

$$Z\{a^n \cos \omega n\} = \frac{1 - a\,z^{-1} \cos \omega}{1 - 2az^{-1} \cos \omega + a^2 z^{-2}} = \frac{z^2 - az \cos \omega}{z^2 - 2az \cos \omega + a^2}$$

10.3.5 Time Reversal Property

$$x(n) \xrightarrow{Z} X(z) \text{ and } x(-n) \xrightarrow{Z} X(z^{-1}) \qquad (10.11)$$

Example 10.21
Determine the Z-transform of $x(n) = u(-n)$.

Solution 10.21

$$x_1(n) = u(-n)$$

$$\text{We take} \quad x(n) = u(n) \quad or \quad X(z) = \frac{1}{1 - z^{-1}}$$

$$x(n) = u(-n) \quad or \quad X(z) = \frac{1}{1 - z}$$

Example 10.22
Determine the Z-transform of $x(n) = \left(\frac{1}{2}\right)^n u(-n)$.

Solution 10.22

$$X(z) = \sum_{n=0}^{\infty} \left(\frac{1}{2}\right)^n u(-n) z^{-n}$$

$$X(z) = \sum_{n=0}^{\infty} \left(\frac{1}{2}\right)^{=n} z^n = \sum_{n=0}^{\infty} (2z)^n$$

If $2Z < 1$, the sum converges and $X(z) = \frac{1}{1-2z}$
  $2Z < 1$ or $|z| < \left|\frac{1}{2}\right|$ ‖

Example 10.23
Determine the Z-transform of the signal

$$x(n) = -\alpha^n u(-n-1) = \begin{cases} 0, & n \geq 0 \\ -\alpha^n, & n \leq -1 \end{cases}$$

Solution 10.23
From the definition (10.1), we have

$$X(z) = \sum_{n=-\infty}^{-1} (-\alpha^n) z^{-n} = -\sum_{L=1}^{\infty} (\alpha^{-1} z)^L$$

where $L = -n$. Using the formula

$$A + A^2 + A^3 + \ldots = A(1 + A + A^2 + \ldots) = \frac{A}{1 - A}$$

When $|A| < 1$ gives

$$X(z) = -\frac{\alpha^{-1}z}{1 - \alpha^{-1}z} = \frac{1}{1 - \alpha z^{-1}}$$

Provided that $|\alpha^{-1}z| < 1$ or equivalently, $|z| < |\alpha|$. Thus

$$x(n) = -\alpha^n u(-n - 1) \xleftrightarrow{z} X(z) = \frac{1}{1 - \alpha(z)^{-1}}, \quad ROC : |z| < |\alpha|$$

The RoC is now the interior of a circle having radius $|\alpha|$. This is shown in Figure 10.2.

Example 10.24

Given the Z-transform of x(n) as

$$X(z) = \frac{z}{z^2 + 4}$$

Use the Z-transform properties to determine $y(n) = x(-n)$.

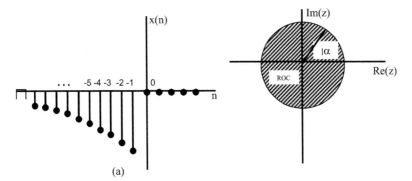

(a)

Figure 10.2 (a) Anticausal signal $x(n) = -\alpha^n u(-n - 1)$ (b) and the ROC of its z-transform.

Solution 10.24

$$x(-n) \xrightarrow{Z} X(z^{-1})$$

Using time reversal property

$$Y(z) = [\frac{z^{-1}}{z^{-2}+4}] = [\frac{z}{1+4z^2}]$$

10.3.6 Differentiation Property

$$x(n) \xrightarrow{Z} X(z)$$

$$nx(n) \xrightarrow{Z} -z\frac{d[X(z)]}{dz} \qquad (10.12)$$

Example 10.25

Determine the Z-transform of $x(n) = nu(n)$.

Solution 10.25

$$x(n) = nu(n), \text{ using the property of derivative}$$

$$nx(n) = -z\frac{d[X(z)]}{dz} \; u(n), \text{ here } x(n) = u(n)$$

$$X(z) = \frac{1}{1-z^{-1}}$$

Substituting X(z) in (10.7) and taking the Z-transform
Using the property of derivative, we have

$$X(z) = -z\frac{d}{dz}[Z\{u(n)\}]$$

Again using the differentiation property for the factor in bracket, we get

$$X(z) = -z\frac{d}{dz}\frac{z}{(z-1)}$$

$$X(z) = -z\frac{d}{dz}\frac{z}{(z-1)} = -z.\frac{(z-1)-z.1}{(z-1)^2}$$

$$X(z) = \frac{z}{(z-1)^2}$$

Example 10.26
A discrete time signal is expressed as $x(n) = n^2 \ u(n)$. Determine its Z-transform.

Solution 10.26
Given that $x(n) = n^2 u(n)$

$$nx(n) = -zX(z) = Z\{x(n)\} = Z\{n^2u(n)\} = Z\{n(n\,u(n)\}$$

Using the property of derivative, we have

$$X(z) = -z\frac{d}{dz}[Z\{nu(n)\}]$$

Again using the differentiation property for the factor in bracket, we get

$$X(z) = -z\frac{d}{dz}\frac{z}{(z-1)^2}$$

$$X(z) = -z\frac{d}{dz}\frac{z}{(z-1)^2} = -z.\frac{(z-1)^2 - 2z^2 + 2z}{(z-1)^4}$$

$$X(z) = z\frac{(z^2-1)}{(z-1)^4}$$

Example 10.27
Given the Z-transform of x(n) as

$$X(z) = \frac{z}{z^2+4}$$

Use the Z-transform properties to determine $y(n) = nx(n)$ in terms of Y(z).

Solution 10.27

$$x(n) \xrightarrow{\ Z\ } X(z) = \frac{z}{z^2+4}$$

$$nx(n) \xrightarrow{\ Z\ } -z\frac{d[X(z)]}{dz}$$

Using the differentiation property

$$Y(z) = -z\frac{d}{dz}[\frac{z}{z^2+4}] = -z[\frac{z^2+4-z(2z)}{(z^2+4)^2}]$$

$$Y(z) = -z[\frac{-z^2+4}{(z^2+4)^2}] = z[\frac{z^2-4}{(z^2+4)^2}]$$

Example 10.28

Determine the Z-transform of the signal

$$x(n) = n[(\frac{1}{2})^n u(n) * (\frac{1}{2})^n u(n)]$$

Solution 10.28

Let $y(n) = [(\frac{1}{2})^n u(n) * (\frac{1}{2})^n u(n)]$ $Z\{y(n)\} = \frac{1}{(1-(\frac{1}{2})z^{-1})^2}$

$$x(n) = ny(n)$$

$$X(z) = -z\frac{d}{dz}\frac{1}{\left[1 - (\frac{1}{2})z^{-1}\right]^2} = \frac{8z^2(6z-1)}{(2z-1)^4}$$

10.3.7 Convolution Property

If $x_1(n) \xrightarrow{Z} X_1(z)$ and $x_2(n) \xrightarrow{Z} X_2(z)$

Then   (10.13)

$$x(n) = x_1(n) * x_2(n) \xrightarrow{Z} X(z) = X_1(z).X_2(z)$$

Example 10.29

Compute the convolution of two signals.

$$x_1(n) = \{1, -2, 1\}$$

$$x_2(n) = \begin{cases} 1, & 0 \le n \le 5 \\ 0, & elsewhere \end{cases}$$

Solution 10.29

$$x_1(n) = \{\underset{\uparrow}{1}, -2, 1\}$$

$$x_2(n) = \{\underset{\uparrow}{1}1, 1, 1, 1, 1\}$$

$$X_1(z) = 1 - 2z^{-1} + z^{-2} \quad X_2(z) = 1 + z^{-1} + z^{-2} + z^{-3} + z^{-4} + z^{-5}$$

Using the property of convolution in Z-domain, we carry out the multiplication:

$$X(z) = X_1(z).X_2(z)$$

$$X(z) = (1 - 2z^{-1} + z^{-2})(1 + z^{-1} + z^{-2} + z^{-3} + z^{-4} + z^{-5})$$

$$X(z) = 1 - z^{-1} - z^{-6} + z^{-7}$$

Hence

$$x(n) = \{\underset{\uparrow}{1}, -1, 0, 0, 0, 0, -1, 1\}$$

Example 10.30
A discrete time signal is given by the expression

$$x(n) = n(-\frac{1}{2})^n u(n) \otimes (\frac{1}{4})^{-n} u(-n)$$

Find the Z-transform.

Solution 10.30

$$x(n) = x_1(n) \otimes x_2(n)$$

$$x_1(n) = n(-\frac{1}{2})^n u(n)$$

$$Z\{u(n)\} = \frac{1}{1 - z^{-1}}$$

Using the scaling property, we have

$$Z[(-\frac{1}{2})^n u(n)] = \frac{z}{z + \frac{1}{2}} \qquad ROC \, |z| > \frac{1}{2}$$

Now, using the Z-domain differentiation property

$$Z[n(-\frac{1}{2})^n u(n)] = -z\frac{d}{dz}[\frac{z}{z + \frac{1}{2}}] = \frac{-\frac{1}{2}z}{(z + \frac{1}{2})^2}$$

$$x_2(n) = (\frac{1}{4})^{-n} u(-n)$$

$$Z\{u(n)\} = \frac{1}{1 - z^{-1}}$$

Using the scaling property

$$Z[(\frac{1}{4})^{-n} u(n)] = Z[(4)^n u(n)] = \frac{1}{1-(4)z^{-1}} = \frac{z}{z-4} \quad ROC \, |z| > \frac{1}{4}$$

Using the time reversal property

$$Z[(\frac{1}{4})^{-n} u(-n)] = \frac{z^{-1}}{z^{-1}-4} \quad ROC \, \left|\frac{1}{z}\right| > \frac{1}{4}$$

$$Z[(\frac{1}{4})^{-n} u(-n)] = \frac{1}{1-4z} \quad ROC \, |z| < 1/4$$

We know that convolution in time domain is equal to multiplication in frequency domain:

$$x(n) = n(-\frac{1}{2})^n u(n) \otimes (\frac{1}{4})^{-n} u(-n)$$

$$Z[x(n)] = Z[n(-\frac{1}{2})^n u(n)] Z[(\frac{1}{4})^{-n} u(-n)]$$

$$X(z) = \frac{-\frac{1}{2}z}{(z+\frac{1}{2})^2 (z-\frac{1}{4})} = -\frac{\frac{1}{8}z}{(z+\frac{1}{2})^2(z-\frac{1}{4})} \quad ROC \quad \frac{1}{2} < |z| < \frac{1}{4}$$

Example 10.31
Determine the Z-transform of the signal

$$x(n) = [(\frac{1}{2})^n u(n) * (\frac{1}{2})^n u(n)]$$

Solution 10.31

$$x(n) = [(\frac{1}{2})^n u(n) * (\frac{1}{2})^n u(n)]$$

$$Z\{x(n)\} = \frac{1}{(1-(\frac{1}{2})z^{-1})^2}$$

10.3.8 Correlation Property

Correlation property states that

If $x_1(n) \leftrightarrow X_1(z)$

and $x_2(n) \leftrightarrow X_2(z)$

(10.14)

$$r_{x1x2}(j) = \sum_{n=-\infty}^{\infty} x_1(n)x_2(n-j) \leftrightarrow X_1(z)X_2(z^{-1})$$

Example 10.32

Compute the cross-correlation sequence $r_{x1x2}(j)$ of the sequences

$$X_1(n) = \{1, 2, 3, 4\}$$
$$X_2(n) = \{4, 3, 2, 1\}$$

Solution 10.32

The cross-correlation sequence can be obtained using its correlation property of the Z-transform.

Therefore, for the given $x_1(n)$ and $x_2(n)$, we have

$$X_1(z) = 1 + 2z^{-1} + 3z^{-2} + 4z^{-3} X_2(z)$$

$$= 4 + 3z^{-1} + 2z^{-2} + z^{-3}$$

Thus $X_2(z^{-1}) = 4 + 3z + 2z^2 + z^3$

Using the property of convolution in the Z-domain, we carry out the multiplication:

$$R_{x1x2}(z) = X_1(z)X_2(z^{-1})$$

$$= (1 + 2z^{-1} + 3z^{-2} + 4z^{-2})(4 + 3z + 2z^2 + z^3)$$

$$R_{x1x2}(z) = X_1(z)X_2(z^{-1})$$

$$= (z^3 + 4z^2 + 10z + 20 + 22z^{-1} + 24z^{-2} + 12z^{-3})$$

10.3.9 Initial Value Theorem

Initial value theorem states that if $x(n)$ is causal discrete time signal with the Z-transform $X(z)$, then initial value may be determined by using the following expression:

$$x(0) = \underset{n \to 0}{Lim}\, x(n) = \underset{|z| \to \infty}{Lim}\, X(z) \tag{10.15}$$

10.3.10 Final Value Theorem

Final value theorem states that for a discrete time signal $x(n)$, if $X(z)$ and the poles of $X(z)$ are all inside the unit circle, then the final value of the discrete time signal, x(∞) may be determined using the following expression:

$$x(\infty) = \underset{n\to\infty}{Lim}\, x(n) = \underset{|z|\to 1}{Lim}\, [(1 - z^{-1})X(z)] \qquad (10.16)$$

Example 10.33
Given the Z-transform of any signal $X(z) = 2 + 3z^{-1} + 4z^{-2}$. Determine the initial and final values of the corresponding discrete time signals.

Solution 10.33
The given expression is $X(z) = 2 + 3z^{-1} + 4z^{-2}$.
    We know that the initial value is given as $x(0) = \underset{n\to\infty}{Lim}\, x(n) = \underset{|z|\to\infty}{Lim}\, X(z)$

$$x(0) = \underset{n\to\infty}{Lim}\, x(n) = \underset{|z|\to\infty}{Lim}\, (2 + 3z^{-1} + 4z^{-2}) = 2 + 0 + 0 = 2$$

Also the final value is given as $x(\infty) = \underset{n\to\infty}{Lim}\, x(n) = \underset{|z|\to 1}{Lim}\, [(1 - z^{-1})X(z)]$

$$x(\infty) = \underset{n\to\infty}{Lim}\, x(n) = \underset{|z|\to 1}{Lim}\, [(1 - z^{-1})(2 + 3z^{-1} + 4z^{-2})]$$

$$x(\infty) = \underset{|z|\to 1}{Lim}\, [(2 + z^{-1} + z^{-2} - 4z^{-3})] = 2 + 1 + 1 - 4 = 0$$

10.3.11 Time Delay Property (One Sided Z-Transform)

The time delay property states that

$$\text{If} \quad x(n) \xrightarrow{Z} X(z) \quad \text{then}$$

$$x(n - k) \xrightarrow{Z} z^{-k} [X(z) + \sum_{n=1}^{k} x(-n)z^{-n}]_{k>0} \qquad (10.17)$$

10.3.12 Time Advance Property

The time advance property states that

$$\text{If} \quad x(n) \xrightarrow{Z} X(z) \quad \text{then} \qquad\qquad (10.18)$$

$$x(n + k) \xrightarrow{Z} z^{k} [X(z) - \sum_{n=0}^{k-1} x(n)z^{-n}]$$

Example 10.34

Determine the response of the following system:

$x(n+2) - 3x(n+1) + 2x(n) = \delta(n)$. Assume all the initial conditions are zero.

Solution 10.34

Given system is $x(n+2) - 3x(n+1) + 2x(n) = \delta(n)$

Taking the Z-transform of both the sides of the above equation, we obtain $X(z)[z^2 - 3z + 2] = 1$.

$$X(z) = \frac{1}{(z-1)(z-2)} = \frac{A}{(z-2)} + \frac{B}{(z-1)}$$

$$X(z) = \frac{1}{(z-2)} + \frac{1}{(z-1)}$$

$$x(n) = Z^{-1}\left[\frac{z^{-1}}{(1-2z^{-1})}\right] + Z^{-1}\left[\frac{z^{-1}}{(1-z^{-1})}\right]$$

$$x(n) = 2^{n-1} + (1)^{n-1} = 1 + (2)^{n-1}$$

Example 10.35

Determine the Z-transform of the following signal:

$$x(n) = \left(\frac{1}{2}\right)^n [u(n) - u(n-10)].$$

Solution 10.35

$$X(z) = \sum_{n=-\infty}^{\infty} \left(\frac{1}{2}\right)^n [u(n) - u(n-10)]z^{-n}$$

$$X(z) = \sum_{n=-\infty}^{\infty} \left(\frac{1}{2z}\right)^n - \sum_{n=-10}^{\infty} \left(\frac{1}{2z}\right)^n = \sum_{n=0}^{9} \left(\frac{1}{2z}\right)^n$$

$$X(z) = \frac{1 - \left(\frac{1}{2z}\right)^{10}}{1 - \left(\frac{1}{2z}\right)} = \frac{(2z)^{10} - 1}{2z - 1} \cdot \frac{1}{(2z)^9}$$

$$X(z) = \frac{1 - \left(\frac{1}{2z}\right)^{10}}{1 - \left(\frac{1}{2z}\right)} = \frac{z^{10} - \left[\frac{1}{2}\right]^{10}}{z - \left[\frac{1}{2}\right]} \cdot \frac{1}{(z)^9}$$

10.4 Inverse Z-Transform (Residue Method)

In this method, we obtain the inverse Z-transform x(n) by summing the residues at all poles:

$$Residue = \frac{1}{(m-1)!} \lim_{z \to \alpha} \left[\frac{d^{m-1}}{dz^{m-1}} \{(z-\alpha)^m X(z)\} \right] \qquad (10.19)$$

The function X(z) may be expanded into partial fractions in the same manner as used with the Laplace transforms. The Z-transform tables may be used to find the inverse Z-transform.

We generally expand the function $X(z)/Z$ into partial fractions, and then multiply z to obtain the expansion in the proper form.

10.4.1 When the Poles are Real and Non-repeated

Example 10.36
E(z) is given by the following transfer function, find e(k).

$$E(z) = \frac{z}{z^2 - 3z + 2}$$

Solution 10.36

$$E(z) = \frac{z}{(z-1)(z-2)}$$

We expand $\frac{E(z)}{z}$ into partial fractions, with the result

$$\frac{E(z)}{Z} = \frac{1}{(z-1)(z-2)} = \frac{-1}{z-1} + \frac{1}{z-2}$$

$$\{E(z)\} = \left[\frac{-z}{z-1} \right] + \left[\frac{z}{z-2} \right]$$

$$e(k) = (-1 + 2^k)u(k)$$

Example 10.37
Solve the following difference equation for x(k) using the Z-transform method for k ≤ 4.

$$x(k) - 4x(k-1) + 3x(k-2) = e(k)$$

where $e(k) = \begin{cases} 2, & k = 0,1 \\ 0, & k \geq 2 \end{cases}$ $\qquad x(-2) = x(-1) = 0.$

Solution 10.37

Taking transform of the difference equation

$$X(z) = E(z) = 2(1 + z^{-1})$$

$$X(z) = \frac{2(1 + z^{-1})}{1 - 4z^{-1} + 3z^{-2}} = \frac{2z(z+1)}{z^2 - 4z + 3}$$

$$\frac{X(z)}{z} = \frac{2(z+1)}{(z-1)(z-3)} = \frac{-2}{z-1} + \frac{4}{z-3}$$

$$X(z) = -\frac{2z}{z-1} + \frac{4z}{z-3}$$

$$x(k) = [-2 + 4(3)^k]u(k)$$

Example 10.38
Given the difference equation: $y(k+2) - 0.75\, y(k+1) + 0.125\, y(k) = e(k)$
where $e(k) = 1, k \geq 0$, and $y(0) = y(1) = 0$.

Solution 10.38

$$Y(z) = \frac{E(z)}{(z - 0.25)(z - 0.5)} \qquad E(z) = \frac{z}{z-1}(given)$$

$$y(z) = \frac{z}{(z-1)(z - \frac{1}{4})(z - 1/2)}$$

$$\frac{y(z)}{z} = \frac{1}{(z-1)(z - 1/4)(z - 1/2)} = \frac{8/3}{z-1} + \frac{16/3}{z - 1/4} + \frac{-8}{z - 1/2}$$

$$y(k) = 8/3 + \tfrac{16}{3}(\tfrac{1}{4})^k - 8(+1/2)^k$$

Example 10.39
Find the inverse Z-transform of each given $E(z)$ by two methods and compare
the value of $e(k)$ for $k = 0, 1, 2$ and 3 obtained by the two methods.

$$E(z) = \frac{z}{(z-1)(z - 0.8)}$$

Solution 10.39

$$E(z) = \frac{z}{(z-1)(z-0.8)}$$

(i)
$$z^2 - 1.8z + 0.8 \overline{\smash{\big)}\, z}$$
$$\frac{z^{-1} + 1.8z^{-2} + 2.44z^{-3}}{z - 1.8 + 0.8z^{-1}}$$
$$\frac{1.8 - 0.8z^{-1}}{1.8...}$$

$$e(0) = 0, e(1) = 1, e(2) = 1.8e(3) = 2.44$$

(ii) $\frac{E(z)}{z} = \frac{1}{(z-1)(z-0.8)} = \frac{5}{z-1} + \frac{-5}{z-0.8}$

$$E(z) = \frac{5z}{z-1} - \frac{5z}{z-0.8},$$

$$z^{-1}\{E(z)\} = 5z^{-1}\left\{\frac{z}{z-1}\right\} - 5z^{-1}\left[\frac{z}{z-0.8}\right]$$

$$e(k) = 5 - 5(0.8)^k, e(k) = 5\left[1 - (0.8)^k\right]$$

$$k = 0, \quad e(0) = 5[1 - (0.8)^0] = 0$$
$$k = 1, \quad e(1) = 5[1 - (0.8)^1] = 1$$
$$k = 2, \quad e(2) = 5[1 - (0.8)^2] = 1.8$$
$$k = 3, \quad e(3) = 5[1 - (0.8)^3] = 2.44$$

Example 10.40
Find the inverse Z-transform of each given E(z) by two methods and compare the value of e(k) for k $=$ 0, 1, 2 and 3 obtained by the two methods.

$$E(z) = \frac{z(z+1)}{(z-1)(z-0.8)} = \frac{z^2+z}{z^2-1.8z+0.8}$$

Solution 10.40

(i)
$$z^2 - 1.8z + 0.8 \overline{\smash{\big)}\, z^2 + z}$$
$$\frac{1 + 2.8z^{-1} + 4.24z^{-2} + 5.39z^{-3}}{z^2 - 1.8z + 0.8}$$
$$\frac{z^2 - 1.8z + 0.8}{2.8z - 0.8}$$
$$\frac{2.8z - 5.04 + 2.24z^{-1}}{4.24 - 2.24z^{-1}}$$
$$4.24$$

$$e(0) = 1, e(1) = 2.8, e(2) = 4.24, e(3) = 5.392$$

(ii) $\frac{E(z)}{z} = \frac{z+1}{(z-1)(z-0.8)} \Rightarrow \frac{10}{z-1} + \frac{-9}{z-0.8}$

$$z^{-1}\{E(z)\} = 10z^{-1}\left\{\frac{z}{z-1}\right\} - 9z^{-1}\left[\frac{z}{z-0.8}\right]$$

$$e(k) = 10 - 9(0.8)^k$$

$$k = 0, \quad e(0) = 10 - 10 = 1$$
$$k = 1, \quad e(1) = 10 - 7.2 = 2.8$$
$$k = 2, \quad e(2) = 10 - 9(.8)^2 = 4.24$$
$$k = 3, \quad e(2) = 10 - 9(.8)^3 = 5.392$$

10.4.2 When the Poles are Real and Repeated

Example 10.41
Find the inverse Z-transform of each given E(z) by two methods and compare the value of e(k) for k = 0, 1, 2 and 3 obtained by the two methods.

$$E(z) = \frac{1}{z(z-1)(z-0.8)} = \frac{1}{z^3 - 1.8z^2 + 0.8z}$$

Solution 10.41

(i) $z^3 - 1.8z^2 + 0.8z \overline{)1}^{\displaystyle z^{-3} + 1.8z^{-4} + 2.44z^{-5}}$

$$e(0) = 2.1815 + 5 - 7.813 = 0$$
$$e(1) = 1.25 + 5 - 7.813(0.8) = 0$$
$$e(k) = 5 - 7.813(0.8)^k k \geq 2$$

(ii) $E(z) = \frac{1}{z^3 - 1.8z^2 + 0.8z} = \frac{E(z)}{z} = \frac{1}{z^2(z-1)(z-0.8)}$

$$\frac{E(z)}{z} = \frac{A_{21}}{z^2} + \frac{A_{11}}{z} + \frac{B}{z-1} + \frac{C}{z-0.8}$$

$$A_{21} = z^2 \cdot \frac{1}{z^2(z-1)(z-0.8)}\Big/z=0 = 1.25$$

$$A_{11} = \frac{d}{dz}\left[z^2 \frac{1}{z^2(z-1)(z-0.8)}\right]$$

$$\frac{-(1)(2z-1.8)}{(z^2-1.8z+0.8)^2}\Big/z=0 = \frac{1.8}{0.8^2} = 2.815$$

$$= \frac{1.25}{z^2} + \frac{2.815}{z} + \frac{5}{z-1} + \frac{-7.813}{z-0.8}$$

$$e(k) = z^{-1}\left\{E(z)\right\} = \frac{1.25}{z} + 2.815 + \frac{5z}{z-1} - \frac{7.813\,z}{z-0.8}$$

$$e(k) = 1.25\,\delta(k-1) + 2.815\,\delta(k) + 5u(k) - 7.813(0.8)^k u(k)$$

Example 10.42

Find the inverse Z-transform of each given E(z) by two methods and compare the value of e(k) for k = 0, 1, 2 and 3 obtained by the two methods.

$$E(z) = \frac{1}{(z-1)(z-0.8)} = \frac{1}{z^2 - 1.8z + 0.8}$$

Solution 10.42

(i) $z^2 - 1.8z + 0.8\overline{)1}$ $\dfrac{z^{-2} + 1.8z^{-3} + 2.44z^{-4}}{}$

$$\frac{1 - 1.8z^{-1} + 0.8z^{-2}}{1.8z^{-1} - 0.8z^{-2}}$$

$$\frac{1.8z^{-1} - 3.24z^{-2} + 1.44z^3}{2.44z^{-2} - 1.44z^{-3}}$$

$$e(0) = 0, e(1) = 0, e(2) = 1, e(3) = 1.8, e(4) = 2.44.$$

(ii) $\dfrac{E(z)}{z} = \dfrac{1}{z(z-1)(z-0.8)} = \dfrac{A}{z} + \dfrac{B}{z-1} + \dfrac{C}{z-0.8}$

$$A = \underset{z\to 0}{Lim}\ z.\frac{1}{z(z-1)(z-0.8)} = \frac{1}{(-1)(-0.8)} = 1.25$$

$$B = \underset{z\to 1}{Lim}\ (z-1).\frac{1}{z(z-1)(z-0.8)}\Big/ = \frac{1}{1(0.2)} = 5$$

$$C = \underset{z\to 0.8}{Lim}\ (z-0.8)\frac{1}{z(z-1)(z-0.8)} = \frac{1}{(0.8)} = -6.25$$

$$\frac{E(z)}{z} = \frac{1.25}{z} + \frac{5}{z-1} - \frac{6.25}{z-0.8}$$

$$E(z) = 1.25 + 5\frac{z}{z-1} - 6.25\frac{z}{z-0.8}$$

$$e(k) = 1.255\,\delta(k) + 5u(k) - 6.25(0.8)^k u(k)$$

$$k = 0,\quad e(0) = 1.25 + 5 - 6.25 = 0$$
$$k = 1,\quad e(1) = 0 + 5 - 6.25(0.8) = 0$$
$$k = 2,\quad e(2) = 0 + 5 - 6.25(0.8)^2 = 1.00$$

Example 10.43

Find the inverse Z-transform x(n) of given X(z).

$$X(z) = \frac{3}{z-2}$$

Solution 10.43

X(z) may be written as $X(z) = 3z^{-1}\frac{z}{z-2}$

$Z\{2^n u(n)\} = \frac{z}{z-2}$, using the time-shifting property

$$Z\{2^{n-1}u(n-1)\} = z^{-1}[\frac{z}{z-2}] = \frac{1}{z-2}$$

Thus, we conclude that $x(n) = 3\{2^{n-1}u(n-1)\}$.

10.4.3 When the Poles are Complex

So far the inverse transform only for functions that have real poles have been considered. The same partial-fraction procedure applies for complex poles; however, the resulting inverse transform contains complex functions:

$$y[n] = Ae^{\Sigma n}\cos(\Omega n + \theta) = \frac{Ae^{\Sigma n}}{2}[e^{j\Omega n}e^{j\theta} + e^{-j\Omega n}e^{-j\theta}]$$

$$= \frac{A}{2}[e^{(\Sigma + j\Omega)n}e^{j\theta} + e^{(\Sigma - j\Omega)n}e^{-j\theta}] \tag{10.20}$$

where Σ and Ω are real. The Z-transform of this function is given by

$$Y(z) = \frac{A}{2}\left[\frac{e^{j\theta}z}{z - e^{\Sigma+j\Omega}} + \frac{e^{-j\theta}z}{z - e^{\Sigma-j\Omega}}\right] \tag{10.21}$$

$$Y(z) = \left[\frac{(A\,e^{j\theta}/2)z}{z - e^{\Sigma+j\Omega}} + \frac{(Ae^{-j\theta}/2)z}{z - e^{\Sigma-j\Omega}}\right] = \frac{k_1}{z - p_1} + \frac{k^*_1}{z - p^*_1} \tag{10.22}$$

where the asterisk indicates the complex conjugate.

$$p_1 = e^{\Sigma}e^{j\Omega} = e^{\Sigma}\angle\Omega$$
$$\Sigma = \ln|p_1| \quad \Omega = \arg p_1 \tag{10.23}$$

$$k_1 = \frac{Ae^{j\theta}}{2} = \frac{A}{2}\angle\theta \Rightarrow A = 2|k_1|; \quad \theta = \arg k_1 \tag{10.24}$$

Hence we calculate Σ and Ω from the poles, and A and θ from the partial-fraction expansion. We can then express the inverse transform as the sinusoid of (10.22).

$$y[n] = Ae^{\Sigma n} \cos(\Omega n + \theta) \tag{10.25}$$

An illustrative example is given next.

Example 10.44

Find the inverse Z-transform of the following transfer function:

$$Y(z) = \frac{-3.894z}{z^2 + 0.6065}$$

Solution 10.44

$$Y(z) = \frac{-3.894z}{z^2 + 0.6065} = \frac{-3.894z}{(z - j0.7788)(z + j0.7788)}$$

$$= \frac{k_1 z}{z - j0.7788} + \frac{k_1^* z}{z + j0.7788}$$

Dividing both sides by z, we calculate k_1:

$$k_1 = (z - j0.7788) \left[\frac{-3.894}{(z - j0.7788)(z + j0.7788)} \right]_{z=j0.7788}$$

$$= \frac{-3.894}{z + j0.7788} \bigg|_{z=j0.7788} = \frac{-3.894}{2(j0.7788)} = 2.5\angle 90°$$

$$p_1 = e^{\Sigma}e^{j\Omega} = e^{\Sigma}\angle\Omega$$

$$\Sigma = \ln |p_1| \quad \Omega = \arg p_1$$

Here, $p_1 = j0.7788$

$$\Sigma = In|p_1| = \ln(0.7788) = -0.250; \quad \Omega = \arg p_1 = \pi/2$$
$$A = 2|k_1| = 2(2.5) = 5; \quad \theta = \arg k_1 = 90°$$

Hence

$$y[n] = Ae^{\Sigma n} \cos(\Omega n + \theta) = 5e^{-0.25n} \cos\left(\frac{\pi}{2}n + 90°\right)$$

This result can be verified by finding the Z-transform of this function using the Z-transform table.

Example 10.45

Find the inverse Z-transform of the following transfer function:

$$X(z) = \frac{z}{z^2 + 1\,z + 0.5}$$

Solution 10.45

$$X(z) = \frac{z}{z^2 + 1\,z + 0.25 + 0.25} = \frac{z}{(z + 0.5)^2 + (0.5)^2}$$

$$= \frac{k_1 z}{z + 0.5 - j0.5} + \frac{k_1^* z}{z + 0.5 + j0.5}$$

Dividing both sides by z, we calculate k_1:

$$k_1 = (z + 0.5 - j0.5) \left[\frac{1}{(z + 0.5 - j0.5)(z + 0.5 + j0.5)} \right]_{z=-0.5+j0.5}$$

$$= \frac{1}{z + 0.5 + j0.5} \bigg|_{z=-0.5+j0.5} = \frac{1}{2(j0.5)} = 1\angle - 90°$$

with $p_1 = -0.5 + j0.5$

$$\Sigma = In|p_1| = \ln(0.707) = -0.34; \quad \Omega = \arg p_1 = 3\pi/4$$

$$A = 2|k_1| = 2(1) = 2; \theta = \arg k_1 = -90°$$

Hence

$$y[n] = Ae^{\Sigma n} \cos(\Omega n + \theta) = 2e^{-0.34n} \cos\left(\frac{3\pi}{4} - 90°\right)$$

11

Analog Filters Design

This chapter begins with general consideration of analogue filters, such as lowpass, highpass, bandpass and band stop, problems associated with passive filters, theory of analogue filter design, The Butterworth approximation, Chebyshev approximation, comparison between Butterworth and Chebyshev filters, examples, problems, solutions and practice problems.

11.1 Introduction

Filters are frequency-dependent entities that are used to remove frequencies lying in certain specific range(s), and pass frequencies lying in certain other specific ranges(s). Based on this definition, we can classify filter into four fundamental categories:

- Low-pass filters (LPF)
- High-pass filters (HPF)
- Band-pass filters (BPF)
- Band-reject filters (BRF)

By response characteristic of a given system, we mean the magnitude (amplitude) vs frequency plot of the transfer function of that system. In mathematical notations, we may denote the transfer function in terms of the Laplace (s-domain), this may be written as

$$H(s) = \frac{V_0(s)}{V_i(s)} \tag{11.1}$$

where $V_0(s)$ = output voltage (in s-domain), and $V_i(s)$ = input voltage (in s-domain). It may be remembered that

$$s = j\omega \tag{11.2}$$

We know that transfer function denotes the ratio of the output to the input of the system, expressed in the frequency domain as

$$H(j\omega) = \frac{V_0(j\omega)}{V_i(j\omega)} \qquad (11.3)$$

where $V_0(j\omega)$ = output voltage (in frequency domain), and $V_i(j\omega)$ = input voltage (in frequency domain).

In terms of z-transformation, we may write

$$H(z) = \frac{V_0(z)}{V_i(z)} \qquad (11.4)$$

From (11.1), (11.3) and (11.4), we find that transfer function can be expressed in different frequency domains. The choice of the domain to be used in a given problem depends on the transformation (such as Fourier, Laplace, z-transform etc.) used to solve the problem. For example, in analogue signal-processing applications, we use continuous-time, Fourier transformation and continuous-time Laplace transformation, while in digital signal processing (DSP) we use discrete-Fourier transformation and z-transformation.

As stated earlier, in the analogue domain, we will be mainly using Fourier transforms for the conversion of time-domain signals into frequency-domain signals. In such applications, the definition of the transfer function, given by (11.3), will be used.

Figure 11.1a shows the frequency-response characteristic of an ideal LPF and Figure 11.1b shows that of an ideal high-pass filter. Similarly Figure 11.1c shows the frequency-response characteristic of an ideal band-pass filter, and Figure 11.1d shows that of an ideal band-reject filter. It may be noted that these are not plotted on log scales.

11.2 LP Filters

As shown in Figure 11.1a, a LPF will allow all the frequencies from $\omega = 0$ to $\omega = \omega_{CL}$ to pass through it, and attenuate (stop, or cut-off) all other frequencies greater than ω_{CL}. We call ω_{CL} as the cut-off frequency of the LPF.

In the ideal LPF characteristic, shown in Figure 11.1a, we find that the magnitude $H(\omega)$ remains constant at unit magnitude [i.e., $H(\omega) = 1$] up to $\omega = \omega_{CL}$, and then vertically falls off to 0.

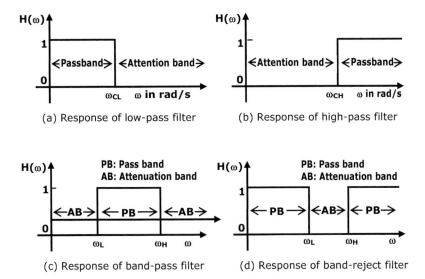

Figure 11.1 Ideal frequency-response characteristics of various types of filters.

It may further be noted that $H(\omega) = 1$ represents the condition that the output $(V_0) =$ input (V_i). This further means that the system remains lossless up to ω_{CL}, and produces infinite loss for frequencies above ω_{CL}.

The band of frequencies that is allowed to pass through the filter is called the pass band (PB), and the band of frequencies that is attenuated is called attenuation band (AB). Further, the AB is also known as stop or cut-off band.

11.2.1 First Order RC LPF Circuit

We start from the transfer function of a pure integrator in $j\omega$ form for understanding the slope per decade:

$$H(s) = \frac{1}{s} \text{ or in } j\omega \text{ form } H(j\omega) = \frac{1}{j\omega}$$

ω	2o log H(jω)	Phase H(jω)(degree)
0.1	20 dB	-90
1	0 dB	-90
10	-20 dB	-90

Thus, here we see that if $H(\omega)$ is pure integrator for a decade of frequency the magnitude comes down at the rate of -20 dB/decade or gets decreased (attenuated) as -20dB per decade (one decade means $10 \times$ frequency).

Now consider Figure 11.2, which shows a simple RC circuit, the proof is also reproduced here for convenience.

Let V_i be the input voltage, V_0 be the output voltage and I be the circulating current in the circuit shown. Then, using Kirchhoff's voltage low, we have

$$V_i = I \left(R + \frac{1}{j\omega C} \right) \quad V_0 = I \frac{1}{j\omega C} \tag{11.6}$$

Therefore, the transfer function is now given by

$$H(\omega) = \frac{1/(j\omega C)}{R + 1/(j\omega C)} \tag{11.7}$$

Simplifying (11.7), we get

$$H(\omega) = \frac{1}{1 + j\omega C R} = \frac{1}{1 + j(\omega/\omega_C)} \tag{11.8}$$

where $\omega_C = 1/(CR)$ is called the upper cut-off frequency.

Figure 11.3 shows the plot of the logarithm of $H(\omega)$ in decibels against the relative frequency (ω/ω_C). As stated earlier, this plot is called the Bode plot.

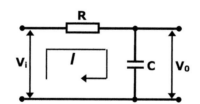

Figure 11.2   RC low-pass filter circuit.

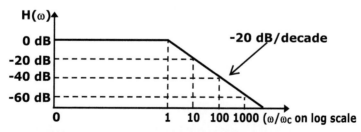

Figure 11.3   Frequency-response characteristics of a low-pass circuit.

Bode plots show variation of the amplitude function H(ω) frequencies. To plot the frequency-response characteristic, we use (11.8), and for understanding the slope of -20 dB/decade we make the following calculation:

$(\omega/\omega_C) \ll 1$

$$H(\omega) \text{in dB} = 20 \log \left(\frac{1}{1 + j\omega/\omega_C} \right) = 20 \log \left[\frac{1}{1^2 + (\omega/\omega_C)^2} \right]^{1/2}$$

$$= 20 \log 1 = 0 \ dB$$

It means the imaginary factor does not dominate at lower few, and the magnitude of the given filter circuit starts from 0 dB/ decade that is in straight line.

$(\omega/\omega_C) \gg 1$

$$H(\omega) \text{in dB} = 20 \log \left(\frac{1}{(\frac{\omega}{\omega_C})^2} \right)^{1/2} = has \ a \ slope \ of \ -20 \ db \ per \ decade$$

It means the imaginary factor dominates at higher frequency and the transfer function is looking like a pure integrator, whose slope is already determined, that is, -20 dB/decade:

$(\omega/\omega_C) = 1$

$$H(\omega) \text{in dB} = 20 \log \left(\frac{1}{1^2 + 1^2} \right)^{1/2} = 20 \log(1/\sqrt{2}) = -3 \ db$$

From the calculations given above, we find that up to $\omega = \omega_C$, H(ω) has a value of -3 dB, which is also called as 3 dB correction.

Figure 11.3 shows the plot of the frequency response based on the calculations given above. The term $(\omega/\omega_C) = 1$ represents the cut-offcondition, and ω_C, as stated before, is called the cut-off frequency.

Now, let us consider (11.8) once again. Substituting s $= j\omega$ in (11.8), we get

$$H(s) = \frac{1}{1 + sCR} = \frac{1/(CR)}{s + 1/(CR)} = \frac{\omega_C}{s + \omega_C} \tag{11.9}$$

In (11.9), we find that the power of the frequency term s $= 1$. Hence, we call this simple LP RC filter as a first-order filter. We now conclude that the rate of attenuation of a first-order filter is -20 dB per decade. Similarly, it can be proved that a second-order filter (with power of s = 2) produces an attenuation of −40 dB per decade, a third-order filter (with power of s = 3) produces an attenuation of −60 dB per decade and so on.

We find that the higher the order of the filter, the better its cut-off characteristics, and the more closely does it approximate the ideal characteristic (please note that the ideal characteristic can be achieved only with a filter having an order of infinity).

11.2.2 Second Order StocktickerRLC LPF Circuit

Figure 11.4 shows a series RLC, with the output taken across the capacitor. From the figure, using Kirchhoff's voltage law (KVL) and Laplace transform, we write the expression for the input voltage as:

$$V_i = I\left(R + sL + \frac{1}{sC}\right) \tag{11.10}$$

Similarly, the output voltage is obtained as

$$V_0 = I\frac{1}{sC} \tag{11.11}$$

Dividing (11.11) by (11.10), we get the expression for the transfer function as

$$H(s) = \frac{V_0}{V_i} = \frac{1/(sC)}{R + sL + (1/sC)} \tag{11.12}$$

Simplifying Equation (11.12), we have

$$H(s) = \frac{1}{sCR + s^2LC + 1} \tag{11.13}$$

Equation (11.13) can be simplified to get the standard expression of H(s) as

$$H(s) = \frac{1/(LC)}{s^2 + (R/L)s + 1/(LC)} = \frac{\omega_C^2}{s^2 + 2\zeta\omega_C s + \omega_C^2} \tag{11.14}$$

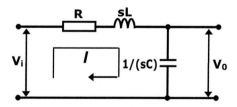

Figure 11.4 Second-order RLC low-pass filter circuit.

where $\omega_C = 1/(\mathrm{LC})$ is the resonant frequency of the series stocktickerRLC circuit, and ξ is the damping factor of the circuit. The term $2\xi\omega_C$ represents the losses in the system. Since the power of s in (11.14) is 2, we say that this is a second-order system. As stated earlier, for a second-order system, the frequency-response curve droops at the rate of -40 dB/decade. This can be proved by drawing the Bode plot of the system.

We now make a comparison of the frequency-response curves of a second-order system and a first-order system. The plot for this is shown in Figure 11.5.

From the plot, shown in Figure 11.5, we conclude that the second-order LPF circuit produces more attenuation than the first-order LPF circuit.

11.2.3 Second Order RC LPF Circuit

Using mesh analysis, we find that the transfer function of the circuit shown in Figure 11.6 is given by

$$H(s) = \frac{V_0}{V_i} = \frac{\omega_C^2}{s^2 + 2\xi\omega_C s + \omega_C^2} \qquad (11.15)$$

where $\omega_C = 1/(\mathrm{RC})$. Comparing (11.14) and (11.15), we find that they are similar expressions, representing the transfer function of the second order passive LPF.

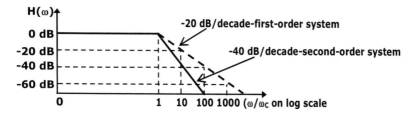

Figure 11.5 Frequency-response characteristics of first- and second-order LPF circuit.

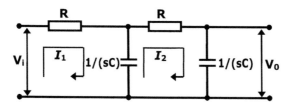

Figure 11.6 6 Second-order RC low-pass filter circuit.

It can be seen that the circuit, shown in Figure 11.6, is obtained by cascading the two first-order circuits, shown in Figure 11.1. We can similarly cascade N (where N is an integer) such first-order circuits to get a Nth-order filter.

11.3 High-Pass Filters

Figure 11.1(b) shows the ideal characteristic of a HP filter. The HPFs pass all frequencies above the cut-off frequency ω_C, and attenuate all those below it. By interchanging R and C in Figure 11.2, we get a first order HP filter. Similarly, by interchanging the positions of R and C in Figure 11.2, we get a second order HP filter. The circuits of the first- and second-order HP filter are shown in Figures 11.7(a) and 11.7(b), respectively.

Proceeding as in the case of the LPF circuits, we can find the transfer functions of the HP filter circuits shown in Figures 11.7a and 11.7b. The transfer function of the first-order HP filter is given by

$$H(s) = \frac{R}{R + 1/(sC)} = \frac{sCR}{sCR + 1} = \frac{s}{s + \omega_{CH}} = \frac{s}{s + \omega_C} \qquad (11.16)$$

where $\omega_{CH} = \omega_C = 1/(CR)$ is the lower cut-off frequency of the filter system.

Similarly, the transfer function of the second-order HP filter system comprising of the stocktickerRLC circuit shown in Figure 11.7b is given by

$$H(s) = \frac{V_0}{V_i} = \frac{sL}{R + sL + 1/(sC)} = \frac{s^2 LC}{sCR + s^2 LC + 1}$$

$$= \frac{s^2}{s^2 + (R/L)s + 1/(LC)} \qquad (11.17)$$

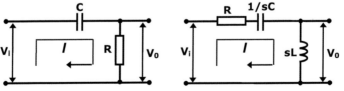

(a) First-order RC high-pass filter   (b) Second-order RLC high-pass filter

Figure 11.7   High-pass filter circuit.

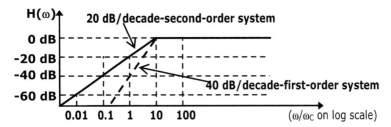

Figure 11.8 Frequency-response characteristics of first and second-order LPF circuit.

The frequency-response characteristics of the two types of filters described above are shown in Figure 11.8.

It can be seen that the practical frequency-response characteristic is far from the ideal situation, shown in Figure 11.1b, as in the case of the LP filter. Here also, to get an approximately ideal characteristic, we must use higher-order filters. This can be generalized in the case of other filters also, i.e., to obtain an approximately ideal characteristic; we must use higher-order structures for all types of filters.

The second-order filter is shown in Figure 11.8. From (11.14), we have the cut-off (resonant) frequency of the circuit given by

$$f_C = \frac{1}{2\pi\sqrt{LC}} \tag{1}$$

as before, choosing the value of C $= 0.01 \ \mu F$, from Equation (5), we get

$$L = \frac{1}{2 \times \pi \times 1000^2 \times 0.01 \times 10^{-6}} = 2.5 \ H \tag{2}$$

This is a high value of L. Hence, to reduce it, we may increase the value of C to 0.1 μF. In that case, on recalculation, we have

$$L = 0.25 \ H \tag{3}$$

It may be remembered that R may be taken as the resistance of the inductor coil, and hence, we need not design it. The completely designed circuit of the second-order LPF is shown in Figure 11.11b.

11.4 Band Pass Filters

The ideal response of a band-pass filter (BPF) is shown in Figure 11.1c. This filter is designed to pass a desired band of frequencies that lie between

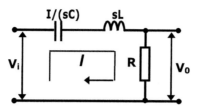

Figure 11.9   Second-order RLC band-pass filter circuit.

two cut-off frequencies called, respectively, the upper and the lower cut-off frequencies. Tuned amplifiers are BPFs in that they permit only a certain range of frequencies to pass through them: all other frequencies, which lie outside this range, are attenuated.

BPF-response characteristic may be thought as the combination of an LPF-response characteristic, with cut-off at ω_H, and an HPF-response characteristic with cut-off at ω_L, as shown in Figure 11.1c. The pass band is given by $\omega_H - \omega_L$.

Practical BPF-response characteristic can also be stated as first order, second order, etc., as in the cases of LPF and HPF. A second-order BPF may be constructed by interchanging R and C in the circuit shown in Figure 11.4. The resultant configuration of the second-order BPF is shown in Figure 11.9.

From Figure 11.9, we get the transfer function of the second-order BPF as

$$H(s) = \frac{V_o}{V_i} = \frac{R}{R + sL + 1/(sC)} = \frac{sCR}{sCR + s^2 LC + 1}$$

$$= \frac{(R/L)s}{s^2 + (R/L)s + 1/(LC)} \tag{18}$$

11.5 Band Reject Filters

Band-reject (elimination or stop) filters (BRFs) have a response characteristic just opposite to that of the BPFs. From Figure 11.1d, we find that the BRF will reject a given range of frequencies, and pass all frequencies that lie outside this range.

The practical response curve of the BRF, as in the cases of LPF, HPF and BPF, can also be stated in terms of the order to the filter. Figure 11.10 shows a typical second-order characteristic of a BRF.

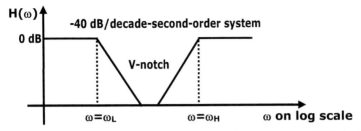

Figure 11.10 Frequency-response characteristics of a second-order BRF circuit.

It can be seen from Figure 11.10 that the practical response curve looks like a V-notch; hence, BRF filters are also called notch filters. Our RLC circuit cannot be used as a notch filter. Instead, other suitable methods are used for constructing the notch filter. Notable among these is the twin-T filter. However, no attempt is made in this section to give the circuit diagram of a twin-T notch filter.

Example 11.1
Design an LPF for a cut-off frequency of 1000 Hz.

Solution 11.1
Because the order of the filter is not specified in the given problem, we shall design both the first- and the second-order filters.

The desired first-order LPF circuit is shown in the given figure.

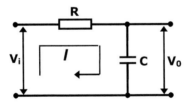

Now, we find that the cut-off angular frequency

$$H(\omega) = \frac{1}{1 + j\omega CR} = \frac{1}{1 + j(\omega/\omega_C)} \tag{1}$$

and the cut $-$ off frequency f_C is given by $\omega_C = \dfrac{1}{RC}$ $f_C = \dfrac{1}{2\pi RC}$ (2)

In the given problem, we have $f_C = 1000$ Hz. Therefore, from Equation (2), we get by substitution

$$1000 = \frac{1}{2\pi RC} \tag{3}$$

In Equation (3), we have two unknown parameters, viz. R and C. We can choose either R or C arbitrarily and then find the other. It is quite common to choose C first, as C is available only in selected values. Here, we choose C = 0.01 μF. Then, from Equation (3), we get

$$R = \frac{1}{2 \times \pi \times 1000 \times 0.01 \times 10^{-6}} \approx 16k\Omega \qquad (4)$$

To get an exact (accurate) value of R, we use a 15-kΩ resistor in series with a 2.2-kΩ potentiometer. The completely designed first-order LPF is shown in Figure 11.11a.

11.6 Designing Higher-Order Filters

As stated earlier, to get an approximate ideal attenuation characteristic at cut-off, we require the order of the filter to be very high. However, it is easy to design filters of the first- and second order. Hence, to get a higher-order filter, the common practice is to cascade filters of the first- and second orders. For example, a third-order filter can be obtained by cascading a first-order filter with a second-order filter, a fourth-order filter can be obtained by cascading two second-order filters and so on. Figure 11.12 shows the block diagram of a third-order filter, based on the above principle.

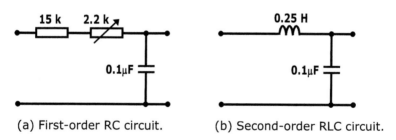

(a) First-order RC circuit.   (b) Second-order RLC circuit.

Figure 11.11 11 Low-pass filter circuits.

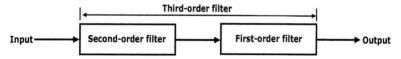

Figure 11.12 Block diagram of a third-order filter.

In designing the first- and second-order filters, we use the transfer functions also. They are repeated here for reference [see Equations (11.9) and (11.14)]:

$$H(s) = \frac{1}{1 + sCR} = \frac{1/(CR)}{s + 1/(CR)} = \frac{\omega_C}{s + \omega_C}$$

$$H(s) = \frac{1/(LC)}{s^2 + (R/L)s + 1/(LC)} = \frac{\omega_C^2}{s^2 + 2\xi\omega_C s + \omega_C^2}$$

(11.9) and (11.14) represent the first- and second-order filters, respectively. The design of the first-order filter is rather simple. However, the design of the second-order filter involves the following steps:

To start with, rewrite (11.14) as a product of poles, then we have

$$H(s) = \frac{\omega_C^2}{s^2 + 2\xi\omega_C s + \omega_C^2} = \frac{\omega_C^2}{(s + a + jb)(s + a - jb)} \tag{8}$$

where (a + jb) and (a − jb) represent the poles of the transfer function. It may be noted that

$$s^2 + 2\zeta\omega_C s + \omega_C^2 = (s + a + jb)(s + a - jb) = s^2 + 2as + a^2 + b^2 \tag{9}$$

Comparing the left-hand side (LHS) and right-hand side of Equation (9), we get

$$a^2 + b^2 = \omega_C^2 \tag{10}$$

and

$$a = \zeta\omega_C \tag{11}$$

Equations (8), (10) and (11) are the basic equations that are used in the design of second-order filters in analogue and DSP.

11.7 Problems Associated with Passive Filters

The following problems are associated with passive filters:

- Cascading several stages of first-order filters produces loading effect on the circuit. This results in the output getting drastically reduced.
- To increase the output, external amplifiers have to be incorporated in the filter circuit.
- Impedance matching becomes a big problem with passive filters.

- If inductors are used in filters, they are bound to produce noise of their own.

Hence it is quite common, nowadays to use active filters, which employ operational amplifiers (op-amps) as the active element. Active filters remove most of the defects of passive filters listed earlier. The design of active filters is discussed later in this chapter.

11.8 Filters Using Operational Amplifiers

Consider the equation of the first-order filter, given by

$$H(s) = \frac{1}{s+1} \tag{11.19}$$

where we have chosen $\omega_C = 1$. (11.19) can be written as

$$H(s) = \frac{Y(s)}{X(s)} = \frac{1}{s+1} \tag{11.20}$$

where we have used Y(s) as the output (instead of V_0) and X(s) as the input (instead of V_i). From Equation (11.20), we get

$$(s+1)\,Y(s) = X(s) \tag{11.21}$$

Taking the inverse Laplace transform of (11.21), and rearranging, we have

$$\frac{dy(t)}{dt} = -y(t) + x(t) \tag{11.22}$$

This can be implemented by using the op-amp setup, shown in Figure 11.13.

The setup consists of a summing integrator, and an inverter, as shown. The output of the summing integrator itself is feedback as one of its inputs, and its second input is the external signal x(t).

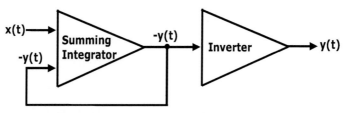

Figure 11.13   Set-up to solve Equation (11.22).

To develop the setup, initially, we assume that we have with us the differential term [dy(t)/dt]. From Equation (11.22), we find that this is the sum of the signal components x(t) and −y(t). If we integrate [dy(t)/dt] using the standard op-amp summing integrator, we get −y(t) as its output.

As stated above, we then feed −y(t) back to the input of the integrator, along with x(t). The output −y(t) of the integrator is then inverted to get the required output function y(t), as shown in Figure 11.13.

It may appear that we are generating the desired (and required) functions from nothing, but this not true. In fact, the functions are generated from the random noise prevalent in the system. We know that thermal noise is present in every system, and that it contains frequencies from 0 Hz to about 10^{13} Hz.

Using feedback connections, one of these frequencies actually produces resonance in the system, which depends on the configurations of the feedback system and develops the desired functions. This type of network, which makes use of feedback connections to produce the desired functions, is known as a recursive (feedback) network. Recursive networks play an important role in filter theory, both in the analogue and the digital regimes. Another kind network structure, which does not use feedback connections at all, is also used for filters. Such networks are called non-recursive filter.

We now conclude our discussions so far as follows:

- Transfer functions of the first- and second-order filters are respectively, first- and second-order differential equations.
- We can use the appropriate op-amp setups for the construction of circuits that simulate these differential equations, and produce the desired (filtered) outputs.
- The op-amp setups used for simulating and solving differential equations are called analogue computers.

We now consider another example to further illustrate the theory of analogue computation. This theory will be extended later to the design of digital HR filters.

Example 11.2

Develop a setup to solve the differential equation:

$$8\frac{d^3y}{dt^3} + 3\frac{d^2y}{dt^2} + 5\frac{dy}{dt} + 8y = 2x \tag{1}$$

Solution 11.2

We first rewrite the given differential equation in the form

$$\frac{d^3y}{dt^3} = -\frac{3}{8}\frac{d^2y}{dt^2} - \frac{5}{8}\frac{dy}{dt} - y + \frac{1}{4}x \qquad (2)$$

Assuming that (d^3y/dt^3) is available to us, we draw the block schematic of the analogue computer setup, as shown in Figure 11.14.

As shown in Figure 11.14, integrator I_1, which is a summing integrator, integrates the sum of the terms

$$-\frac{3}{8}\frac{d^2y}{dt^2}, \quad -\frac{5}{8}\frac{dy}{dt}, -y, \quad \text{and} \quad +\frac{1}{4}x$$

which is equal to (d^3y/dt^3) to produce $(-d^2y/dt^2)$. We then integrate $(-d^2y/dt^2)$ in integrator I_2 to give (dy/dt). This term is then integrated by integrator I_3 to give $-v_0$. We then invert $-y$ using inverter 2 to give y, which is the required solution. Since the outputs of the integrators are terms with '1' as their respective coefficients, we use multipliers, as shown in Figure 11.14, to get the required coefficients of 3/8, 5/8 and 1/4. Figure 11.15 shows how the multiplication is effected in analogue integration. It may be remembered that Figure 11.15 represents a summing integrator.

From the basic op-amp integrator theory, we know that output voltage

$$-V_0 = \frac{1}{R_1C}\int V_1dt + \frac{1}{R_2C}\int V_2dt + \frac{1}{R_3C}\int V_3dt \qquad (3)$$

Let in Equation (3), $V_1 = d^2y/dt^2$, $V_2 = dy/dt$ and $V_3 = y$. Assume also that we have chosen $C = 1\ \mu F$. Then comparing coefficients of d^2y/dt^2

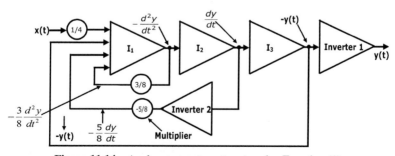

Figure 11.14 Analog computer set-up to solve Equation (1).

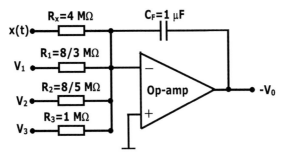

Figure 11.15 Summing integrator using op-amp.

etc., in Equations (2) and (3), we find that

$$\frac{1}{R_1 C} = \frac{3}{8}, \quad \frac{1}{R_2 C} = \frac{5}{8}, \quad \frac{1}{R_3 C} = 1$$

from which we get

$$R_1 = \frac{3}{8} M\Omega, \quad R_2 = \frac{5}{8} M\Omega. \quad R_3 = 1 \ M\Omega$$

From the above calculations, we observe that the passive components of the integrator circuit will take care of the required multiplicands. In fact, in actual analogue computer setups, resistors are variable potentiometers, whose values can be precisely adjusted to get the desired value of R. It may be noted that the input x(t) also has a multiplying factor of 1/4, which may be generated as discussed earlier. This is also shown in Figure 11.14.

Referring back to Figure 11.14, we find all the individual terms for the summation to get $(d^3 y/dt^3)$ are fed back to the input of integrator I_1 to form a set of feedback loops. The resultant structure, as stated earlier, is a recursive network.

It may be noted that the main purpose of presenting the theory of the analogue computer setup in the preceding paragraphs is to illustrate how digital filter structures can be produced from it.

11.9 Representing Structure of Analogue Computers

In developing Figure 11.14, we had assumed the availability of the highest order differential term available to us, and then used successive integrations using the op-amp integrators to produce the desired function.

We can also look at it in another way. Suppose, we know that we have the desired output y available with us. Then, we can use successive differentiations to get the highest order differential term to us. We illustrate this idea by choosing Equation (1) of example again.

Example 11.3
Rewriting Equation (1) of Example 11.2, we have

$$8\frac{d^3y}{dt^3} + 3\frac{d^2y}{dt^2} + 5\frac{dy}{dt} + 8y = 2x$$

Solution 11.3
We can modify this as

$$y = \frac{1}{4}x - \frac{d^3y}{dt^3} - \frac{3}{8}\frac{d^2y}{dt^2} - \frac{5}{8}\frac{dy}{dt} \tag{2}$$

It is a well-known fact (from the basic circuit theory) that an integrator introduces a delay to functions that are applied at its input. This delay is specified in some unit of time. Usually, we choose this 'some' as unity. Hence, we can represent integrators as unit delays. The block diagram of the analogue computer setup using this modification is shown in Figure 11.16.

In Figure 11.16, the unit delays are represented as small square blocks, instead of representing them as the op-amp integrators. The reason for this representation is that in DSP, these blocks need not be the op-amp integrators,

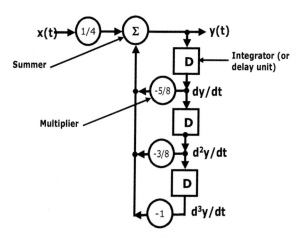

Figure 11.16   Modified analog computer set-up to solve Equation (1).

but can be any circuit such as D flip-flop that produces a delay action on its input signal.

On closer observation, we find that the analogue setups shown in Figures 11.14 and 11.16 are not different at all, as all the blocks used in both the cases are one and the same. However, this will be different in digital processing, where normally op-amps are not used at all as integrators. Figure 11.16 is used in the DSP filter structures.

11.10 Step-By-Step Design of Analogue Filters

The squared magnitude response $H(j\omega)^2$ [power transfer characteristic] of the filter is shown in Figure 11.17 (below).

It can be seen that ideal characteristics can never be achieved in practice due to limitations of the devices used. Also, in most cases, we can tolerate the undesired frequencies, if their amplitude levels are low and do not interfere with the desired frequencies. The aim of the designer, therefore, is to design a filter that will pass maximum amount of desired frequencies, and introduce enough amount of attenuation to undesired frequencies. This suggests that the ideal characteristic can be achieved only approximately in a practical

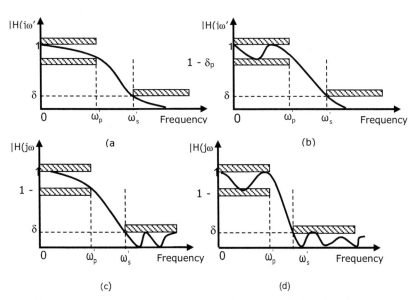

Figure 11.17 Sketches of frequency responses of some classical analog filters: (a) Butterworth (b) Chebyshev type I; (c) Chebyshev type II; (d) Elliptic.

situation. Currently, we use four types of mathematical functions to achieve the required approximations in the response characteristics of filters. These functions, which are used for the design of filters of the recursive type, are listed hereunder:

- The Butterworth approximation function
- The Chebyshev approximation function
- The Elliptical approximation function
- The Bessel's approximation function

Each approximation has certain plus and minus points. In this text, we mainly discuss the Butterworth and the Chebyshev filters only.

11.11 Butterworth Approximation Function

We had developed the transfer function of a first-order LP RC filter, which is given by

$$H(\omega) = \frac{1}{1 + j\omega CR} = \frac{1}{1 + j(\omega/\omega_C)} \tag{11.23}$$

Representing Equation (11.23) in its squared-magnitude form and N = 1, we have

$$|H(\omega)|^2 = \frac{1}{1 + (\omega/\omega_C)^2} \tag{11.24}$$

Taking the square root of (11.24), we get the magnitude of H(ω) as

$$|H(\omega)| = \left[\frac{1}{1 + (\omega/\omega_C)^2}\right]^{1/2} \tag{11.25}$$

When N stages are cascaded, we get the nth-order filter. The cut-off will become sharper and sharper, as we increase the order of the filter used. (11.26) may be generalized to develop an approximate expression or the transfer function of an nth-order filter. Thus, we find

$$|H(\omega)| = \left[\frac{1}{1 + (\omega/\omega_C)^{2N}}\right]^{1/2} \tag{11.26}$$

(11.26) represents a Butterworth polynomial of order N. We can design analogue and digital filters based on this approximation function.

It is found that the pass-band gain remains more or less constant and flat up to the cut-off frequency, as shown in Figure 11.18. Hence, this is

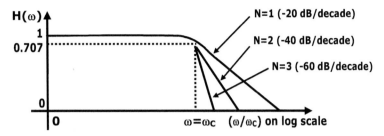

Figure 11.18   Butterworth frequency-response characteristics of an LPF circuit.

also known as a maximally flat response. Ripples may be permitted in the frequency-response characteristic but their amplitude must be very low ($< \pm 0.1$). The responses using Butterworth approximation for various orders of filters are also shown in Figure 11.18. In many cases, H(ω) is expressed in decibels. For this, we take 20 x logarithm of both sides of (11.26) to get

$$20 \log |H(\omega)| = 20 \log \left[\frac{1}{1 + (\omega/\omega_C)^{2N}} \right]^{1/2} = -10 \log[1 + (\omega/\omega_C)^{2N}]$$

(11.27)

That is, $|H(\omega)|$ *in dB* $= -10 \log \left[1 + \left(\dfrac{\omega}{\omega_C} \right)^{2N} \right]$ (11.28)

(11.28) is a design equation, which will be used in our design examples on filters.

11.11.1 Step-By-Step Design of Butterworth Filter

Filters are designed to meet certain specifications or requirements, which are specified in the given design problem. Mostly, prototype filters are designed as LPFs. Then, by using frequency-transformation techniques, we convert this design to the desired (i.e., high-pass, band-pass or band-reject) type of filter design.

Figure 11.19 shows the characteristic of an LPF. To design the filter, we must have, or be given, information on

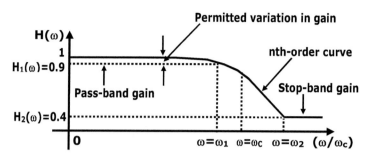

Figure 11.19   Butterworth characteristics showing specifications of an LPF.

- The amount of gain required in the pass band
- The frequency up to which pass-band gain must remain more of less constant
- The amount of attenuation required
- The frequency from which the attenuation must start

Based on the specifications given earlier, we determine the order and cut-off frequency of the filter required to meet them. Using the information so obtained, we develop the filter transfer function, which is then used to construct the desired filter.

For convenience, the specifications may be written in a systematic form as given below:

1. Pass-band gain required $H_1(\omega)$
2. Frequency up to which the pass-band gain must remain more or less constant: ω_1
3. Amount of attenuation required: $H_2(\omega)$
4. Frequency from which the attenuation must start: ω_2

As stated earlier, from these specifications, we determine the cut-off frequency ω_C and the order of the filter n.

11.11.2 Design Procedure for Butterworth Filter

The transfer function $H(\omega)$ will be usually specified either as a mere number or in units of decibels. Accordingly, there are two ways through which the design steps may progress initially to determine N and ω_C.

After the determination of these factors the two will merge into one and design steps will progress through one direction. Let us first determine N and ω_C when $H(\omega)$ is specified as a mere number.

11.11.3 Design Procedure when H(ω) is Specified as a Mere Number

Step 1: Determination of N and ω_C

We have, from (11.29), the transfer function given by

$$|H(\omega)| = \left[\frac{1}{1 + (\omega/\omega_C)^{2N}}\right]^{1/2} \tag{11.29}$$

Now, at $\omega = \omega_1$, we have

$$|H_1(\omega)| = \left[\frac{1}{1 + (\omega_1/\omega_C)^{2N}}\right]^{1/2} \tag{11.30}$$

Similarly, at $\omega = \omega_2$, we have

$$|H_2(\omega)| = \left[\frac{1}{1 + (\omega_2/\omega_C)^{2N}}\right]^{1/2} \tag{11.31}$$

Equations (11.29) and (11.30) can be solved to get the values of N and ω_C.

Step 2: Determination of the poles of H(s)

It is found that the poles of the Butterworth polynomial lie on a circle, whose radius is ω_C. The poles of a second-order filter are shown in Figure 11.20, which always lie on LHS plane for fulfilling the stability criterion.

The LP Butterworth filter is characterized by the following magnitude-squared-frequency response:

$$|H(\omega)|^2 = \frac{1}{1 + \left(\frac{\omega}{\omega_c}\right)^{2N}}$$

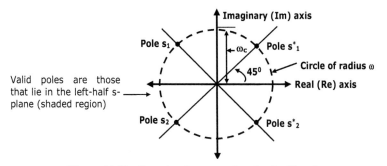

Figure 11.20 Location Butterworth poles for N = 2.

Where N is the filter order and ω_c is the 3 dB cut-off frequency of the LPF (for the normalized prototype filter, ω_c is always equal to 1). The magnitude-frequency response of a typical Butterworth LPF is depicted in Figure 11.17, and is seen to be monotonic in both the passband and stopband. The response is said to be maximally flat because of its initial flatness. The method of calculating poles of the Butterworth filter, the Butterworth polynomials are discussed from the following formula, which has been derived for understanding:

$$\left(\frac{\omega}{\omega_c}\right)^{2N} + 1 = 0 \qquad \left(\frac{j\omega}{j\omega_c}\right)^{2N} = -1 \qquad \left(\frac{s}{j\omega_c}\right)^{2N} = -1 \qquad (11.32)$$

$$(s)^{2N} = (-1)(j\,\omega_c)^{2N} \quad \text{or} \quad (s)^{2N} = (e^{-j\pi})\left(e^{j\pi/2}\right)^{2N} \omega_c^{2N}$$

As the value of a complex number is unchanged when multiplied by $e^{j(2\pi m)}$ for integer value of m, i.e., $m = 1, 2, 4, \ldots$

$$(s)^{2N} = (e^{-j\pi})\left(e^{j\pi/2}\right)^{2N} \omega_c^{2N} e^{j(2\pi m)} \qquad (11.33)$$

$$(s)^{2N} = \omega_c^{2N} e^{j(2m+N-1)\pi}$$

therefore, distinct location of poles is found by

$$\left(\frac{2m + N - 1}{2\,N}\right) \pi \text{ to give } (s_m) = \omega_c e^{j(\theta)}$$

The transfer function of the normalized analogue Butterworth filter, H(s), contains zeros at infinity and poles which are uniformly spaced on a circle using the following relationship:

$$s_m = e^{j\pi(2m+N-1)/2N}$$

$$= \omega_C \left[\cos\frac{(2m + N - 1)\pi}{2N} + j\sin\frac{(2m + N - 1)\pi}{2N}\right] \quad m = 1, 2, \ldots, N$$

If ω_C is of radius 1 in the s-plane then above equation simplifies to

$$s_m = e^{j\pi(2m+N-1)/2N}$$

$$= \cos\left[\frac{(2m + N - 1)\pi}{2N}\right] + j\sin\left[\frac{(2m + N - 1)\pi}{2N}\right] \quad m = 1, 2, \ldots, N$$

$$(11.34)$$

The poles occur in complex conjugate pairs and lie on the LHS of the s-plane. The normalized Butterworth polynomial up to order 5 has been calculated using the above equation and been placed below in a tabular form. Students are advised to verify themselves that how it has been derived.

Step 3: To find the expression for H(s)
Once we get the location of the poles (see Figure 11.21), we can obtain the transfer function of a second-order filter by using the expression

Using (11.34)
N = 1;
m = 1;
$s = \cos \pi + j \sin \pi = -1$

$$H(s) = \frac{1}{(s+1)}$$

Using (11.34)
N = 2;
m = 1;
$s1 = \cos(3\pi/4) + j \sin(3\pi/4) = -0.707 + j0.707$
m = 2;
$s2 = \cos(3\pi/4) + j \sin(3\pi/4) = -0.707 - j0.707$

$$H(s) = \frac{1}{(s+0.707 - j0.707)(s+0.707 - j0.707)}$$

$$= \frac{1}{(s+0.707)^2 + (0.707)^2}$$

$$H(s) = \frac{1}{(s+1.414+0.5)+(0.5)} = \frac{1}{(s+1.414\,s+1.0)}$$

Using (11.34)
N = 3;
m = 1;
$s1 = \cos(4\pi/6) + j \sin(4\pi/3) = -0.5 + j0.866$
m = 2;
$s2 = \cos(6\pi/6) + j \sin(6\pi/6) = -1.0$
m = 3;
$s2 = \cos(8\pi/6) + j \sin(8\pi/6) = -0.5 - j0.866$

$$H(s) = \frac{1}{(s + 0.5 - j0.866)(s + 0.5 + j0.866)(s + 1)}$$

$$= \frac{1}{(s + 0.5)^2 + (0.866)^2(s + 1)}$$

$$H(s) = \frac{1}{(s + 1.0 + 0.25) + (0.75)} = \frac{1}{(s + 1\ s + 1.0)(s + 1)}$$

Using (11.34)

N = 4;

m = 1;

s1 = $\cos(5\pi/8) + j\sin(5\pi/8) = -0.3826 + j0.923$

m = 2;

s2 = $\cos(7\pi/8) + j\sin(7\pi/8) = -0.9238 + j0.383$

m = 3;

s3 = $\cos(9\pi/8) + j\sin(9\pi/8) = -0.9238 - j0.383$

m = 4;

s4 = $\cos(11\pi/8) + j\sin(11\pi/8) = -0.3826 - j0.923$

$$H(s) = \frac{1}{(s2 + 0.765s + 1)(s2 + 1.848s + 1)}$$

Similarly for N = 5, the Butterworth filter transfer function can be derived, which is listed in Table 11.1.

and using the above calculation with $\omega_C = 1.5$, we have

$$H(s) = \frac{1}{\left[s + 1.5/\sqrt{2} + j(1.5/\sqrt{2})\right]\left[s + 1.5/\sqrt{2} - j(1.5/\sqrt{2})\right]}$$

$$= \frac{2.25}{s^2 + 1.5\sqrt{2}s + 2.25} \tag{11.35}$$

Table 11.1   Normalized Butterworth polynomial

Order of N	Factors for the Denominator Polynomial
1	$s + 1$
2	$s^2 + 1.414s + 1$
3	$(s + 1)(s^2 + s + 1)$
4	$(s^2 + 0.765s + 1)(s^2 + 1.848s + 1)$
5	$(s + 1)(s^2 + 0.618s + 1)(s^2 + 1.618s + 1)$

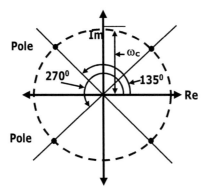

Figure 11.21 Diagrams showing the positions of poles.

Once the transfer function is obtained, the design of the filter is easy. Let us now choose a specific example to illustrate our design steps.

Example 11.4
Design a Butterworth LPF for the following specifications:

1. Pass-band gain required: 0.9
2. Frequency up to which pass-band gain must remain more or less steady ω_1 : 100 rad/s
3. Gain in the attenuation band: 0.4
4. Frequency from which the attenuation must start ω_2: 200 rad/s

Solution 11.4
The design steps are given hereunder:

Step 1: Determination of N and ω_C
Substituting the value of $H(\omega) = 0.9$ for $\omega = 100$ rad/s in Equation (11.31) yields

$$0.9 = \left[\frac{1}{1 + (100/\omega_C)^{2N}} \right]^{1/2} \tag{1}$$

Similarly at $\omega = 200$ rad/s, we have

$$0.4 = \left[\frac{1}{1 + (200/\omega_C)^{2N}} \right]^{1/2} \tag{2}$$

Squaring Equation (1), we get

$$0.81 = \frac{1}{1 + (100/\omega_C)^{2N}} \tag{3}$$

Inverting Equation (3), we get

$$1 + \left(\frac{100}{\omega_C}\right)^{2N} = \frac{1}{0.81} \tag{4}$$

Equation (3) may be simplified to get

$$\left(\frac{100}{\omega_C}\right)^{2N} = 1.234 - 1 = 0.234 \tag{5}$$

Similarly, Equation (2) may be simplified as

$$\left(\frac{200}{\omega_C}\right)^{2N} = \frac{1}{0.16} - 1 = 5.25 \tag{6}$$

Dividing Equation (6) with Equation (5), we get

$$\frac{5.25}{0.234} = \left(\frac{200}{100}\right)^{2N} = 2^{2N} \tag{7}$$

Taking logarithm of both sides of Equation (7), we have

$$2N \log 2 = \log 22.436 \tag{8}$$

From Equation (8), we find that

$$N = 2.244 \tag{9}$$

Equation (9) shows that to get the desired pass-band and attenuation-band characteristics, we must have a filter of order 2.244. However, it is quite customary to specify the order of a given filter in integers or whole numbers. We must, therefore, choose the value of N as the immediate next higher integer. This is chosen because the specifications given insist that we must have an attenuation of at least 0.2. This can be achieved only if we choose the immediate next higher integer as the value of N. It can be easily seen that, if we choose the immediate lower integer as N, then we will not be able to meet the desired attenuation.

Based on a similar argument, we choose the order of the filter $N = 3$

$$\tag{10}$$

With this new value of N, we now calculate the value of cut-off frequency ω_C by using either Equation (5) or Equation (6). It is seen that value of ω_C, obtained by using Equation (6), gives better attenuation characteristics. It may be remembered here that the filter is designed mainly to cut-off frequencies higher than ω_C, and hence a better attenuation characteristics means a better cut-off of undesired frequencies.

Now, using Equation (6) and (10), we get

$$\left(\frac{200}{\omega_C}\right)^6 = 5.25 \tag{11}$$

Solving Equation (11), we find that cut-off frequency

$$\omega_C = 151.7 \text{ rad/s} \tag{12}$$

Step 2: Determination of the poles and expression for H(s)
Using (11.34)

$N = 3;$
$m = 1;$
$s1 = \cos(4\pi/6) + j\sin(4\pi/3) = -0.5 + j0.866$
$m = 2;$
$s2 = \cos(6\pi/6) + j\sin(6\pi/6) = -1.0$
$m = 3;$
$s2 = \cos(8\pi/6) + j\sin(8\pi/6) = -0.5 - j0.866$

$$H(s) = \frac{1}{(s + 0.5 - j0.866)(s + 0.5 + j0.866)(s + 1)}$$

$$= \frac{1}{[(s + 0.5)^2 + (0.866)^2](s + 1)} \tag{13}$$

$$H(s) = \frac{1}{(s + 1.0 + 0.25) + (0.75)} = \frac{1}{(s + 1\,s + 1.0)(s + 1)}$$

The poles are located as shown in Figure 11.22.

It should be noted that the poles of the transfer function complex poles always occur in conjugate.

Modifying H(s), when $\omega_C \gg 1$
In Equation (14), we have used $\omega_C = 1$. However, in most practical cases, $\omega_C \gg 1$. In our case, we had obtained $\omega_C = 151.7$ rad/s. So, we must

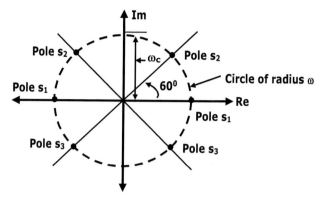

Figure 11.22 Location Butterworth poles for N = 3.

modify the poles by multiplying with ω_C. Thus, the new poles are

$$\text{Pole } s_1 = (-1) \times 151.7 = -151.7$$

$$\text{Pole } s_2 = \left(-\frac{1}{2} + j\frac{\sqrt{3}}{2}\right)(151.7) = -75.85 + j131.376$$

$$\text{Pole } s_3 = \left(-\frac{1}{2} - j\frac{\sqrt{3}}{2}\right)(151.7) = -75.85 - j131.376 \quad (14)$$

Using the above values in Equation (14), we get

$$H(s) = \frac{151.7^3}{(s + 75.85 + j131.376)(s + 75.85 - j131.376)(s + 1)} \quad (15)$$

Equation (15) can be simplified

$$H(s) = \frac{3.491 \times 10^6}{(s^2 + 151.7s + 0.02301 \times 10^6)(s + 151.7)} \quad (16)$$

11.11.4 Design Procedure when H(ω) is Specified in Decibels

Consider now the situation when H(ω) is expressed in decibels. We proceed as follows:

Step 1: Determination of N and ω_C: We have Equation (11.28) giving

$$|H(\omega)| \; in \; dB = -10\log\left[1 + \left(\frac{\omega}{\omega_C}\right)^{2N}\right]$$

Now, at $\omega = \omega_1$, we have

$$|H(\omega)| \; in \; dB = -10 \log \left[1 + \left(\frac{\omega_1}{\omega_C} \right)^{2N} \right] \qquad (11.36)$$

Similarly, at $\omega = \omega_2$, we have

$$|H(\omega)| \; in \; dB = -10 \log \left[1 + \left(\frac{\omega_2}{\omega_C} \right)^{2N} \right] \qquad (11.37)$$

Equations (11.36) and (11.37) can be solved to get the values of N and ω_C. Let us take an example to illustrate this.

Example 11.5
Design a Butterworth LPF for the following specifications:

1. Pass-band gain: -0.5 dB
2. Frequency up to which pass-band gain must remain more or less steady, ω_2: 100 rad/s.
3. Gain in the attenuation band: -20 dB
4. Frequency from which the attenuation must start, ω_2: 200 rad/s.

Solution 11.5
The design steps are given hereunder:

Step 1: Determination of N and ω_C: We have, at $\omega = 100$ rad/s

$$-0.5 = -10 \log \left[1 + \left(\frac{100}{\omega_C} \right)^{2N} \right] \qquad (1)$$

Equation (1) can be solved as shown hereunder to get one of a pair of equations for determining the values of N and ω_C. Rearranging Equation (1), we get

$$0.05 = \log \left[1 + \left(\frac{100}{\omega_C} \right)^{2N} \right] \qquad (2)$$

On further simplification, we get

$$10^{0.05} = 1.122 = 1 + \left(\frac{100}{\omega_C} \right)^{2N} \qquad (3)$$

Reducing Equation (3) yields

$$0.122 = \left(\frac{100}{\omega_C}\right)^{2N} \tag{4}$$

Now, at $\omega = 200$ rad/s, (11.39) gives

$$-20 = -10 \log\left[1 + \left(\frac{200}{\omega_C}\right)^{2N}\right] \tag{5}$$

Simplifying Equation (5), we find

$$99 = \left(\frac{200}{\omega_C}\right)^{2N} \tag{6}$$

Now, we have with us Equation (3) and (6), the two desired equations, to find N and ω_C. Solving them as in case (a), we get the values of N and ω_C. Carrying out the calculations, we find that

Order of the filter required, $N = 5$

Cut $-$ off frequency, $\omega_C = 126.58$ rad/s

Once N and ω_C are obtained, the rest of the procedure is as the same as that in part (a).

We now proceed to design the required 'active' analogue filters. The filter structures used here are known as the Sallen and Key structures, named after their originators. Figures 11.23(a) and 11.23(b) show the second-order and first-order Sallen and Key filter structures. By cascading these basic structures, filters of any order can be obtained.

Based on the above argument, we can construct the fifth-order filter by cascading two second-order filters with a first-order filter. Figure 11.24 shows the structure of the fifth-order filter, based on this theory.

Step 2: Determination of filter components R and C
From the analysis of the second-order filter, we have cut-off frequency given by

$$\omega_C = \frac{1}{RC} \tag{7}$$

It is quite common to assume the value of C at first, since C is available in certain finite values in the open market. We fix C usually at a suitable

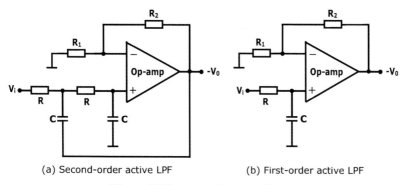

(a) Second-order active LPF   (b) First-order active LPF

Figure 11.23   Active low-pass filters.

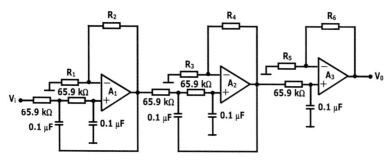

Figure 11.24   Fifth-order active LPF.

value in between 0.01 μF and 1 μF. Here, we choose C = 0.1 μF. Then from Equation (7), we get

$$R = \frac{1}{\omega_C C} \qquad (8)$$

Once ω_C is given, R can be calculated. In Example 11.4, we had found $\omega_C = 151.7$ rad/s. We use this value to find out the value of R. Substituting C = 0.1 μF, and $\omega_C = 151.7$ rad/s in Equation (8), we get

$$R = \frac{1}{0.1 \times 10^{-6} \times 151.7} = 65.9 \ k\Omega \qquad (9)$$

Usually, in low-precision circuits, it its quite common to obtain the value of R by using a standard resistor of 56 kΩ, and then using a pot of 15 kΩ in series with it. However, active filters require precise values R and C to get accurate cut-off frequency. Hence, in these applications, we must use precision components for R and C.

Step 3: Determination of amplifier elements R_1, R_2, R_3, R_4, R_5 and R_6
From analysis and Table 11.1, we have, for a fifth-order filter, the coefficients of the terms given by

$$3 - A_1 = 0.618 \tag{10}$$

$$3 - A_2 = 1.618 \tag{11}$$

where A_1 and A_2 are the gains of the respective amplifiers of the second-order filters used. The gain of the first-order filter can be chosen arbitrarily, as it is not dependent on any factor in particular. We also know that for the non-inverting amplifiers used in our design, the gains are given by

$$A_1 = 1 + \frac{R_2}{R_1} \tag{12}$$

$$A_2 = 1 + \frac{R_4}{R_3} \tag{13}$$

$$A_3 = 1 + \frac{R_6}{R_5} \tag{14}$$

From Equation (10), we get

$$A_1 = 2.382 \tag{15}$$

$$A_2 = 1.382 \tag{16}$$

The gain of the first-order amplifier, as stated earlier, can be arbitrarily chosen. Let us choose $A_3 = 50$. If the cut-off frequency of the filter is high, then due to the limitations of the op-amp gain at higher frequencies, the gain chosen must be low. This is the only restriction we impose on the gain of the amplifier of the first-order filter. From Equations (12) and (15), we find

$$\frac{R_2}{R_1} = 1.382 \tag{17}$$

Fixing $R_1 = $ kΩ (arbitrarily chosen), we get $R_2 = 13.82$ kΩ from Equation (17). It may be noted that it is customary to choose the values of R between 1 kΩ and 1 MΩ in practical circuits.

In a similar fashion, from Equations (13) and (16), we find $R_3 = 10$ kΩ, and $R_4 = 3.82$ kΩ. Finally, we choose the elements of the first-order filters, for a gain of 50, as $R_5 = 10$ kΩ and $R_6 = 49$ kΩ. This completes our design of the fifth order active filter. The completely designed filter circuit with values incorporated is shown in Figure 11.24.

Example 11.6
This example is used here to summarize the two approaches given in parts (a) and (b) of the design procedure. Given the desired specifications as

1. Pass-band gain required: 0.89 or ± 1 dB
2. Frequency up to which pass-band gain must remain more or less steady: 25 Hz
3. Amount of attenuation required: 0.215 or -13.35 dB
4. Frequency from which the attenuation must start: 75 Hz

Solution 11.6
In this case, we have carefully chosen the values of H(ω) so that the end results of the two approaches will be the same.

We proceed as given in parts (a) and (b) of the design procedure. First, we start as in part (a).

Step 1: Determination of N and ω_C
We have $H_1(\omega) = 0.89$ (given). Now

$$\omega_1 = 2\pi f_1 = 2\pi \times 25 = 157 \ rad/s$$

and

$$\omega_2 = 2\pi f_2 = 2\pi \times 75 = 471 \ rad/s$$

For $\omega_1 = 157$ rad/s in Equation (8.9) yields

$$0.89 = \left[\frac{1}{1 + (157/\omega_C)^{2N}} \right]^{1/2} \tag{1}$$

Similarly, at $\omega = 471$ rad/s, we have

$$0.215 = \left[\frac{1}{1 + (471/\omega_C)^{2N}} \right]^{1/2} \tag{2}$$

Squaring Equation (1), we get

$$0.792 = \frac{1}{1 + (157/\omega_C)^{2N}} \tag{3}$$

Inverting Equation (3), we get

$$1 + \left(\frac{157}{\omega_C} \right)^{2N} = \frac{1}{0.792} \tag{4}$$

Equation (3) may be simplified to get

$$\left(\frac{157}{\omega_C}\right)^{2N} = 1.262 - 1 = 0.262 \tag{5}$$

Similarly, rearranging Equation (5)

$$1 + \left(\frac{471}{\omega_C}\right)^{2N} = \frac{1}{0.046} = 21.63 \tag{6}$$

Further simplification yields

$$1 + \left(\frac{471}{\omega_C}\right)^{2N} = 20.63 \tag{7}$$

Now, divide Equation (5) with Equation (7) to get

$$\left(\frac{471}{157}\right)^{2N} = \frac{20.63}{0.262} \tag{8}$$

Simplifying Equation (8), we have

$$3^{2N} = 78.74 \tag{9}$$

Taking logarithm of both sides and simplifying, we get

$$N = 1.98 \tag{10}$$

Choosing the next higher value, we get

$$N = 2 \tag{11}$$

To find the corresponding ω_C, we use Equation (7). Substituting for N = 2 in Equation (17) of example 11.2, and solving, we get

$$\omega_C = 221 \text{ rad/s} \tag{12}$$

Next, we proceed to solve for N and ω_C as in part (b).
We have, at $\omega = 157$ rad/s

$$-1 = -10\log\left[1 + \left(\frac{157}{\omega_C}\right)^{2N}\right] \tag{13}$$

Rearranging Equation (13), we have

$$0.01 = \log \left[1 + \left(\frac{157}{\omega_C}\right)^{2N}\right] \tag{14}$$

On further simplification, we get

$$10^{0.1} = 1.259 = 1 + \left(\frac{157}{\omega_C}\right)^{2N} \tag{15}$$

Reducing Equation (15) yields

$$0.1259 = \left(\frac{157}{\omega_C}\right)^{2N} \tag{16}$$

Now, when $= 471$ rad/s, Equation (9.6) gives

$$-13.35 = -10 = \log \left[1 + \left(\frac{471}{\omega_C}\right)^{2N}\right] \tag{17}$$

Simplifying Equation (17), we find

$$21.62 = \left(\frac{471}{\omega_C}\right)^{2N} \tag{18}$$

Now, we have with us Equations (16) and (18), the desired equations, to find N and ω_C. it can be seen that they are almost the same as Equations (5) and (7), respectively. Solving then we get the same values of N and ω_C as in part (a).

Steps 2 and 3: Determination of the poles and valid poles of H(s)
Once we get N and ω_C, we proceed to find the valid poles of the transfer function. For order 2, and $\omega_C = 1$, we have already found the poles and valid poles of H(s). Multiplying these values with ω_C, we get the valid poles of H(s). The poles are

$$s_1, s_2 = \omega_C \left(\frac{1}{\sqrt{2}} \pm j\frac{1}{\sqrt{2}}\right) = 221 \left(\frac{1}{\sqrt{2}} \pm j\frac{1}{\sqrt{2}}\right) = 56.27 \pm j56.27 \tag{19}$$

The transfer function H(s) can now be written as

$$H(s) = \frac{6332.626}{(s + 56.27 + j56.27)(s + 56.27 + j56.27)}$$

$$= \frac{6332.626}{s^2 + 11.54 + 6332.626} \tag{20}$$

11.12 Chebyshev Approximation

When pass-band gain contains ripples, we use the Chebyshev approximation. Usually, Chebyshev polynomials are used in filter design, when the pass-band gain contains ripples greater than $\pm$ 1 dB. The frequency-response characteristics of a Chebyshev LPF are shown in Figure 11.25. Figure 11.25(a) shows the characteristic, when $H(\omega)$ is expressed as a mere number, and Figure 11.25(b) shows the characteristic, when $H(\omega)$ is expressed in decibels.

In the case of the Butterworth design, from the given specifications, we determine the order N and the cut-off frequency ω_C of the desired filter. However, in the case of Chebyshev design, from the given specifications, we

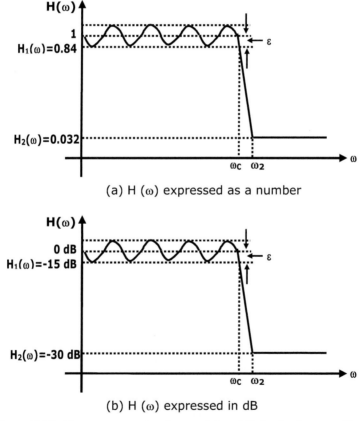

(a) H (ω) expressed as a number

(b) H (ω) expressed in dB

Figure 11.25   Frequency-response characteristics of Chebyshev low-pass filter.

determine only N, since the specifications contain ω_C. Using N, we determine the poles, and hence the transfer function of the desired Chebyshev filter.

It is found that the poles of the Butterworth polynomial lie on a circle, whose radius is ω_C, while the poles of Chebyshev filter lies on the ellipse. The Chebyshev makes use of the Chebyshev polynomial, which, in equation form, is given by

$$|H(\omega)|^2 = \frac{1}{1 + \varepsilon^2 C_N^2 \left(\frac{\omega}{\omega_C}\right)} \tag{11.38}$$

For finding polynomial of the normalized Chebyshev, LPF is described by a relation that looks trigonometric in nature where $C_N(x)$ is actually a polynomial of degree N.

$$C_N(x) = \cos[N\phi] = \text{where} \quad \phi = \left[\cos^{-1}(x)\right] \tag{11.39}$$

$$C_N(x) = \cos\left[N\cos^{-1}(x)\right] \tag{11.40}$$

$$C_0(x) = \cos[0] = 1 \quad \text{and} \quad C_1(x) = \cos\left[\cos^{-1}(x)\right] = x \tag{11.41}$$

$$\text{Where } x = \frac{\omega}{\omega_c} \tag{11.42}$$

Using trigonometric identities

$$\cos(\alpha \pm \beta) = \cos\alpha \, \cos\beta \mp \sin\alpha \, \sin\beta$$

$$\cos[(N+1)\phi] + \cos[(N-1)\phi] = 2\cos[N\phi]\cos[\phi] \tag{11.43}$$

$$C_{N+1}(x) = 2\cos[N\phi]\cos[\phi] - C_{N-1}(x) \quad \text{for} \quad N \geq 1 \tag{11.44}$$

N = 1

$$C_2(x) = 2\cos[\phi]\cos[\phi] - C_0(x) = C_2(x) = 2\,x\,x - 1 = 2\,x^2 - 1$$

The normalized Chebyshev polynomial up to order 5 has been calculated using (11.43) and (11.44) and been placed below in tabular form without the ripple factor (ϵ) *and given in tabular form for different values of N.* Students are advised to verify themselves that how it has been derived.

Normalized Chebyshev Polynomial without value of ϵ

Order of N	$C_N(x)$, $x = \omega/\omega_C$
0	1
1	x
2	$2\,x^2 - 1$
3	$4\,x^3 - 3\,x$
4	$8\,x^4 - 8\,x^2 + 1$
5	$16\,x^5 - 20$
	$x^3 + 5\,x$

$$|H(\omega)|^2 = \frac{1}{1 + \varepsilon^2 C_N^2(x)} \qquad (11.45)$$

Using (11.45), the normalized Chebyshev polynomial, with the ripple factor (ϵ), is derived and is given in tabular form for different values of ϵ. The formula for the pass-band ripple (A_P) and the ripple factor (ϵ) is given in (11.46).

$$A_P(dB) = 10\log(1 + \varepsilon^2) \qquad (11.46)$$

$$\varepsilon = \sqrt{10^{A_P} - 1} \qquad (11.47)$$

This method of finding polynomial of the normalized Chebyshev LPF, which lies on an ellipse in the s-plane can be found with the help of the following formula of coordinates:

$$\alpha = \frac{1}{N}\sinh^{-1}\left(\frac{1}{\varepsilon}\right) \qquad (11.48)$$

$$s_m = -\sinh(\alpha)\sin[\tfrac{2m-1}{2N}]\pi + j\cosh(\alpha)\cos[\tfrac{2m-1}{2N}]\pi \quad m = 1, 2, \ldots, N \qquad (11.49)$$

Using (11.48) and (11.49)

N = 1

$\epsilon = 0.3493$

m = 1

$\alpha = (1/N) * (a\sinh(1/\epsilon))$

$\alpha = 1.774$

$s1 = -\sinh(1.774) * \sin(\pi/2) + j\cosh(1.774)\cos(\pi/2)$

$s1 = -2.863$

$$H(s) = \frac{1}{s + 2.863}$$

Using (11.48) and (11.49)

$N = 2$

$\epsilon = 0.3493$

$m = 1, 2$

$\alpha = (1/N) * (a \sinh(1/\epsilon))$

$\alpha = 1.774/2 = 0.887$

$s1 = -\sinh(0.887) * \sin(\pi/4) + j\cosh(0.887)\cos(\pi/4)$

$s1 = -0.7128 - j1.0041$

$s2 = -\sinh(0.887)\sin(3\pi/4) + j\cosh(0.887)\cos(3\pi/4)$

$s2 = -0.7128 + j1.0041$

$$H(s) = \frac{1}{s^2 + 1.426s + 1.5164}$$

Normalized Chebyshev Polynomial *Pass-band 0.5 dB Ripple ($\epsilon = 0.3493$)*

Order of N	Factors of Polynomial ($s = j\omega$)
1	$s + 2.863$
2	$s^2 + 1.426s + 1.5164$
3	$(s + 0.626)(s^2 + 0.626s + 1.42453)$
4	$(s^2 + 0.350s + 1.062881)(s^2 + 0.846s + 0.35617)$

Normalized Chebyshev Polynomial *Pass-band 3.0 dB Ripple ($\epsilon = 0.9953$)*

Order of N	Factors of Polynomial ($s = j\omega$)
1	$s + 1.002$
2	$s^2 + 0.2986s + 0.8395$
3	$(s + 0.299)(s^2 + 0.2986s + 0.8395)$
4	$(s^2 + 0.17s + 0.90214)(s^2 + 0.412s + 0.1961)$

where $\epsilon =$ amount of ripple in the magnitude, and $C_N =$ Chebyshev coefficient, given by

$$C_N = \cosh\left(N \cosh^{-1}\frac{\omega}{\omega_C}\right), \quad if \frac{\omega}{\omega_C} \geq 1 \qquad (11.50)$$

and

$$C_N = \cos\left(N \cos^{-1}\frac{\omega}{\omega_C}\right), \quad if \frac{\omega}{\omega_C} \leq 1 \qquad (11.51)$$

We usually will have to use the first of the two conditions in our designs, as in most of the design, $(\omega/\omega_C)>1$. The Chebyshev polynomial may be expressed in decibels also as

$$|H(\omega)| \; in \; dB = 20 \log \left(\frac{1}{1 + \varepsilon^2 C_N^2} \right)^{1/2} \tag{11.52}$$

Simplifying Equation (11.50), we get

$$|H(\omega)| \; in \; dB = -10 \log(1 + \varepsilon^2 C_N^2) \tag{11.53}$$

11.12.1 Step-by-step Design of Chebyshev LPF

As in the case of the Butterworth design, we have to first give the specifications desired of the Chebyshev LPF. The specifications may be stated as follows:

1. Pass-band gain required: $H_1(\omega)$
2. Frequency up to which pass-band gain must remain more or less steady: ω_1
3. Amount of attenuation required: $H_2(\omega)$
4. Frequency from which the attenuation must start: ω_2

Let us take a practical example for illustrating the design.

Example 11.7

Design a Chebyshev LPF for the following specifications:

1. Pass-band gain required: 0.84 (or -1.5 dB)
2. Frequency up to which pass-band gain must remain more or less steady state 150 rad/s
3. Amount of attenuation required: 0.0316 (or -30 dB)
4. Frequency from which the attenuation must start, ω_2: 300 rad/s

Solution 11.7

As in the case of the Butterworth design, in this case also, $H(\omega)$ may be specified as a number, or in decibels. We illustrate both these designs. First, we discuss the design steps, when $H(\omega)$ is specified as a number.

(a) **H (ω) Specified as a Number Step 1: Determination of N**
    Given the expression

$$|H(\omega)|^2 = \frac{1}{1 + \varepsilon^2 C_N^2} \tag{1}$$

at $\omega_1 = \omega_C = 150$ rad/s, we have

$$|H_1(\omega)|^2 = \frac{1}{1 + \varepsilon^2} \tag{2}$$

since $C_N = 1$, for $\omega = \omega_C$. Substituting for $H_1(\omega) = 0.84$ in Equation (2), we get

$$0.84^2 = \frac{1}{1 + \varepsilon^2} \quad or \quad \varepsilon = 0.646 \tag{3}$$

Rearranging Equation (3), and solving, we get

$$\varepsilon^2 = 0.417 \tag{4}$$

Now, using the second condition, we have at $\omega_2 = 300$ rad/s

$$0.0316^2 = \frac{1}{1 + 0.417C_N^2} \quad or \quad C_N = 48.98 \tag{5}$$

Substituting $C_N = 48.98$ in Equation (11.50a), we get

$$48.98 = \cosh\left(N \cosh^{-1} \frac{\omega}{\omega_C}\right) \tag{6}$$

In Equation (6), we substitute the values of $\omega_2 = 300$ rad/s, and $\omega_C = 150$ rad/s to give

$$\cosh^{-1}(49.98) = N \cosh^{-1} \frac{300}{150} \quad or \quad N = \frac{\cosh^{-1}(48.98)}{\cosh^{-1} 2} = 3.48 \tag{7}$$

Choosing the next higher value, the order of the desired filter N $= 4$.

(b) $H(\omega)$ Specified in Decibels

For this design, we use Equation (11.52). Substituting in it for $H_1(\omega) = -1.5$ dB, we get

$$-1.5 = -10 \log(1 + \varepsilon^2) \quad or \quad \varepsilon^2 = 0.413 \tag{8}$$

Similarly, using the second condition, we have at $\omega_2 = 300$ rad/s

$$-30 = -10 \log(1 + 0.413C_N^2) \quad or \quad C_N = 49.18 \tag{9}$$

This is very close to the value of C_N obtained. Therefore, we find N $= 4$.

Step 2: Determination of the poles of H(s)

The polynomial of the normalized Chebyshev LPF, which lies on an ellipse in the s-plane can be found as follows:

$$s_m = \sigma_m + j\omega_m$$

$$\alpha = \frac{1}{N}\sinh^{-1}\left(\frac{1}{\varepsilon}\right) \tag{10}$$

$$s_m = -\sinh(\alpha)\sin[\tfrac{2m-1}{2N}]\pi + j\cosh(\alpha)\cos[\tfrac{2m-1}{2N}]\pi \quad m = 1, 2, ..., N \tag{11}$$

Rule 2: Determine the factor α from the expression

$$\alpha = \frac{1}{N}\sinh^{-1}\frac{1}{\varepsilon}$$

Rule 2: We now find α from Equation (11.41) as

Using (11.39) and (11.40)

N = 4

$\epsilon = 0.3493$

m = 1

$\alpha = (1/N) * (a\sinh(1/\epsilon))$

$\alpha = 0.4435$

s1 $= -\sinh(0.4435) * \sin(\pi/8) + j\cosh(0.4435)\cos(\pi/8)$

s1 $= -0.1754 - j1.0163$

s2 $= -\sinh(0.4435) * \sin(3\pi/8) + j\cosh(0.4435)\cos(3\pi/8)$

s2 $= -0.4233 - 0.4209j$

s3 $= -\sinh(0.4435) * \sin(5\pi/8) + j\cosh(0.4435)\cos(5\pi/8)$

s3 $= -0.4233 + 0.4209i$

s4 $= -\sinh(0.4435) * \sin(7\pi/8) + j\cosh(0.4435)\cos(7\pi/8)$

s4 $= -0.1754 + j1.0163$

$$H(s) = \frac{1}{(s^2 + 0.350s + 1.062881)(s^2 + 0.846s + 0.35617)}$$

$$\omega_C = 150\text{rad/s}$$

Substituting the values obtained in Equations (1) to (11), we get the required transfer function as

$$H(s) = \frac{12.269.10^6}{(s^2 + 79.944s + 5215.783)(s^2 + 35.608s + 5352.255)} \tag{12}$$

The expression obtained in Equation (12) can be used to design the desired filter.

11.13 Butterworth and Chebyshev Filters' Comparison

We can now make a comparison between the Butterworth and the Chebyshev filters. As pointed out earlier, Butterworth filters show maximally flat pass-band gain. Their frequency response characteristics show no ripples in the pass- and stop-bands.

The Chebyshev filters permit certain finite amount of ripples in the pass-band, but no ripples in the stop-band. However, for the inverse Chebyshev filter, the ripples are in the stop-band, and there are no ripples in the pass-band.

If ripples are permissible in the pass-band, then to meet the same specifications, the order of the desired Chebyshev filter can be seen to be lower than that of the desired Butterworth filter. For example, assume that, to meet a given set of specifications, we require a fifth order Butterworth filter. Then it can be seen that these same specifications can be met with a Chebyshev filter of the fourth or still lower order. This means that we require smaller number of components for the construction of Chebyshev filters than Butterworth filters.

As stated above, even though the design specifications can be met with a lower-order Chebyshev filter, their design procedure is more complex than that of the Butterworth filter. However, the design steps of the two types of filters differ only up to the determination of the transfer function H(s). Once H(s) is determined, the remaining procedure for the actual (physical) implementation of the two filters merges into one common procedure. This statement is true for other types of filters also. So, we conclude that the difference in the design procedures among various types of filters exist only up to the determination of the transfer function H(s). Once this is fixed, the physical implementation of the filter follows the same procedures for all types of filters. It may be noted that the major difference among various types of filters lies in the amount of ripples permissible in the pass- and stop-bands. The higher the amount of ripples permitted, the lower the order of the filter required. We now consider a design example to prove that, given the same set of specifications, the order of the desired Chebyshev filter will be lower than that of the desired Butterworth filter.

Example 11.8
Design a Butterworth/Chebyshev LPF for the following specification.

1. Pass-band gain required: 0.84
2. Frequency up to which pass-band gain must remain more or less steady, ω:150 rad/s

3. Amount of attenuation required: 0.0316
4. Frequency from which the attenuation must start, w_2: 300 rad/s.

Solution 11.8

As in the case of the Butterworth design, H(w) may be specified as a number in decibels. We illustrate both these designs. Here, we discuss the design steps, when H(w) is specified as a number.

(a) **Butterworth design-determination of N**
   Given

$$|H(w)| = \left[\frac{1}{1 + (w/w_C)^{2N}}\right]^{1/2}$$

   Substituting values, and assuming $w_1 \neq w_C$, we get

$$0.84^2 = \frac{1}{1 + (150/w_C)^{2N}}$$

   where we have used $w_1 = 150$ rad/s. Solving (1), we find

$$\left(\frac{150}{w_C}\right)^{2N} = 0.417 \tag{1}$$

   Similarly, we have at $w_2 = 300$ rad/s

$$0.0316^2 = \frac{1}{1 + (300/w_C)^{2N}} \tag{2}$$

   Solving Equation (2) yields

$$\left(\frac{300}{w_C}\right)^{2N} = 1000 \tag{3}$$

   Dividing Equation (3) with Equation (1) and solving, we get the order of the filter as N = 6.

(b) **Chebyshev design determination of N**
   From the specifications given in Example 11.7, we have already determined the value of N for the Chebyshev design, and this is obtained as

$$N = 4 \tag{3}$$

   Comparing Equations (3) and (3), we find that the order of the desired Chebyshev filter is lower than that of the desired Butterworth filter. This proves our previous argument.

11.14 Practice Problems

Practice Problem 11.1
Develop an analogue computer setup to solve the differential equation

$$10\frac{d^3y}{dt^3} + 5\frac{d^2y}{dt^2} + 2\frac{dy}{dt} + y = x$$

Practice Problem 11.2
Develop an analogue computer setup to solve the differential equation

$$\frac{d^5y}{dt^5} - 6\frac{d^4y}{dt^4} + \frac{d^3y}{dt^3} + 3\frac{d^2y}{dt^2} - 2\frac{dy}{dt} + 7y = 12x$$

Practice Problem 11.3
Develop an analogue computer setup to solve the differential equation

$$\frac{d^6y}{dt^6} - 6\frac{d^4y}{dt^4} + \frac{d^3y}{dt^3} + 7\frac{d^2y}{dt^2} = 12x$$

Practice Problem 11.4
Develop an analogue computer setup to obtain the function $y(t) = A\cos\omega t$.
[Hint: Develop the relevant differential equation by differentiating the given function twice.]

Practice Problem 11.5
Design a passive first-order (a) LP RC filter for a cut-off frequency of 5 kHz. Specify the values of the components used.

Practice Problem 11.6
Design a passive first-order (a) band-pass and (b) band-reject RC filter for cut-off frequencies of 2 and 4 kHz, respectively. Specify the values of the components used.

Practice Problem 11.7
Design a passive second-order (a) LP and (b) RLC filter for a cut-off frequency of 5 kHz. Specify the values of the components used.

Practice Problem 11.8
Design a passive second-order (a) band-pass and (b) band-reject RLC filter for cut-off frequencies of 2 and 4 kHz, respectively. Specify the values of the components used.

Practice Problem 11.9

Design a Butterworth filter for the following specifications:

1. Pass-band gain required: 0.85
2. Frequency up to which pass-band gain must remain more or less steady, ω_1: 1000
3. Amount of attenuation required: 0.10
4. Frequency from which attenuation must start, ω_2: 3000 rad/s

Practice Problem 11.10

Design a Butterworth filter for the following specifications:

1. Pass-band gain required: 0.95
2. Frequency up to which pass-band gain must remain more or less steady, ω_1: 500 Hz
3. Amount of attenuation required: 0.20
4. Frequency from which attenuation must start, ω_2: 3000 Hz

Practice Problem 11.11

Design a Butterworth filter for the following specifications:

1. Pass-band gain required: -1 dB
2. Frequency up to which pass-band gain must remain more or less steady, ω_1: 200 Hz
3. Amount of attenuation required: -40 dB
4. Frequency from which attenuation must start, ω_2: 600 Hz

Practice Problem 11.12

Design a Butterworth filter for the following specifications:

1. Pass-band gain required: -0.5 dB
2. Frequency up to which pass-band gain must remain more or less steady, ω_1: 100 Hz
3. Amount of attenuation required: -30 dB
4. Frequency from which attenuation must start, ω_2: 500 Hz

Practice Problem 11.13

Design a Butterworth filter for the following specifications:

1. Pass-band gain required: -1.5 dB
2. Frequency up to which pass-band gain must remain more or less steady, ω_1: 250 Hz
3. Amount of attenuation required: -35 dB
4. Frequency from which attenuation must start, ω_2: 800 Hz

Practice Problem 11.14

Design a Chebyshev filter for the following specifications:

1. Pass-band gain required: -2 dB
2. Frequency up to which pass-band gain must remain more or less steady, ω_1: 250 Hz
3. Amount of attenuation required: -40 dB
4. Frequency from which attenuation must start, ω_2: 800 Hz

Practice Problem 11.15

Design a Chebyshev filter for the following specifications:

1. Pass-band gain required: 0.85
2. Frequency up to which pass-band gain must remain more or less steady, ω_C: 1000 Hz
3. Amount of attenuation required: 0.10
4. Frequency from which attenuation must start, ω_2: 300 Hz

Practice Problem 11.16

Design a Chebyshev filter for the following specifications:

1. Pass-band gain required: -1 dB
2. Frequency up to which pass-band gain must remain more or less steady, ω_1: 200 Hz
3. Amount of attenuation required: -40 dB
4. Frequency from which attenuation must start, ω_2: 600 Hz

Practice Problem 11.17

Design a Chebyshev filter for the following specifications:

1. Pass-band gain required: -2.5 dB
2. Frequency up to which pass-band gain must remain more or less steady, ω_1: 100 Hz
3. Amount of attenuation required: -50 dB
4. Frequency from which attenuation must start, ω_2: 500 rad/s

Practice Problem 11.18

Design a Chebyshev filter for the following specifications:

1. Pass-band gain required: -3 dB
2. Frequency up to which pass-band gain must remain more or less steady, ω_1: 50 rad/s
3. Amount of attenuation required: -20 dB
4. Frequency from which attenuation must start, ω_2: 250 rad/s

Practice Problem 11.19

Design Butterworth and Chebyshev filters for the following specifications [Consider ω_1 (Butterworth design) $= \omega_C$ (Chebyshev design)]:

1. Pass-band gain required: -2 dB.
2. Frequency up to which pass-band gain must remain more or less steady, ω_1: 100 rad/s
3. Amount of attenuation required: -20 dB
4. Frequency from which attenuation must start, ω_2: 250 rad/s

Practice Problem 11.20

Design Butterworth and Chebyshev filters for the following specifications:

1. Pass-band gain required: -1.5 dB
2. Frequency up to which pass-band gain must remain more or less steady, ω_1: 200 Hz
3. Amount of attenuation required: -30 dB
4. Frequency from which attenuation must start, ω_2: 400 Hz

Practice Problem 11.21

Design Butterworth and Chebyshev filters for the following specifications:

1. Pass-band gain required: 0.75
2. Frequency up to which pass-band gain must remain more or less steady, ω_1: 1500 Hz
3. Amount of attenuation required: 0.4
4. Frequency from which attenuation must start, ω_2: 4000 Hz

Practice Problem 11.22

Design Butterworth and Chebyshev filters for the following specifications:

1. Pass-band gain required: -0.88
2. Frequency up to which pass-band gain must remain more or less steady, ω_1: 300 Hz
3. Amount of attenuation required: 0.2
4. Frequency from which the attenuation must start, ω_2: 600 Hz

Practice Problem 11.23

Design Butterworth and Chebyshev filters for the following specifications:

1. Pass-band gain required: -0.85
2. Frequency up to which pass-band gain must remain more or less steady, ω_1: 500 Hz
3. Amount of attenuation required: 0.15
4. Frequency from which attenuation must start, ω_2: 600 Hz

12

Future Trends

This chapter provides applications in the form of presented and published articles in well-known journals.

12.1 Skin Lesion Segmentation from Dermoscopic Images using Convolutional Neural Network

Kashan Zafar[1], Syed Omer Gilani[1*], Asim Waris[1], Ali Ahmed[1], Mohsin Jamil[2] Muhammad Nasir Khan[3] and Amer Sohail Kashif[1]

[1]Department of Biomedical Engineering and Sciences, School of Mechanical and Manufacturing Engineering, National University of Sciences and Technology, Islamabad 44000, Pakistan
[2]Department of Electrical and Computer Engineering, Memorial University of Newfoundland, Canada
[3]Department of Electrical Engineering, University of Lahore, Pakistan
*Corresponding author: omer@smme.nust.edu.pk;

Clinical treatment of skin lesion is primarily dependent on timely detection and delimitation of lesion boundaries for accurate cancerous region localization. Prevalence of skin cancer is on the higher side, especially that of melanoma, which is aggressive in nature due to its high metastasis rate. Therefore, timely diagnosis is critical for its treatment before the onset of malignancy. To address this problem, medical imaging is used for the analysis and segmentation of lesion boundaries from dermoscopic images. Various methods have been used, ranging from visual inspection to the textural analysis of the images. However, accuracy of these methods is low for proper clinical treatment because of the sensitivity involved in surgical procedures or drug application. This presents an opportunity to develop an automated model with good accuracy so that it may be used in a clinical setting. This paper proposes an automated method for segmenting lesion boundaries that combines two architectures, the U-Net and the ResNet, collectively called Res-Unet. Moreover, we also used image inpainting for hair removal, which improved the segmentation results significantly. We trained our model on the ISIC 2017 dataset and validated it on the ISIC 2017 test set as well as the PH^2 dataset. Our proposed model attained a Jaccard Index of 0.772 on the ISIC 2017 test set and 0.854 on the PH^2 dataset, which are comparable results to the currently available state-of-the-art techniques.

12.1.1 Introduction

Computer-aided technologies for the diagnostic analysis of medical images have received significant attention from the research community. These are efficiently designed and modified for the purposes of inter-alia segmentation and classification of the region of interest (ROI) [1], which in this instance involves cancerous regions. Needless to mention, the effective treatment of cancer is dependent on early detection and delimitation of lesion boundaries, particularly during its nascent stages because cancer generally has the characteristic tendency of delayed clinical onset [2]. Every year, nearly 17 million people are affected by cancer and about 9.6 million people die due to delayed diagnosis and treatment [3]. This makes cancer the leading cause of death worldwide [4]. In the case of skin cancer, it is one of the most prevalent types of the disease in both adults and children [5] and occurs or originates in the epidermal tissue. Various computer-aided techniques have been proposed for cancer boundary detection from dermoscopic images [3].

Among the different types of skin cancers, melanoma is not only the most dangerous and aggressive in nature due to its high metastasis rate, but also has the greatest prevalence [4]. Melanoma is a malignant type of skin cancer that develops through the irregular growth of pigmented skin cells called melanocytes [5]. It can develop anywhere on the epidermal layer of the skin and presumably may also affect the chest and back and propagate from the primary site of the cancer [6]. Its incidence rate has risen up to 4–6% annually and has the highest mortality rate among all types of skin cancers [7]. Early diagnosis is crucial as it increases the 5-year survival rate up to 98% [8,9].

From the above details pertaining to the incidence and mortality rate associated with melanoma, timely diagnosis becomes all the more necessary for providing effective treatment to the affected. Insofar as the detection and segmentation of lesion boundaries, there are two streams of methodologies: first, traditional methods that usually resort to visual inspection by the clinician, and second, semi-automated and automated methods, which mostly involve point-based pixel intensity operations [10,11], pixel clustering methods [12–15], level set methods [16], deformable models [17], deep-learning based methods [18–20], etc.

Be that as it may, most of the methods being used today are not semi-automated because the accuracy associated therewith is generally prone to errors due to the following reasons: the inherent limitations of the methods [21], and changing character of dermoscopic images induced due to the florescence and brightness inhomogeneities [22]. For this very reason, the world has shifted towards more sophisticated methods, inter-alia, the convolutional neural networks (CNNs) [23].

In this paper, we intend to exploit the properties and model architectures based on CNN for skin lesion boundary delimitation and segmentation. In addition, we propose our own novelty within the already available techniques which greatly increases the segmentation accuracy, which is image inpainting. Image inpainting, together with other image pre-processing techniques such as morphological operations, is used to remove the hair structures contained within the dermoscopic images that otherwise handicap the architecture because of complexities present in the images.

This research examines the accuracy of the proposed technique together with the adoption of the proposed pre-processing method. We also benchmark our proposed scheme with other available methods by way of results through network accuracy, Jaccard Index, Dice score and other performance metrics that aid us in comparison.

12.1.1.1 Literature Review

This section delineates and chalks-out the relevant work done on the issue of segmentation of skin lesions. It is done with an added emphasis and focus on the recent studies that have incorporated deep-learning methods for the aforementioned purpose of lesion segmentation.

At the outset, it is contended that accurate segmentation and delimitation of skin lesion boundaries can aid and assist the clinician in the detection and diagnosis process, and may later also help towards classification of the lesion type. There has been a gamut of studies done for the purposes of segmentation and classification of skin lesions, and for a general survey of these, the reader can refer the following two papers authored by Oliveira et al. [24] and Rafael et al. [25].

We hereinafter present a review of the literature vis-à-vis two aspects (i.e., pre-processing and segmentation techniques, respectively). Both aspects have a direct effect on the outcome of the results (the prediction) and therefore, both are catered into the broader scheme of methodology presented in this paper. Additionally, since dermoscopic images have varying complexities and contain different textural, intensity and feature inhomogeneity, it becomes necessary to apply prior pre-processing techniques so that inhomogeneous sections can be smoothened out.

12.1.1.1.1 Pre-processing techniques

Researchers encounter complications while segmenting skin lesions due to low brightness and the noise present in the images. These artefacts affect the accuracy of segmentation. For better results, Celebi et al. [26] proposed a technique that enhances image contrast by searching for idyllic weights for converting RGB images into greyscale by maximizing Otsu's histogram bimodality measure. Optimization resulted in a better adaptive ability to distinguish between tumour and skin and allowed for accurate resolution of the regions, whereas Beuren et al. [27] described the morphological operation that can be applied on the image for contrast enhancement. The lesion is highlighted through the colour morphological filter and simply segmented through binarization. Lee et al. [28] proposed a method to remove hair artefacts from dermoscopic images. An algorithm based on morphological operations was designed to remove hair-like artefacts from skin images. Removing hair, characterized as noise, from skin images has a noteworthy effect on segmentation results. A median filter was found to be effective on noisy images. A non-linear filter was applied to images to smooth them [29]. Celebi et al. [30] established a concept where the size of the filter to be applied must be proportional to the size of the image for effective smoothing.

Image inpainting is a pre-processing technique used for both removing parts from an image and for restoration purposes, so that the missing and damaged information in images is restored. It is of vital importance in the field of medical imaging and through its application, unnecessary structures or artefacts from the images (i.e., hair artefacts in skin lesions images) can be removed [31–33].

12.1.1.1.2 Segmentation techniques

Most image segmentation tasks use traditional machine learning processes for feature extraction. The literature explains some of the important techniques used for accurate segmentation. Jaisakthi et al. [34] summarizes a semi-supervised method for segmenting skin lesions. Grab-cut techniques and K-means clustering are employed conjunctively for segmentation. After the former segments the melanoma through graph cuts, the latter fine-tunes the boundaries of the lesion. Pre-processing techniques such as image normalization and noise removal measures are used on the input images before feeding them to the pixel classifier. Mohanad Aljanabi et al. [35] proposed an artificial bee colony (ABC) method to segment skin lesions. Utilizing fewer parameters, the model is a swarm-based scheme involving pre-processing of the digital images, followed by determining the optimum threshold value of the melanoma through which the lesion is segmented, as done by Otsu thresholding. High specificity and Jaccard Index are achieved by this algorithm.

Pennisi et al. [36] introduced a technique that segments images using the Delaunay triangulation method (DTM). The approach involves parallel segmentation techniques that generate two varying images that are then merged to obtain the final lesion mask. Artefacts are removed from the images after which one process filters out the skin from the images to provide a binary mask of the lesion, and similarly, the other technique utilizes Delaunay triangulation to produce the mask. Both of these are combined to obtain the extracted lesion. The DTM technique is automated and does not require a training process, which is why it is faster than other methods. Emre Celebi et al. [37] provide a brief overview of the border detection techniques (i.e., edge based, region based, histogram thresholding, active contours, clustering, etc.) and especially pays attention to evaluation aspects and computational issues. Lei Bi et al. [38] suggested a new automated method that performed image segmentation using image-wise supervised learning (ISL) and multiscale super pixel based cellular automata (MSCA). The authors used probabilistic mapping for automatic seed selection that removes user-defined seed selection; afterwards, the MSCA model was employed for segmenting skin lesions. Ashnil Kumar et al. [39] introduced a fully convolutional network (FCN) based method for segmenting dermoscopic images. Image features were learned from embedded multistages of the FCN and achieved an improved segmentation accuracy (than previous works) of skin lesion without employing any pre-processing part (i.e., hair removal, contrast improvement, etc.). Yading Yuan et al. [40] proposed a convolution deconvolutional neural network (CDNN) to automate the process of the segmentation of skin lesions. This paper focused on training strategies that makes the model more efficient, as opposed to the use of various pre- and post-processing techniques. The model generates probability maps where the elements correspond to the probability of pixels belonging to the melanoma. Berseth et al. [41] developed a U-Net architecture for segmenting skin lesions based on the probability map of the image dimension where the tenfold cross validation technique was used for training the model. Mishra [42] presented a deep learning technique for extracting the lesion region from dermoscopic images.

This paper combines Otsu's thresholding and CNN for better results. U-Net based architecture was used to extract more complex features. Chengyao Qian et al. [43] proposed an encoder–decoder architecture for segmentation inspired by DeepLab [44] and ResNet 101 was adapted for feature extraction. Frederico Guth et al. [45] introduced a U-Net 34 architecture that merged insights from U-Net and ResNet. The optimized learning rate was used for fine-tuning the network and the slanted triangular learning rate strategy (STLR) was employed.

12.1.2 Materials and Methods

12.1.2.1 Dataset Modalities

We trained and tested our CNN model on dermoscopic skin images acquired from two publicly accessible datasets (i.e., PH^2 [46] and ISIC 2017 [47]), the latter provided by the 'International Skin Imaging Collaboration' (ISIC). Example images from both datasets are shown in Figure 12.1.

We compared our model in the task of Lesion Segmentation, part 1 of the 2017 ISBI Skin Lesion Analysis Toward Melanoma Detection challenge. We evaluated our model on the ISIC-17 test data consisting of 600 images to compare its performance with state-of-the-art pipelines. Additionally, our model was also tested on the PH^2 dataset with its 200 dermoscopic images including 40 melanoma, 80 common nevi and 80 atypical nevi images.

12.1.2.2 Proposed Methodology

In this section, we introduce our devised methodology, which was trained and tested on the datasets (details presented later), and the subsequent results are reported and discussed. At the outset, it is pertinent to mention that we proposed a method that out-performed other similar available methods, both in terms of model accuracy and in pixel-by-pixel similarity measure, also called the IOU overlap (sometimes also referred to as the Jaccard Index).

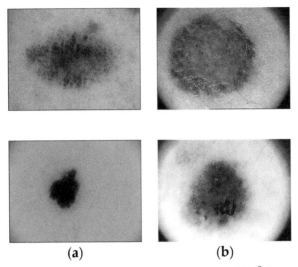

(a)   **(b)**

Figure 12.1   Examples of the ISIC-17 dataset (a) and PH^2 dataset (b).

We herein proceed to describe, point by point, the various subsections of the proposed method.

12.1.2.2.1 Image pre-processing

Images are pre-processed using resizing, scaling, hair removal and data centering techniques before being given as input to the CNN model. For noise removal, morphological operations are applied. We obtained promising results by applying pre-processing practices, which are as follows:

Image Resizing: It is good practice to resize images before they are fed into the neural network. It allows the model to convolve faster, thereby saving computational power and dealing with memory constraints. Dermoscopic images vary in size and to overcome such individual differences, the images and their corresponding ground truths are down sampled to 256×256 resolution. All the RBG images are in the JPEG file format while the respective labels are in the PNG format.

Image Normalization and Standardization: Images are normalized before training to remove poor contrast issues. Normalization changes the range of pixel values, rescaling the image between 0 and 1 so that the input data is centred around zero in all dimensions. Normalization is obtained by subtracting the image from its mean value, which is then divided by the standard deviation of the image.

Hair Removal: Dermoscopic images contain hair-like artefacts that cause issues while segmenting lesion regions. A series of morphological operations are applied to the image to remove these hair-like structures. The inpainting algorithm [48] is then applied to replace the pixel values with the neighbouring pixels, explained as follows:

- Conversion of RGB to greyscale image;
- Black top-hat filter [49,50] is applied to the greyscale image;
- Inpainting algorithm is implemented on the generated binary mask and
- Inpainting of the hair-occupied regions with neighbouring pixels.

A 17×17 cross shaped structuring element is defined, as shown in Figure 12.2c. Black top-hat (or black hat filter) filtering is obtained by subtracting closing of image from original image. If A is the original input image and B is the closing of the input image, then black top-hat filter is defined by Equation (12.1):

$$\text{Black Hat}(A) = A_{BH} = (AB) - A \qquad (12.1)$$

Closing morphological operation is the erosion of the dilation of set A and B. Closing fills small holes in the region while keeping the initial region sizes intact. It preserves the background pixels that are like the structuring element, while eliminating all other regions of the background.

The image obtained after applying the closing operation on a greyscale image is subtracted from the image itself to obtain hair-like structures. Binary mask of the hair elements is obtained by applying a threshold value of '10'

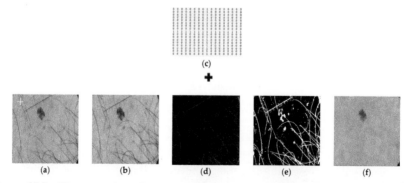

Figure 12.2 Representation of a single image as it is passed through the hair removal algorithm. Left to right: (a) Input image; (b) Greyscale image; (c) Cross shaped structuring element employed during morphological operations; (d) Image obtained after applying black top-hat filter; (e) Image after thresholding; (f). Final image obtained as output.

on the image obtained from the black top-hat filter. Images obtained from the black top-hat filter and after thresholding, respectively, are highlighted in Figures 12.2d and e.

The image based on the fast marching method was employed [51]. The inpainting algorithm replaces the hair structures with the bordering pixels of the image to restore the original image. This technique is commonly used in recovering old or noisy images. The image to be inpainted and the mask obtained after thresholding was used to inpaint those hairy regions that were extracted with the neighbouring pixels and output was achieved (Figure 12.2f).

12.1.2.2.2 Model architecture

Deep learning architectures are currently being used to solve visual recognition and object detection problems. The CNN models have shown good impact over semi-automated methods for semantic segmentation. The U-Net architecture, which is based on an encoder–decoder approach, has revealed significant results in medical image segmentation. The outputs of these networks are binary segmentation masks.

In general, the CNN models are the combination of layers (i.e., convolutional, max pooling, batch normalization and activation layer). The CNN architectures have been widely used in computer assisted medical diagnostics.

For this purpose, a CNN architecture was trained on an ISIC 2017 dataset. The network architecture (Figure 12.3) takes insight from both the U-Net and

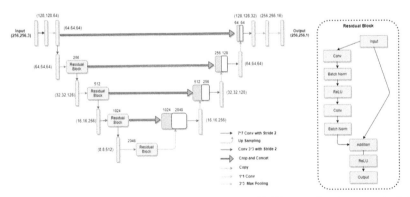

Figure 12.3 Schematic diagram representing UResNet-50. The encoder shown on the left is the ResNet-50, while the U-Net decoder is shown on the right. Given in the parenthesis are the channel dimensions of the incoming feature maps to each block. Arrows are defined in the legend.

ResNet. The contracting path (convolutional side) is based on the ResNet architecture, and the expansive path (deconvolutional side) is based on the U-Net pipeline. Overall, the network performs in an encoder–decoder fashion and is composed of 50 layers (ResNet-50). Input images of resolution 256×256 are fed into the model. The convolutional network architecture is shown in Table 12.1.

On the contracting side, after the first convolutional layer, a max pooling layer is defined with a kernel of 3×3 and a stride of 2 that halves the input dimension. Repetitive blocks are introduced with three convolutional layers per block; the 1×1 convolutional layer is defined before and after each 3×3 convolutional layer. It reduces the number of channels in the input before the 3×3 convolutional layer and again, the 1×1 is defined to restore dimensions. This is called the 'bottleneck' design, which reduces the training time of the network.

After 5 units of downsampling, the dimension ranges to 8×8 and 2048 filters. In contrast, the deconvolutional side or expansive path (Table 12.2) consists of 10 layers that perform deconvolution.

12.1.2.2.3 Network training
We trained our model for 100 epochs and applied data augmentation during runtime, which enhances the performance as more data increases the predictability of the model so that it can classify better, thereby producing

Table 12.1   Convolutional network architecture based on ResNet-50

Layer Name	Output Size	Kernel Size and No. of Filters
Conv 1	128×128	7×7, 64, Stride 2
Max Pooling	64×64	3×3, Stride 2
Conv 2	64×64	$\begin{bmatrix} 1 \times 1 \times 64 \\ 3 \times 3 \times 64 \\ 1 \times 1 \times 256 \end{bmatrix} \times 3$
Conv 3	32×32	$\begin{bmatrix} 1 \times 1 \times 128 \\ 3 \times 3 \times 128 \\ 1 \times 1 \times 512 \end{bmatrix} \times 4$
Conv 4	16×16	$\begin{bmatrix} 1 \times 1 \times 256 \\ 3 \times 3 \times 256 \\ 1 \times 1 \times 1024 \end{bmatrix} \times 6$
Conv 5	8×8	$\begin{bmatrix} 1 \times 1 \times 512 \\ 3 \times 3 \times 512 \\ 1 \times 1 \times 2048 \end{bmatrix} \times 3$

Table 12.2　The deconvolution architecture based on U-Net

Layer Name	Kernel	Output Size and No. of Filters
U1	2×2	$16 \times 16 \times 2048$
D1	3×3	$16 \times 16 \times 256$
D2	3×3	$16 \times 16 \times 256$
U2	2×2	$32 \times 32 \times 256$
D3	3×3	$32 \times 32 \times 128$
D4	3×3	$32 \times 32 \times 128$
U3	2×2	$64 \times 64 \times 128$
D5	3×3	$64 \times 64 \times 64$
D6	3×3	$64 \times 64 \times 64$
U4	2×2	$128 \times 128 \times 64$
D7	3×3	$128 \times 128 \times 32$
D8	3×3	$128 \times 128 \times 32$
U5	2×2	$256 \times 256 \times 32$
D9	3×3	$256 \times 256 \times 16$
D10	3×3	$256 \times 256 \times 16$
Output	1×1	$256 \times 256 \times 1$

Table 12.3　Hyperparameters maintained during training

Name	Value
Input Size	$256 \times 256 \times 3$
Batch Size	16
Learning Rate	$1e^{-3}$
Optimizer	Adam
Epoch	100
Loss Function	Binary Crossentropy

a significant effect on the segmentation results. We rotated images in three dimensions, which increased the dataset threefold.

Early stopping is defined and the learning rate is reduced if the model loss does not decrease for 10 epochs. Our model stopped after approximately 70 epochs. Transfer learning was employed for training the model on our dataset, utilizing pre-trained weights obtained through training on the ImageNet dataset. Table 12.3 shows the hyperparameters used to train our model.

12.1.3 Results

12.1.3.1 Model Evaluation

Our model was evaluated on images obtained from the International Skin Imaging Collaboration ISIC 2017. We trained our CNN model on the training

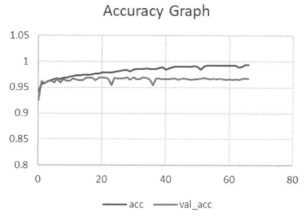

Figure 12.4   Training and validation accuracy of the proposed convolutional neural network model for 70 epochs.

group of ISIC 2017, which consisted of 2000 skin lesion images. During this process, a total training accuracy of 0.995 was obtained for 70 epochs. The variations of accuracy between the training and validation group during training are highlighted in Figure 12.4.

The model was tested on the validation and test set taken from the ISIC 2017 dataset. Furthermore, the model was also tested on the PH^2 dataset comprising of 200 dermoscopic images. The ground truths were also available in order to check the performance of the proposed CNN model. All images went through the pre-processing step before being fed into the CNN architecture as described earlier. Parameters of convolutional layers were set during the training process. During the evaluation process, the model parameters were not changed in order to assess our model's performance on the pre-set parameters. The results of multiple subjects are shown in Figure 12.5.

The receiver operative characteristics (ROC) curve was used to evaluate the performance binary classifiers. The ROC is a plot between the true positive rate (sensitivity) as a function of false positive rate (specificity) at different thresholds. This study emphasizes segmenting the lesion region, with 1 representing the lesion region and 0 representing the black region of the image. The ROC curve is the best evaluation technique that defines separability between classes. Each data point in a curve shows the values at a specific threshold. Figure 12.6 shows the ROC curve of the model on the ISIC test set.

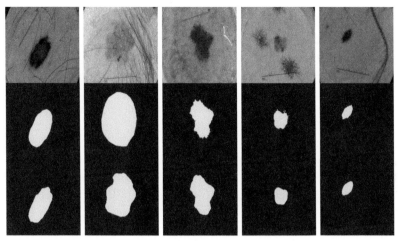

Figure 12.5 Example results of multiple patients. The first row contains the original images of five patients from the test set. The second row contains corresponding ground truths as provided. The third row contains predicted masks from the proposed method.

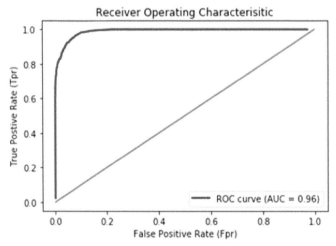

Figure 12.6 Receiver operative characteristics (ROC) curve generated on the ISIC-17 test set.

The ROC curve dictates the model's capability to distinguish between classes accurately. The higher the area under the curve, the higher the network's ability to distinguish two classes more precisely (i.e., either 0 or 1). The AUC of our proposed model was 0.963, illustrating the model's remarkable competence of differentiability.

12.1.4 Benchmarks

12.1.4.1 Comparison with Different Frameworks

The model was tested both with and without pre-processing on the ISIC-17 dataset to ascertain the efficiency of the hair-removing algorithm. A Jaccard of 0.763 (Table 12.4) was achieved when the inpainting algorithm was not employed to remove the hair structures from the images, which improved considerably to 0.772 with the implementation of the pre-processing technique.

For evaluation, we compared our results with the existing deep learning frameworks (enlisted in Table 12.5) that had been tested on the ISIC-17 dataset. FCN-8s [52] achieved a JI (0.696) and DC (0.783), respectively. Although our proposed method was the deepest among the below listed frameworks, we improved the results by balanced data augmentation and reduced overfitting. Simple U-Net obtained a JI of 0.651 and a DC of 0.768.

Our proposed method is a combination of the ResNet50 based encoder and U-Net based decoder, which achieved a Jaccard index of 0.772 and Dice coefficient of 0.858.

12.1.4.2 Comparison with Top 5 Challenge Participants of Leaderboard

The intent was that this research would segment the lesion regions with higher accuracy when compared to other methods. Three different groups of images were used to validate our network: (1) the ISIC 2017 test group; (2) ISIC 2017 validation group and the (3) PH^2 dataset. The test group consisted of 600 dermoscopic images and the validation group was composed of 150 images.

Table 12.4   Model performance on the ISIC-2017 test set

Methods	Jaccard Index
No Pre-processing	*Pre-processed*
0.763	0.772

Table 12.5   Comparison with different frameworks on the ISIC-2017 test set

Methods	Jaccard Index	Dice Coefficient
FCN-8s [52]	0.696	0.783
U-Net [53]	0.651	0.768
II-FCN [54]	0.699	0.794
Auto-ED [55]	0.738	0.824
FCRN [56]	0.753	0.839
Res-Unet (Proposed)	**0.772**	**0.858**

The PH^2 dataset is a renowned dataset and was used for further evaluation of our network and benchmarking our results with existing methods and participants in the challenge. Table 12.6 depicts our results in terms of the Jaccard Index as per the challenge's demand, in comparison with the top five participants from the ISIC-17 Challenge. The top ranked participant Yading Yaun et al. [57] obtained a Jaccard index of 0.765.

Different methods have been employed for segmentation purpose. The second ranked participant Matt Berseth et al. [58] obtained a JA (0.762) by employing a U-Net framework. Our technique achieved a Jaccard index of 0.772 with the proposed technique stated earlier. Based on our results, our model performed better than the existing techniques used in the associative field of study.

12.1.4.3 Evaluation of Model on the PH^2 Dataset

To evaluate the robustness of our proposed model, we further tested the architecture on the PH^2 dataset and compared our segmentations with the existing state-of-the-art techniques. The results are listed in Table 12.7. Our method achieved promising results. FCN-16s achieved a JA of 0.802 and DC of 0.881, respectively. Another framework, Mask-RCNN attained a JA of 0.839 and a DC of 0.907 on the PH^2 dataset.

Table 12.6   Comparison of results with the challenge participants

Authors	Model	Jaccard Index
Yading Yuan et al. [57]	CDNN	0.765
Matt Berseth et al. [58]	U-Net	0.762
Popleyi et al. [59]	FCN	0.760
Euijoon Ahn et al. [60]	ResNet	0.758
Afonso Menegola et al. [61]	VGG16	0.754
Proposed Method	**Res-Unet**	**0.772**

Table 12.7   Comparison with different frameworks on the PH^2 Dataset

Methods	Jaccard Index	Dice Coefficient
FCN-16s	0.802	0.881
DeeplabV3+	0.814	0.890
Mask-RCNN	0.830	0.904
Multistage FCN	–	0.906
SSLS	–	0.916
Res-Unet (proposed)	**0.854**	**0.924**

12.1.5 Conclusions

Skin lesion segmentation is a vital step in developing a computer aided diagnosis system for skin cancer. In this paper, we successfully developed a skin lesion segmentation algorithm using the CNN with an advanced hair-removal algorithm that effectively removed hair structures from the dermoscopic images, improving the accuracy considerably. We tested our model architecture on the ISIC-2017 dataset and PH^2 dataset, and the Jaccard index obtained thereof was 0.772 and 0.854, respectively. Our proposed method achieved promising results compared with the state-of-the-art techniques in terms of the Jaccard index. Furthermore, our CNN model was tested on a PH^2 dataset along with the ISIC-17 test set and produced better segmentation and performed better than the existing methods in the literature. Empirical results show that the combination of the U-Net and ResNet shows impressive results.

The limited training data used requires extensive augmentation to prevent the model from overfitting. A large dataset is therefore needed for better accuracy and generalization of the model. Furthermore, for it to achieve state-of-the-art results, the model was made to be complex and efficient, which takes more time to train as opposed to the conventional U-Net.

Our future work includes using a larger dataset to reduce overfitting problems and hypertuning the parameters for more effective training. Additionally, a conditional random field (CRF) application can also be applied to refine the model output.

Acknowledgements

In this section you can acknowledge any support given which is not covered by the author contribution or funding sections. This may include administrative and technical support, or donations in kind (e.g., materials used for experiments).

Conflicts of Interest

The authors declare no conflicts of interest.

References

[1] Pham, D.L.; Xu, C.; Prince, J.L. Current methods in medical image segmentation. *Annu. Rev. Biomed. Eng.* **2000**, *2*, 315–337.

[2] Macià, F.; Pumarega, J.; Gallén, M.; Porta, M. Time from (clinical or certainty) diagnosis to treatment onset in cancer patients: The choice of diagnostic date strongly influences differences in therapeutic delay by tumor site and stage. *J. Clin. Epidemiol.* **2013**, *66*, 928–939.

[3] Oliveira, R.B.; Filho, M.E.; Ma, Z.; Papa, J.P.; Pereira, A.S.; Tavares, J.M.R.S. Computational methods for the image segmentation of pigmented skin lesions: A review. *Comput. Methods Programs Biomed.* **2016**, *131*, 127–141.

[4] Matthews, N.H.; Li, W.-Q.; Qureshi, A.A.; Weinstock, M.A.; Cho, E. Epidemiology of melanoma. In *Cutaneous Melanoma: Etiology and Therapy*; Codon Publications, 2017, pp. 3–22.

[5] Graham, C. *The SAGE Encyclopedia of Cancer and Society*, 2nd ed.; 2015.

[6] Colorectal Cancer Facts & Figures 2017–2019. Atlanta, American Cancer Society, 2017.

[7] Matthews, N.H.; Li, W.-Q.; Qureshi, A.A.; Weinstock, M.A.; Cho, E. Epidemiology of melanoma. In *Cutaneous Melanoma: Etiology and Therapy*; Codon Publications, 2017, pp. 3–22.

[8] Mohan, S.V.; Chang, A.L.S. Advanced basal cell carcinoma: Epidemiology and therapeutic innovations. *Curr. Dermatol. Rep.* **2014**, *3*, 40–45.

[9] Guy, G.P.; Machlin, S.R.; Ekwueme, D.U.; Yabroff, K.R. Prevalence and costs of skin cancer treatment in the U.S., 2002–2006 and 2007–2011. *Am. J. Prev. Med.* **2015**, *48*, 183–187.

[10] Barata, C.; Ruela, M.; Francisco, M.; Mendonca, T.; Marques, J.S. Two systems for the detection of melanomas in Dermoscopy images using texture and color features. *IEEE Syst. J.* **2014**, *8*, 965–979.

[11] Schaefer, G.; Rajab, M.I.; Celebi, M.E.; Iyatomi, H. Color and contrast enhancement for improved skin lesion segmentation. *Comput. Med. Imaging Graph.* **2011**, *35*, 99–104.

[12] Gómez, D.D.; Butakoff, C.; Ersbøll, B.K.; Stoecker, W. Independent histogram pursuit for segmentation of skin lesions. *IEEE Trans. Biomed. Eng.* **2008**, *55*, 157–161.

[13] Maeda, J.; Kawano, A.; Yamauchi, S.; Suzuki, Y.; Marçal, A.R.S.; Mendonça, T. Perceptual image segmentation using fuzzy-based hierarchical algorithm and its application to images Dermoscopy. In Proceedings of the 2008 IEEE Conference on Soft Computing on Industrial Applications, Muroran, Japan, 25–27 June 2008; pp. 66–71.

[14] Yüksel, M.E.; Borlu, M. Accurate segmentation of dermoscopic images by image thresholding based on type-2 fuzzy logic. *IEEE Trans. Fuzzy Syst.* **2009**, *17*, 976–982.

[15] Xie, F.Y.; Qin, S.Y.; Jiang, Z.G.; Meng, R.S. PDE-based unsupervised repair of hair-occluded information in Dermoscopy images of melanoma. *Comput. Med. Imaging Graph.* **2009**, *33*, 275–282.

[16] Silveira, M.; Nascimento, J.C.; Marques, J.S.; Marcal, A.R.S.; Mendonca, T.; Yamauchi, S.; Maeda, J.; Rozeira, J. Comparison of segmentation methods for melanoma diagnosis in dermoscopy images. *IEEE J. Sel. Top. Signal Process.* **2009**, *3*, 35–45.

[17] Ma, Z.; Tavares, J.M.R.S. A novel approach to segment skin lesions in dermoscopic images based on a deformable model. *IEEE J. Biomed. Heal. Inform.* **2016**, *20*, 615–623.

[18] Mishra, R.; Daescu, O. Deep learning for skin lesion segmentation. In Proceedings of the 2017 IEEE International Conference on Bioinformatics and Biomedicine, BIBM, Kansas City, MO, USA, 13–16 November 2017.

[19] Li, Y.; Shen, L. Skin lesion analysis towards melanoma detection using deep learning network. *Sensors* **2018**, *18*, 556, doi:10.3390/s18020556.

[20] Jafari, M.H.; Karimi, N.; Nasr-Esfahani, E.; Samavi, S.; Soroushmehr, S.M.R.; Ward, K.; Najarian, K. Skin lesion segmentation in clinical images using deep learning. In Proceedings of the International Conference on Pattern Recognition, Cancun, Mexico, 4–8 December 2016.

[21] Tamilselvi, P.R. Analysis of image segmentation techniques for medical images. *Int. Conf. Emerg. Res. Comput. Information, Commun. Appl.* **2014**, *2*, 73–76.

[22] Schaefer, G.; Rajab, M.I.; Celebi, M.E.; Iyatomi, H. Color and contrast enhancement for improved skin lesion segmentation. *Comput. Med. Imaging Graph.* **2011**, *35*, 99–104.

[23] Yuan, Y.; Chao, M.; Lo, Y.C. Automatic skin lesion segmentation using deep fully convolutional networks with jaccard distance. *IEEE Trans. Med. Imaging* **2017**, *36*, 1876–1886.

[24] Oliveira, R.B.; Filho, M.E.; Ma, Z.; Papa, J.P.; Pereira, A.S.; Tavares, J.M.R.S. Computational methods for the image segmentation of pigmented skin lesions: A review. *Comput. Methods Programs Biomed.* **2016**, *131*, 127–141.

[25] Korotkov, K.; Garcia, R. Computerized analysis of pigmented skin lesions: A review. *Artif. Intell. Med.* **2012**, *56*, 69–90.

[26] Celebi, M.E.; Iyatomi, H.; Schaefer, G. Contrast enhancement in dermoscopy images by maximizing a histogram bimodality measure. In Proceedings of the International Conference on Image Processing ICIP, Cairo, Egypt, 7–10 November 2009; pp. 2601–2604.

[27] Beuren, A.T.; Janasieivicz, R.; Pinheiro, G.; Grando, N.; Facon, J. Skin melanoma segmentation by morphological approach. In Proceedings of the ACM International Conference Proceeding Series, August 2012, pp. 972–978.

[28] Lee, T.; Ng, V.; Gallagher, R.; Coldman, A.; McLean, D. DullRazor: A software approach to hair removal from images. *Comput. Biol. Med.* **1997**, *27*, 533–543.

[29] Chanda, B.; Majumder, D.D. *Digital Image Processing and Analysis*, 2nd ed.; PHI Learning Pvt. Ltd.: Delhi, India, 2011.

[30] Celebi, M.E.; Kingravi, K.A.; Iyatomi, H.; Aslandogan, Y.A.; Stoecker, W.V.; Moss, R.H.; Malters, J.M.; Grichnik, J.M.; Marghoob, A.A.; Rabinovitz, H.S.; et al. Border detection in Dermoscopy images using statistical region merging. *Skin Res. Technol.* **2008**, *14*, 347–353.

[31] Guillemot, C.; Le Meur, O. Image Inpainting: Overview and recent advances. *IEEE Signal Process. Mag.* **2014**, *31*, 127–144.

[32] Shen, H.; Li, X.; Cheng, Q.; Zeng, C.; Yang, G.; Li, H.; Zhang, L. Missing information reconstruction of remote sensing data: A technical review. *IEEE Geosci. Remote Sens. Mag.* **2015**, *3*, 61–85.

[33] Li, X.; Shen, H.; Zhang, L.; Zhang, H.; Yuan, Q.; Yang, G. Recovering quantitative remote sensing products contaminated by thick clouds and shadows using multitemporal dictionary learning. *IEEE Trans. Geosci. Remote Sens.* **2014**, *52*, 7086–7098.

[34] Jaisakthi, S.M.; Mirunalini, P.; Aravindan, C. Automated skin lesion segmentation of Dermoscopic images using grabcut and kmeans algorithms. *IET Comput. Vis.* **2018**, *12*, 1088–1095.

[35] Aljanabi, M.; Özok, Y.E.; Rahebi, J.; Abdullah, A.S. Skin lesion segmentation method for Dermoscopy images using artificial bee colony algorithm. *Symmetry* **2018**, *10*, 347, doi:10.3390/sym10080347.

[36] Pennisi, A.; Bloisi, D.D.; Nardi, D.; Giampetruzzi, A.R.; Mondino, C.; Facchiano, A. Skin lesion image segmentation using Delaunay Triangulation for melanoma detection. *Comput. Med. Imaging Graph.* **2016**, *52*, 89–103.

[37] Celebi, M.E.; Iyatomi, H.; Schaefer, G.; Stoecker, W.V. Lesion border detection in Dermoscopy images. *Comput. Med. Imaging Graph.* **2009**, *33*, 148–153.

[38] Bi, L.; Kim, J.; Ahn, E.; Feng, D.; Fulham, M. Automated skin lesion segmentation via image-wise supervised learning and multi-scale Superpixel based cellular automata. In Proceedings of the International Symposium on Biomedical Imaging, Prague, Czech Republic, 13–16 April 2016; pp. 1059–1062.

[39] Bi, L.; Kim, J.; Ahn, E.; Kumar, A.; Fulham, M.; Feng, D. Dermoscopic image segmentation via multistage fully convolutional networks. *IEEE Trans. Biomed. Eng.* **2017**, *64*, 2065–2074.

[40] Yuan, Y. Automatic skin lesion segmentation with fully convolutional-deconvolutional networks. *arXiv* **2017**. arXiv:1703,05165.

[41] Berseth, M. ISIC 2017-Skin Lesion Analysis Towards Melanoma Detection, International Skin Imaging Collaboration, 2017. Available online: https://arxiv.org/abs/1703.00523 (accessed on 9 September 2019).

[42] Mishra, R.; Daescu, O. Deep learning for skin lesion segmentation. In Proceedings of the 2017 IEEE International Conference on Bioinformatics and Biomedicine, BIBM, Kansas City, MO, USA, 13–16 November 2017; pp. 1189–1194.

[43] Qian, C.; Jiang, H.; Liu, T. *ISIC 2018-Skin Lesion Analysis. 2018. ISIC – Skin Image Analysis Workshop and Challenge @ MICCAI 2018 Hosted by the International Skin Imaging Collaboration (ISIC);* Springer: Berlin, Germany, 2018.

[44] Chen, L.-C.; Papandreou, G.; Kokkinos, I.; Murphy, K.; Yuille, A.L. IEEE Transactions on Pattern Analysis and Machine Intelligence DeepLab: Semantic Image Segmentation with Deep Convolutional Nets, Atrous Convolution, and Fully Connected CRFs. *IEEE Trans. Pattern Anal. Mach. Intell.* **2016**, *40*, 834–848.

[45] Guth, F.; Decampos, T.E. Skin Lesion Segmentation Using U-Net and Good Training Strategies. Available online: https://arxiv.org/abs/1811.11314 (accessed on 9 September 2019).

[46] Mendonca, T.; Ferreira, P.M.; Marques, J.S.; Marcal, A.R.S.; Rozeira, J. PH^2 – A dermoscopic image database for research and benchmarking. In Proceedings of the 35th Annual International Conference of the IEEE Engineering in Medicine and Biology Society (EMBC), Osaka, Japan, 3–7 July 2013.

[47] Codella, N.C.F.; Gutman, D.; Celebi, M.E.; Helba, B.; Marchetti, M.A.; Dusza, S.W.; Kalloo, A.; Liopyris, K.; Mishra, N.; Kittler, H.; et al. Skin lesion analysis toward melanoma detection: A challenge at the 2017 international symposium on biomedical imaging (ISBI), hosted by the international skin imaging collaboration (ISIC). In Proceedings of the

2018 IEEE 15th International Symposium on Biomedical Imaging (ISBI 2018), Washington, DC, USA, 4–7 April 2017.

[48] Telea, A. An image inpainting technique based on the fast marching method. *J. Graph. Tools.* **2004**, *9*, 23–34.

[49] Thapar, S. Study and implementation of various morphology based image contrast enhancement techniques. *Int. J. Comput. Bus. Res.* 2012, 2229–6166.

[50] Wang, G.; Wang, Y.; Li, H.; Chen, X.; Lu, H.; Ma, Y.; Peng, C.; Tang, L. Morphological background detection and illumination normalization of text image with poor lighting. *PLoS ONE.*, **2014**, *9*, e110991, doi:10.1371/journal.pone.0110991.

[51] Telea, A. An image inpainting technique based on the fast marching method. *J. Graph. Tools.* **2004**, *9*, 23–34.

[52] Shelhamer, E.; Long, J.; Darrell, T. Fully convolutional networks for semantic segmentation. In Proceedings of the IEEE Conference on Computer Vision and Pattern Recognition, Boston, MA, USA, 7–12 June 2015; pp. 3431–3440.

[53] Ronneberger, O.; Fischer, P.; Brox, T. U-Net: Convolutional networks for biomedical image segmentation. In Proceedings of the Medical Image Computing and Computer Assisted Interventions, Munich, Germany, 5–9 October 2015; pp. 234–241.

[54] Wen, H. II-FCN for skin lesion analysis towards melanoma detection. *arXiv* **2017**. arXiv:1702,08699.

[55] Attia, M.; Hossny, M.; Nahavandi, S.; Yazdabadi, A. Spatially aware melanoma segmentation using hybrid deep learning techniques. *arXiv* **2017**. arXiv:1702,07963.

[56] Li, Y.; Shen, L. Skin lesion analysis towards melanoma detection using deep learning network. *Sensors.* **2018**, *18*, 556, doi:10.3390/s18020556.

[57] Yuan, Y.; Chao, M.; Lo, Y.C. Automatic skin lesion segmentation using deep fully convolutional networks with jaccard distance. *IEEE Trans. Med. Imaging.* **2017**, *36*, 1876–1886.

[58] Berseth, M. ISIC 2017-skin lesion analysis towards melanoma detection. *arXiv* **2017**. arXiv:1703,00523.

[59] Bi, L.; Kim, J.; Ahn, E.; Feng, D. Automatic Skin Lesion Analysis using Large-scale Der-moscopy Images and Deep Residual Networks. Available online: https://arxiv.org/ftp/arxiv/papers/1703/1703.04197.pdf (accessed on).

[60] Bi, L.; Kim, J.; Ahn, E.; Feng, D.; Fulham, M. Automated skin lesion segmentation via image-wise supervised learning and multi-scale

superpixel based cellular automata. In Proceedings of the International Symposium on Biomedical Imaging, Prague, Czech Republic, 13–16 April 2016; pp. 1059–1062.

[61] Menegola, A.; Tavares, J.; Fornaciali, M.; Li, L.T.; Avila, S.; Valle, E. Recod Titans at ISIC Challenge 2017. Available online: https://arxiv.org/abs/1703.04819 (accessed on).

12.2 Photodetector based Indoor Positioning Systems Variants: New Look

Muhammad Nasir Khan, Omer Gilani and Mohsin Jamil

Indoor positioning systems (IPSs) using photodetectors at receiving end are revealed. In the first part, an overview of the visible light communication (VLC) technology is presented. The second part of this paper briefs the VLC channel model and its basic parameters and their effect on light propagation. In the third portion the light emitting diode (LED) based positioning algorithms are discussed in detail and further protocols are investigated that are used for position calculation exploiting various techniques. The types of different photodetectors based IPSs are discussed along with the metrics. A table is drawn providing a comparison of the different IPSs. A new look of the IPS is proposed with an extended feature and better accuracy. A conclusion is drawn in this field for researchers to enhance the VLC-based systems towards the future research.

12.2.1 Introduction

Visible light communication (VLC) is a branch of optical wireless communication (OWC) engineering. In the VLC, the same light that is used for illumination can be used for communication purpose. The VLC receiver's purpose is to convert the received modulated light signals back into electrical signals. Intensity of transmitter output is dimmed up and down with respect to current (digital 1 and 0) flowing through light sources [1]. This process is achieved by varying current, e.g., photodiode but imperceptible to the human eyes. Figure 12.7 show a simple VLC based communication model. Light emitting diode (LED) is a semiconductor device used to transmit high speed data using the technique given in Refs. [2–3].

For positioning, mostly global positioning system (GPS) is used as it provides accurate results [4]. In indoor environments the positioning is somewhat difficult as satellite signals degrade and in these areas other systems based on indoor wireless signals (e.g., Wi-Fi, Bluetooth, radio frequency identification [RFID] and ZigBee) have been proposed and deployed over the last few decades [5, 6]. LED based positioning systems have high bandwidth, high data rates and very low cost which make it better than all other systems used before for positioning. In Ref. [7] it is shown that the LEDs-based IPS performs better than the systems developed for position calculation using Bluetooth and Wi-Fi. A novel concept for using the VLC

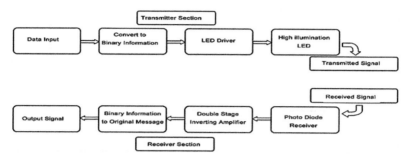

Figure 12.7   VLC based communication system.

with three-dimensional (3D) indoor positioning is presented in Ref. [8]. A VLC transceiver using experimental measurements was modelled. Proposed is a 3D-positioning algorithm using received signal strength indication, which changes based on the angle and distance of the location-based service. The simulations are shown using three LED arrays in a room of dimensions 1.5×1.5 m $\times 2$ cm of height. Receiver is placed at 0.9 m away. The results in the 2D system have accuracy with 3 cm error. Receive-signal strength (RSS) trilateration algorithm was used for positioning. This provides a low-cost solution but noise from the other light sources and eavesdropper nodes were not considered so there is a need for improvement from noise and security point of view.

In Ref. [9], an IPS based on VLC is proposed. This system measures the location of receiver in 3D without prior knowledge of receiver height with respect to ground and its alignment. Light sensor and accelerometer, which can be found in most smartphones, are used at the receiver's side. The experiments are performed in a space of $5 \times 3 \times 3$ m. All transmitters have to be at the same height. Experiment results positioning error of 0.25 m is observed when 1.22 rad of field of view (FoV) at the receiver side is used. The approach using single photodetector would be simpler but less accurate and robust than one which uses camera [10]. Approach that uses camera is not energy efficient and also aggravates positioning latency. The positioning technique adopted in Ref. [11] using camera as detector and a single source gives accurate positioning estimation. In Ref. [12] a highly accurate 3D IPS system based on the LED is proposed which is based on the frequency division multiplexing (FDM) and time-difference-of-arrival (TDOA) methods.

12.2.2 Characteristics of Led-Based IPS

This section describes the prominent features regarding optical wireless system which is used for localization in VLC based positioning systems.

12.2.2.1 Channel Model

The LEDs transmit the signals according to Lambert's emission law and behave as a Lambertian source [13]. The VLC channel can be described by the following equation [14]:

$$P_i\left(\theta,\varphi\right) = \frac{\left(m_i+1\right)A_R P_{T_i}}{2\pi} \cdot \frac{\cos^{m_i}\left(\theta\right)\cos^M\left(\varphi\right)T_s\left(\varphi\right)g(\varphi)}{D_i^2} \qquad (12.1)$$

where D_i denotes the distance between ith transmitter from the receiver, A_R is the photodetector's effective area, P_{T_i} is the luminance source power, θ is the irradiance angle, φ is the incidence angle, $T_s(\varphi)$ is the optical filter gain, M denotes Lambertian orders of the receiver, m is Lambertian orders of the transmitter and $g(\varphi)$ is the optical concentrator's gain.

Equations (12.2) and (12.3) can be used for calculating the values of m and M by

$$m = -\frac{\ln 2}{\ln\left(\cos\theta_{1/2}\right)}. \qquad (12.2)$$

$$M = -\frac{\ln 2}{\ln\left(\cos\varphi_{1/2}\right)}. \qquad (12.3)$$

In the absence of lens, the gain of the optical filter is equal to the gain of the optical concentrator. We can increase the A_R for better sensitivity but this will decrease the bandwidth. In Ref. [15] influence on IPSs of various sizes of photodetectors was discussed [15]. Figure 12.8 describes the scenario as for calculation of M light intensity. Some researchers assumed 60 to be the half-power semi-angle and set M = 1 [16]. However, to obtain Lambertian orders of receiver and transmitter we measure intensity of light at different incident angles keeping $\theta = 0$ and $\varphi = 0$ accordingly as shown in Figure 12.8 [17]. The relationship of m to the positioning accuracy is studied in Ref. [18]. When m is increased from 30 to 80, an increase in signal concentration is observed; this increase results in increased signal-to-noise ratio (SNR) as is seen in (12.2) so the estimated distance error is minimized from 0.05 m to 0.01 m at an irradiance angle of 0, and optical power of 0.9 W is used at 3 m distance.

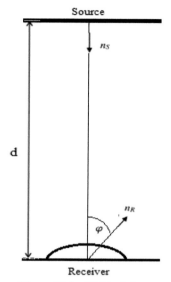

Figure 12.8   Field of view.

The radiation pattern properties of the LED source are affected by the lens parameters, i.e., refractive index, size and arrangement [19]. In Ref. [20], researchers have considered the optical gain for both transmitter and receiver because of lens usage on both ends. As a result, (12.4) is given hereunder as the updated form of (12.1):

$$P_r = (P_r/d^2)\, C_{opt}\, G_t\,(\theta)\, G_r\,(\varphi) \tag{12.4}$$

where $G_t\,(\theta) = \exp\left(\frac{-\theta_s}{K_t}\right)$ and $G_r\,(\varphi) = \exp\left(\frac{-\varphi_s}{k_r}\right)$, S_t and S_r are slope constants, which are determined by lens shapes of transmitters and the receiver, respectively. K_t and K_r are related to $K_t = \frac{(\theta_{1/2})_{S_t}}{\ln(1/2)}$ and $K_r = \frac{(\varphi_{1/2})_{S_r}}{\ln(1/2)}$, respectively.

The propagation of VL signal through the environment is described by the channel model. The received signal at PD is sum of line-of-sight (LOS), and non-LOS signals creating problem is with regard to positioning, which is discussed in the next section.

12.2.2.2 Multiplexing Protocols

Transmission capacity can be increased by using several multiplexing methods to transmit the VLC signals in parallel. Data transmitted from various sources using multiplexing techniques to enhance the data rate must be accommodated properly at receiver so that the data received will be error free.

For that purpose, advance and suitable multiplexing protocols are required for multiplexing. In LED based position detection systems mostly applied multiplexing protocols are time division multiplexing (TDM) and FDM.

In the FDM the data is transmitted by multiple LEDs and each LED transmit its data by using a specific sub-band allocated to that LED with the assurance that the bands have enough guard band so that the signals transmitted by any LED will not be overlapped with any signal from any other LED. Whereas in TDM the LEDs transmit their signal in a specific time slot allocated to them from the whole transmission time such that no more than one LED will transmit its signal at the

same time slot. Important points to be considered while applying multiplexing in the LED-based IPSs are the following:

(i) Keying frequency must be at a rate which will avoid the visible flicker as this disrupts the primary function of the LEDs and causes harmful effects to human eye.

(ii) Mostly the zeros and ones are represented by low and high voltages in modulation techniques applied traditionally for VLC communication and this method can cause problem if more number of zeros should come in transmission causing dimming of light for long instance and this problem can be solved using inverted-LPPM (I-LPPM) which uses inverted pulse position of the PPM modulation.

12.2.2.3 Field of View

The FoV selection is an important aspect in the VLC based positioning systems. Large FoV results in low SNR because the signals coming as result of reflections are also added to the desired signal in NLOS environment [21]. In Ref. [22] the accuracy is compromised with the increase in FoV using a single photodetector but in Ref. [23] multiple photodetectors are used with different FoVs and this research shows better performance in position calculation while using the photodetectors with wider FoV. So it is a trade-off depending on the environment and structure.

12.2.2.4 Noise

The SNR is the major factor influencing the performance of the VLC systems. A very low SNR is observed in outer region because of power degradation due to large distance and irradiance angle. In the VLC mainly shot noise and thermal noise affects the system given by

$$\sigma_{shot}^2 = 2q\gamma\left(P_{rec}\right)B + 2qI_{bg}I_2B \qquad (12.5)$$

$$P_{rec} = \sum_{i=1}^{n} H_i(0)P_i \qquad (12.6)$$

where, q is electronic charge, K denotes the Boltzmann's constant, I_2 is noise bandwidth factor, $H_i(0)$ is channel DC gain, n denotes number of LEDs, P_i is instantaneous emitted power for the ith LED, B is equivalent noise bandwidth, I_{bg} is background current and ??is fixed photodetector's capacitance. Noise produced by feedback resistor and FET channel noise is represented in (12.6). It is seen from (12.4) and (12.6) that B affects the IPS accuracy as it is directly proportional to noise.

The total noise variance is given by

$$\sigma_{noise}^2 = \sigma_{thermal}^2 + \sigma_{shot}^2 \qquad (12.7)$$

$$SNR\,(dB) = 10\log_{10}\frac{\left(R_p\frac{P_{LoS}}{A}\right)^2}{\sigma_{noise}^2} \qquad (12.8)$$

Equation (12.8) describes the SNR of the LOS component. As compared to shot noise, thermal noise is often negligible in case of large area optical receivers.

12.2.2.5 Multipath Effect

Within a room when the LEDs are used for positioning and communication purposes then the multipath light comparatively weaker than the LOS channel also comes from different components like walls, tables, ceilings, mirrors and other surfaces within the indoor environment creating multipath effect, which is a major participant in the degradation of SNR, BER and other performance metrics.

In Ref. [24] time dispersion of different surfaces having different reflectivities is presented. The RSS and TOA based positioning systems use received power and distance from transmitter and receiver for position calculation which becomes erroneous if reflected components and LOS components are mixed before reception. However, under multipath conditions there is no encouraging effect while increasing power of transmitter because localization error achieves a stable value very quickly. In Ref. [25], simulations result shows that localization accuracy was gradually improved by increasing transmitter power but when the NLOS signals were not present in the environment.

12.2.2.6 Error
Error is the crucial parameter in the LED-based designs, which causes system performance degradation. The TDM-based systems are very sensitive to time synchronization. When the LEDs are allocated and the neighbouring slots and time synchronization fails, then this causes error, which can be modelled as

$$\widehat{P_1} = P_1 + aP_2 + \frac{2}{P_T}n(t) \tag{12.9}$$

$$\widehat{P_2} = (1-a)P_2 + \frac{2}{P_T}n(t) \tag{12.10}$$

where a denotes synchronization error percentage. Equations (12.8) and (12.9) show that RSS value always have some error as the received power is always different from the actual value, so whenever the RSS-based IPS algorithm will use TDM protocols then the above defined error must have to be accounted for better performance. Positioning error increases with the increase in the synchronization error.

12.2.3 LED-Positioning Algorithms

For IPSs several positioning algorithms have been developed for RF-based systems. These algorithms can also be deployable in the LED-based IPSs. Using the developed algorithm other techniques are applied which provide good improvement in localization calculations.

12.2.3.1 Received Signal Strength
Visible light system uses mostly the RSS-based method for indoor positioning because this method detects the transmitted signals' characteristics using a single photodetector and needs no extra device which makes the system very inexpensive. According to the channel model the signal strength decreases by increasing the distance between transmitter and receiver. By using the RSS model further some algorithms such as trilateration, fingerprinting and proximity can be used for better calculation of position using visible light with better results as proved by many published research papers.

12.2.3.1.1 Trilateration
Algorithms based on trilateration for positioning use a minimum of three transmitters with known locations. Receiver measures the intensity of light

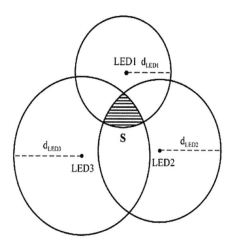

Figure 12.9   Location of receiver.

for each transmitter and calculates the distance between transmitter and specific receiver and uses these measurements to draw circles with calculated distance as radius of these circles. Then the location of the receiver is determined by calculating the intersection point of these circles as shown in Figure 12.9 by solving equations simultaneously.

$$(x - x_1)^2 + (y - y_1)^2 = d_1{}^2$$
$$(x - x_2)^2 + (y - y_2)^2 = d_2{}^2$$
$$(x - x_3)^2 + (y - y_3)^2 = d_3{}^2$$

The real estimation is shown in Figure 12.10 whose ideal case is shown in Figure 12.9. Due to quantization errors in calculations and measured distance, real estimations usually have some uncertainty. So in practice these circles intersect by covering some area rather than a point as shown in Figure 12.10. In 3D localization, a minimum of four transmitters is required and four simultaneous equations have to be solved for position calculation. Major difference of 3D from 2D case is that there are spheres instead of circles and the area upon which the intersects have a volume instead of point.

12.2.3.1.2 Fingerprinting
This system has two steps involved for indoor position calculation. The first step consists of an off-line survey in which information of an indoor environment is recorded after processing in a database. To produce the database,

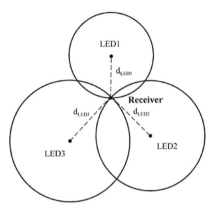

Figure 12.10   Estimation of location of receiver.

the indoor map is distributed into a grid containing numerous locations called fingerprint locations and RSS values observed at each fingerprints location are stored in the database. The RSS-based fingerprinting algorithm is mainly used in the LED based positioning systems [26]. Position is calculated by using fingerprints' locations having the lowest Euler distance. In scenarios where transmitting power of the LEDs is different this affects RSS calculations, and in response produces error in position calculation which can be reduced using normalization.

12.2.3.1.3 Proximity
This positioning algorithm is not much accurate but is easy to implement as it relies upon density of transmitter distribution. In this algorithm when an object receives signals from the LEDs whose position is known, it is considered that they are near to each other and then target location is allotted to the transmitter whose RSS signal strength is maximum among all signals compared.

12.2.3.2 Time of Arrival/Time Difference of Arrival
GPS uses the time a wireless signal takes to travel from transmitter to receiver [33]. This system is very sensitive to time synchronization. To avoid time synchronization often IPSs apply TDOA instead of TOA. The TDOA system uses at least two receivers to calculate the time difference between receivers. If multiple LEDs transmit using FDM then at receiver side single receiver is capable by using BPF to receive signals separately and calculate

the TDOA by using these signals. This makes IPSs application possible by
using the TDOA.

12.2.3.2.1 Trilateration

Trilateration method can also be employed in the systems based on the
TDOA/TOA because the light speed is constant thus causing phase delay
at receiver. The TDOA-based systems require a minimum of three phase
differences. At transmitter side each transmitter is assigned a frequency
identifier making it different from other sources. Assuming there are three
LEDs represented by the radiated power can be expressed as

$$P_i(t) = P_{CONT} + P_{MOD}\cos(2\pi f_i t + \varphi_o) \tag{12.11}$$

where,

$P_{CONT} = $ continuous optical signal
$P_{MOD} = $ modulated signal
$\varphi_o = $ initial phase of the optical signal
$f_i = 1, 3, 5 \ KHz$ is the modulated frequency

The received signal can be given by

$$E(t) = K.\left|R.\sum_{i=1}^{3} P_i(t) \otimes h_i(t)\right|^2 \tag{12.12}$$

where

$R = $ responsivity of the receiver, $K = $ proportionality constant and
$h_i(t) = $ channel impulse response.

Output signals from the BPFs after processing according to specific
frequencies is given by

$$E_{BPFi}(t) = L_i \cdot \cos\{2\pi f_i(t - d_i/c) + \varphi_o\} \tag{12.13}$$

where

$$L_i = 2 \cdot K \cdot R \cdot H(0)_i \cdot P_{CONT} \cdot P_{MOD}$$

$$d_i = \text{distance between LED}_i \text{ and the receiver}$$

The signal's frequency is unified using down converter and as a result we
have

$$E_{SIGi}(t) = K_i \cdot \cos\{\pi f_i t - 2 \cdot \frac{j\pi f_i d_i}{c} + \varphi_{TOT}\} \tag{12.14}$$

where j $= 1, 3, 5$ and φ_{TOT} is the total phase shift. After that, Hilbert transform is used for phase difference detection. From three LEDs, two phase

differences can be calculated. In Ref. [34] a technique is introduced to detect the third phase difference. Finally, the phase differences are extracted as follows:

$$\Delta\varphi_{12} = 2\pi f_1 \frac{d_1 - 3d_2}{c} = \tan^{-1}\left(\frac{I_{12}}{Q_{12}}\right) \tag{12.15}$$

$$\Delta\varphi_{13} = 2\pi f_1 \frac{d_1 - 5d_3}{c} = \tan^{-1}\left(\frac{I_{13}}{Q_{13}}\right) \tag{12.16}$$

$$\Delta\varphi_{21} = 2\pi f_1 \frac{d_2 - 3d_1}{c} = \tan^{-1}\left(\frac{I_{21}}{Q_{21}}\right) \tag{12.17}$$

where f_1 is the reference frequency, I and Q can be calculated using the following equations:

$$I_{12} = E_1(t) \cdot Hilbert\left[E_2(t)\right] - Hilbert[E_1(t)] \cdot E_2(t) \tag{12.18}$$
$$Q_{12} = E_1(t) \cdot E_2(t) + Hilbert\left[E_1(t)\right] \cdot Hilbert[E_2(t)] \tag{12.19}$$

where

$Hilb[\cdot] = $ Hilbert transform
$E = $ Received signal power after BPF and down conversion

After that distances are calculated using the above equations and then position of receiver is calculated using trilateration and the distances calculated above.

12.2.3.2.2 Multilateration

The technique for direction calculation mostly used in the TDOA-based systems is multilateration [35]. This algorithm is based on distance difference between two transmitters whose location is known. The distance difference between each set of two transmitters is given by

$$d_i - d_j = \sqrt{(a_i - a_R)^2 + (b_i - b_R)^2 + (c_i - c_R)^2}$$
$$- \sqrt{(a_j - a_R)^2 + (b_j - b_R)^2 + (c_j - c_R)^2} \tag{12.20}$$

where

$(a_i, b_i, c_i) = $ position of the ith LED in 3D

$(a_R, b_R, c_R) = $ position of the receiver

The calculated difference between distance of each LED and the PD is not absolute. So it forms a hyperbolic curve in Cartesian plane to facilitate the infinite number of locations. Another curve is formed by using a second pair of transmitter. There are several locations where these curves can intersect pointing the possible location of target. By increasing the number of curves the location calculation can be determined easily. To remove the intersection error produced due to noise can be reduced using optimization techniques. Major error produced in the systems that employ multilateration for position calculation is due to thermal noise, shot noise and signals reflected in the NLOS environments [35].

12.2.3.2.3 Angle of Arrival

The angle formed between the normal to photodetector plane and LOS is known as AOA and the methods for AOA calculations are image transformation and modelling. Image transformation uses the photo of light captured and trigonometric relationships to calculate the AOA [36]. MEMS sensors are required for calculations if the camera is not placed horizontally [37]. Modelling approach uses channel model to estimate the calculations [38].

In the AOA the popular algorithm used to determine the position is triangulation [38]. Triangulation has two types namely lateration and angulation. Lateration is also known as trilateration. In angulation we have transmitters whose location is known and angles are measured relative to these sources. Angulation is also referred to as triangulation. After AOA measurements the position is located using the intersection of direction lines as shown in Figure 12.11. The AOA based positioning systems have several benefits over other algorithms that make them more suitable for position calculation.

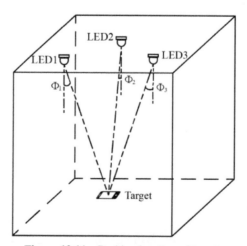

Figure 12.11 Position location of target.

This algorithm is not used much in wireless systems before and in AOA we do not have to account for the noise from reflection if image transformation approaches are used. Time synchronization is not mandatory in these types of systems and they are more stable if calculated using the MEMS sensors.

12.2.3.3 Data Smoothing Filters

All the IPSs algorithms deployed so far are based on the LOS communication pattern. In real environment there are many other sources participating producing noise at the receiver, which disrupts the positioning phenomena. Also, the LOS based positioning systems have rough results even in smooth environments. So in all the above situations and algorithms filtering is employed to improve the results of positioning systems. Two types of filters are mostly deployed: Kalman filter (KF) and particle filter (PF) [22].

The KF filter assumes the system to be linear and Gaussian. In the LED-based IPSs the type of noise distribution is assumed to be Gaussian white noise and the main noise contribution factors are shot noise and thermal noise. The KF algorithm in first phase calculates the present state and covariance based on previous values and in second phase it modifies the predicted values. For a linear system optimal estimation can be achieved by using the KF, however, for non-linear systems extended KF and unscented KF should be used. The KF is applied in many positioning methods such as in Ref. [40]. The KF is used in a system which is based on the AOA, and in Ref. [41] EKF is used with multilateration for position calculation. Where trilateration is used for position calculation, the KF can smooth the navigation route in Ref. [42]. Simulation in Ref. [42] results in reduction of position error from 0.143 m to 0.115 m.

To estimate the previous values, the PF uses a set of weighted particles [43]. Simulation results show that the PF is much complex and time-consuming than the KF but it is more stable. To avoid degeneracy problem a sampling process is applied in this filter [44, 42].

12.2.4 Types of Systems

A photodiode provides current when light falls on it and the current it provides is further processed to calculate the RSS, TDOA or AOA. The photodiode is a very cheap device. As to gain maximum information at receiving end a photodiode array is used instead of a single photodiode. Recent developments in the VLC systems have broaden the scope of its applications. IPSs based on the VLC are divided on the basis of number of transmitters and receivers.

Single receiver and multiple transmitter approach requires a minimum of three LEDs; however, the number of LEDs can be increased to gain more accuracy and for high dimension systems. Single transmitter and receiver approach is implemented in the research [27] with inertial sensors as an auxiliary device. It is a complex and time-consuming approach and is advised to be applied only when there are low numbers of transmitters to implement any system. This system calculates the RSS peak value when phone faced squarely to light and angle at this situation is incident angle when phone is horizontal. From this information, distance is calculated and direction is calculated by cell phone compass and using this distance and direction, location is calculated.

Single transmitter and multiple receiver type of systems use multiple receivers and single transmitter for positioning purpose. In Ref. [43] a receiver is designed with three photodetectors that use the RSS values of optical signals to derive distance between transmitter and each receiver in a 2D environment as the position of each photodetector is at fixed location with respect to other photodetectors. By using the rotation angle of receiver side the device can be located in the middle of the receiving device by estimating the rotation angle between the receiving photodetectors while rotating them with reference to a specific photodetector. This design was further improved in Ref. [38] by separated and tilted photodetectors. The receivers are close enough so that the received signals have same distance from transmitter but intensity was different and incidence angle gain difference between two receivers at each location on the map with fixed azimuth angle of the receivers. The results turned out to be a hyperbolic cluster for each pair [38]. Finally, the location of the receiver was the intersection point of three lines of incidence angle gain difference. In Ref. [30] angular diversity technique is exploited at receiver side using 6 tilted photodetectors.

Other hybrid systems using the RF signals are studied in Refs. [27–31]. Performance of photodetectors' 3D positioning systems was explored in Refs. [16, 22]. This system uses accelerometers to calculate altitude for the AOA triangulation positioning.

12.2.5 Analysis Metrics

A comparison is presented in Table 12.8 of the photodiode-based IPSs. Several photodiodes-based systems are compared according to accuracy, complexity and cost. In Table 12.8, height represents the height of receiver from the ground.

Table 12.8 Photodiode-based IPS comparison

System	Signal Algorithm	Positioning Algorithm	Experiment Type	Testbed Dimensions (m)	Accuracy (cm)	Noise	Cost	Complexity
[22]	RSS	Trilateration	Experiment	5 × 8, 2 × 12, 3.5 × 6.5	30	No	Low	Medium
[19]	RSS	Trilateration	Simulation	6 × 4 × 4 H : 1	<20 2D	Yes	Low	Medium
[34]	RSS	Trilateration	Experiment	0.6×0.6×0.6	2.4 3D	No	Low	Medium
[13]	RSS	Trilateration	Simulation	6 × 6 × 4.2 H = 1	<15 2D	Yes	Medium	Medium
[35]	RSS	Trilateration	Simulation	0.9×0.9×1.5	1.58 2D	Yes	low	Medium
[26]	RSS	Trilateration	Experiment	Large	0.3 − 0.443D	Yes	Medium	High
[36]	RSS	Trilateration+ PDR + Filters	Experiment	2.5 × 2.84 × 2.5	14 2D	Yes	Medium	High
[37]	RSS	Filters	Simulation	Large	High	Yes	Medium	Low
[38]	RSS	Fingerprinting	Experiment	1.8 × 1.2 × 1 Step : 0.1	14.84 2D	Yes	High	High
[10]	RSS	Fingerprinting	Simulation	6 × 6 × 4	9.1 ∼ 26.4 2D	No	High	High

(*Continued*)

Table 12.8 Continued

System	Signal Algorithm	Positioning Algorithm	Experiment Type	Testbed Dimensions (m)	Accuracy (cm)	Noise	Cost	Complexity
[39]	RSS	Fingerprinting	Simulation	30 × 30 Unit : 1	81 2D	Yes	High	High
[40]	RSS	Proximity	Simulation	10 × 2.5 × 3	<130 2D	Yes	Low	Low
[9]	RSS	Modeling	Experiment	5 × 4 × 3	60	Yes	Medium	Medium
[41]	TDOA	Trilateration	Simulation	5 × 5 × 3	0.18 2D	No	Low	Medium
[10]	TDOA	Trilateration	Simulation	5 × 5 × 3	0.002 3D	No	Low	Medium
[42]	TDOA	Multialteration	Simulation	5 × 5 × 3	3.59 3D	Yes	Low	Medium
[43]	AOA	Triangulation	Experiment	5 × 3 × 3	25 3D	Yes	Medium	Medium
[44]	AOA	Triangulation	Simulation	4 × 6	<15 2D	No	Low	Medium
[27]	AOA	Coverage	Experiment	0.08 × 0.08 × 0.4	0.82 2D	Yes	High	Medium
[45]	AOA	Coverage	Simulation	5 × 3 × 3 H : 1	12.9 2D	Yes	High	Medium

12.2.5.1 Accuracy

The localization systems are characterized based on accuracy in Table 12.8. The accuracy takes into account the dimensions, e.g., 2D or 3D and also whether the noise factor is accumulated or not in experiment or simulations. Generally, real-time results slightly differ from the results gained from simulations. From accuracy point of view, the trilateration-based systems perform far better than proximity-based systems and the RSS- and AOA-based systems are merely the same under this metric.

12.2.5.2 Complexity

Complexity increases with the increase in the number of LEDs and addition of auxiliary devices to the system. On design and implementation basis, triangulation and trilateration are most complex. The complexity of fingerprinting is directly proportional to area under observation and inversely proportional to area of fingerprints, and proximity is least complex.

12.2.5.3 Cost

Cost includes time and expenditure on devices, infrastructure and labour required to implement the system. The LED-based IPS are very cheap from all other positioning systems available in competition so far as LED's infrastructure is already deployed for lightening purpose and they are energy efficient. In LED-based IPS options the fingerprinting is most expensive from cost point of view as it requires high labour at its offline mode in which data is collected of locations.

12.2.6 Challenges and Future Concerns

Research has explored many possibilities in IPSs using LEDs. There are many areas where there is need for research and there are many open challenges still in this field. Some of the challenges in this field are listed hereunder for researchers to concentrate for further work. High decoding rate is required at the receiver side if the VLC-based system is transmitting at high rates, so there is need to new decoding techniques, which seems to be the main challenge in the VLC-based communication. Using multiple LEDs for positioning purposes enables multiplexing protocols' implementation at transmitter side. So there is a need to further exploit application of these multiplexing protocols for efficient and accurate position detection and high rate data communication while using the shared medium.

Since the LED based positioning system is difficult to apply for smart-phone users, there is a need to design the LED-based IPSs for smartphone applications so that it will enhance the commercialization of the VLC-based technology for smartphone users. Channel models derived so far have to be further improved to accommodate the multipath effect and several other noises present in indoor environment for the LED based IPSs. Simulations and experiments have to be performed on these models to show their validity and effectiveness. Furthermore, to counter the effects produced by large scale and complex environments it is suggested that the experiments have to be performed in these types of environments. Low accuracy is noticed in a system applying RSS-based systems at corners and edges, so there is a need to find out a solution to these problems as not much work in this area have been done so far.

12.2.7 New Look

After having a full review of the existing technology, the first aspect to be noted is that most of the researchers have neglected the effect of shot noise (e.g., in the sense of signal dependency). As is seen from (12.7), both noises play important role in analysing the system. From the recent research works [29, 31–32], the effect of background noise is simulated and highlight the effect of signal independency. From Refs. [31, 32], it is shown that by considering the effect of signal independency, we better analyse the pragmatic implementation of the future research projects.

The second view point is that most of the current VLC systems have the capability of transmitting small text data at comparatively high speed. The future demands are relatively different: large data size, text documents, audio transmission and live streaming of audio/video files.

The third feature, which should be highlighted, is the working of analogue-to-digital conversion. The aim behind is that to transmit the audio/video files, we must encode them to save bandwidth and enhance data capacity but currently the working of these converters is not feasible in terms of efficiency and complexity. The problem has much importance on its own and should be highlighted.

It is possible to do the conversion efficiently using the latest existing encoding methods. In the near future, the VLC technology is going to be very popular by improving fast and becoming ubiquitous worldwide owing to its enormous benefits. It is therefore highlighted for the new era researcher to investigate further this technology and introduce a new advancement in the

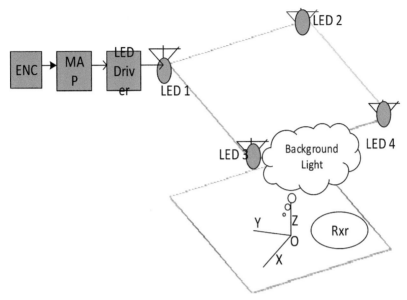

Figure 12.12   New look of VLC-IPS system.

field of wireless communication (as the internet of things is gaining lot of attention of future investors).

Figure 12.12 gives the overview of the new looking project for the VLC-IPS, which could be implemented for further processing and experimental work. It is proposed that the real time compressed video is to be transmitted using Wi-Fi technology with the help of smartphone camera. Then using the decoding procedure IDs should be assigned for the dark and bright images. It is shown in Figure 12.12, the background light, which affects the signal. In previous works [20–23, 32], the background light is ignored which should be highlighted for experimental work. The above given new look project can be simulated using the MATLAB image acquisition toolbox and the algorithm [33] and the pre-set saved database.

Recently [46] a new work based on the VLC-IPS is suggested assuming the compression technique for separation of signal and proximity for estimation. No doubt the solution is better when compared with low computational complexity and low cost but it has its own demerits of transmitting low data rate. It is therefore, (obliging notion behind the proposed review research), a provided review gives full understanding of future demands and current lack of research work, which should be incorporated to reach the future destiny of wireless communication.

12.2.8 Conclusion

A review of the different VLC based positioning algorithms is conducted. Different parameters of channel, signal accessing techniques, positioning algorithms and multiplexing techniques are discussed. Then comparison is made on complexity, cost and accuracy basis. After that challenges and future concerns are discussed with the hope that this review will help the researchers to deploy the VLC-based IPSs in a more accurate and efficient way of fulfilling the future demands.

Acknowledgements

The authors would like to thank Dr. Ishtiaq Ahmad and Dr. Ali Raza for their thorough discussion and helping in completing the research work. We also acknowledge the Punjab Higher Education Commission (PHEC), Lahore, Pakistan for providing the travel grant for this research.

References

[1] Navin Kumar, Nuno Lourenço, Michal Spiez and Rui L. Aguiar, "Visible Light Communication Systems Conception and VIDAS", IETE Technical Review, Vol: 25, Issue: 6, pp: 359–367, Nov-Dec 2008.

[2] Nam-Tuan Le and Yeong Min Jang, "Virtual Cognitive MAC for Visible Light Communication System", International Journal of Smart Home, Vol: 6, Issue: 2, pp: 95–100, 2012.

[3] Pure VLC, "Visible Light Communication: An introductory guide", [online] www.purevlc.net, 2012.

[4] E. D. Kaplan and C. J. Hegarty, Understanding GPS: principles and applications: Artech House Publishers, 2006.

[5] H. Lan, C. Yu, Y. Zhuang, Y. Li, and N. El-Sheimy, "A Novel Kalman Filter with State Constraint Approach for the Integration of Multiple Pedestrian Navigation Systems," Micromachines, vol. 6, p. 926, 2015.

[6] Y. Zhuang and N. El-Sheimy, "Tightly-Coupled Integration of WiFi and MEMS Sensors on Handheld Devices for Indoor Pedestrian Navigation," Sensors Journal, IEEE, vol. 16, pp. 224–234, 2016.

[7] N. U. Hassan, A. Naeem, M. A. Pasha, T. Jadoon, and C. Yuen, "Indoor Positioning Using Visible LED Lights: A Survey," ACM Comput. Surv., vol. 48, pp. 1–32, 2015.

[8] S. H. Yang, E. M. Jeong, D. R. Kim, H. S. Kim, Y. H. Son, and S. K. Han, "Indoor three-dimensional location estimation based on LED visible light communication," Electronics Letters, vol. 49, pp. 54–56, 2013.

[9] M. Yasir, S.-W. Ho, and B. N. Vellambi, "Indoor Positioning System Using Visible Light and Accelerometer," Journal of Lightwave Technology, vol. 32, pp. 3306–3316, 2014/10/01 2014.

[10] U. Nadeem, N. U. Hassan, M. A. Pasha, and C. Yuen, "Highly accurate 3D wireless indoor positioning system using white LED lights," Electronics Letters, vol. 50, pp. 828–830, 2014.

[11] T. Komine and M. Nakagawa, "Fundamental analysis for visible-light communication system using LED lights," IEEE Transactions on Consumer Electronics, vol. 50, pp. 100–107, 2004.

[12] M. F. Keskin, E. Gonendik, and S. Gezici, "Improved Lower Bounds for Ranging in Synchronous Visible Light Positioning Systems," Journal of lightwave technology, vol. 34, pp. 5496–5504, 2016.

[13] W. Gu, W. Zhang, M. Kavehrad, and L. Feng, "Three-dimensional light positioning algorithm with filtering techniques for indoor environments," Optical Engineering, vol. 53, pp. 107107–107107, 2014.

[14] X. Zhang, J. Duan, Y. Fu, and A. Shi, "Theoretical accuracy analysis of indoor visible light communication positioning system based on received signal strength indicator," Journal of Lightwave Technology, vol. 32, pp. 3578–3584, 2014.

[15] S.-H. Yang, H.-S. Kim, Y.-H. Son, and S.-K. Han, "Three-Dimensional Visible Light Indoor Localization Using AOA and RSS with Multiple Optical Receivers," Journal of Lightwave Technology, vol. 32, pp. 2480–2485, 2014/07/15 2014.

[16] Z. Zhou, M. Kavehrad, and P. Deng, "Indoor positioning algorithm using light-emitting diode visible light communications," Optical Engineering, vol. 51, pp. 085009-1-085009-6, 2012.

[17] N. A. Mohammed and M. A. Elkarim, "Exploring the effect of diffuse reflection on indoor localization systems based on RSSI-VLC," Optics express, vol. 23, pp. 20297–20313, 2015.

[18] U. Nadeem, N. U. Hassan, M. A. Pasha, and C. Yuen, "Indoor positioning system designs using visible LED lights: performance comparison

of TDM and FDM protocols," Electronics Letters, vol. 51, pp. 72–74, 2015.

[19] W. Zhang, M. I. S. Chowdhury, and M. Kavehrad, "Asynchronous indoor positioning system based on visible light communications," Optical Engineering, vol. 53, pp. 045105–045105, 2014.

[20] M.-G. Moon, S.-I. Choi, J. Park, and J. Y. Kim, "Indoor Positioning System using LED Lights and a Dual Image Sensor," Journal of the Optical Society of Korea, vol. 19, pp. 586–591, 2015/12/01 2015.

[21] S. h. P. Won, W. W. Melek, and F. Golnaraghi, "A Kalman/Particle Filter-Based Position and Orientation Estimation Method Using a Position Sensor/Inertial Measurement Unit Hybrid System," IEEE Transactions on Industrial Electronics, vol. 57, pp. 1787–1798, 2010.

[22] L. Li, P. Hu, C. Peng, G. Shen, and F. Zhao, "Epsilon: A visible light based positioning system," in 11th USENIX Symposium on Networked Systems Design and Implementation (NSDI 14), 2014, pp. 331–343.

[23] Q. Xu, R. Zheng, and S. Hranilovic, "IDyLL: indoor localization using inertial and light sensors on smartphones," Proceedings of the 2015 ACM International Joint Conference on Pervasive and Ubiquitous Computing, Osaka, Japan, 2015.

[24] D. Taniuchi, X. Liu, D. Nakai, and T. Maekawa, "Spring model based collaborative indoor position estimation with neighbor mobile devices," *IEEE Journal of Selected Topics in Signal Processing,* vol. 9, pp. 268–277, 2015.

[25] Z. Luo, W. Zhang, and G. Zhou, "Improved spring model-based collaborative indoor visible light positioning," *Optical Review,* vol. 23, pp.479–486, 2016.

[26] B. Xie, K. Chen, G. Tan, M. Lu, Y. Liu, J. Wu, *et al.*, "LIPS: A Light Intensity–Based Positioning System for Indoor Environments," *ACM Trans. Sen. Netw.,* vol. 12, pp. 1–27, 2016.

[27] J. Herrnsdorf, M. J. Strain, E. Gu, R. K. Henderson, and M. D. Dawson, "Positioning and space-division multiple access enabled by structured illumination with light-emitting diodes," *Journal of Lightwave Technology,* vol. 35, pp. 2339–2345, 2017.

[28] M. N. Khan and M. Jamil, "Performance Improvement in Lifetime and Throughput of LEACH Protocol," Indian Journal of Science and Technology, vol. 9, no. 21, pp. 1–6, 2016.

[29] M. N. Khan and M. Jamil, "Maximizing Throughput of Free Space Communication Systems Using Puncturing Technique," Arabian Journal for Science and Engineering, vol. 39, no. 12, pp. 8925–8933, 2014.

[30] M. N. Khan, H. K. Hasnain and M. Jamil, "Digital Signal Processing: A Breadth-first Approach", Rivers Publishers, Denmark, pp. 221–247, 2016.

[31] Muhammad Nasir Khan, and M. Jamil, "Maximizing Throughput of Hybrid FSO-RF Communication System: An Algorithm," IEEE Access, vol. 6, pp. 30039–30048, 2018.

[32] Y. Li, Z. Ghassemlooy, X. Tang, B. Lin and Y. Zhang, "A VLC Smartphone Camera Based Indoor Positioning System," in *IEEE Photonics Technology Letters*, vol. 30, no. 13, pp. 1171–1174, 1 July1, 2018. doi: 10.1109/LPT.2018.2834930

[33] C. Fu, C. W. Cheng, W. H. Shen, Y. L. Wei and H. M. Tsai, "Lightbib: marathoner recognition system with visible light communications," *IEEE Int. Conf. Data Science and Data Intensive Sys.*, 2015, pp. 572–578.

[34] H. S. Kim, D. R. Kim, S. H. Yang, Y. H. Son, and S. K. Han, "An Indoor Visible Light Communication Positioning System Using a RF Carrier Allocation Technique," Journal of Lightwave Technology, vol. 31, pp. 134–144, 2013.

[35] S.-H. Yang, D.-R. Kim, H.-S. Kim, Y.-H. Son, and S.-K. Han, "Visible light based high accuracy indoor localization using the extinction ratio distributions of light signals," *Microwave and Optical Technology Letters,* vol. 55, pp. 1385–1389, 2013.

[36] Z. Li, A. Yang, H. Lv, L. Feng, and W. Song, "Fusion of visible light indoor positioning and inertial navigation based on particle filter," *IEEE Photonics Journal,* 2017.

[37] D. Zheng, K. Cui, B. Bai, G. Chen, and J. A. Farrell, "Indoor localization based on LEDs," in 2011 IEEE International Conference on Control Applications (CCA), 2011, pp. 573–578.

[38] J. Vongkulbhisal, B. Chantaramolee, Y. Zhao, and W. S. Mohammed, "A fingerprinting-based indoor localization system using intensity modulation of light emitting diodes," Microwave and Optical Technology Letters, vol. 54, pp. 1218–1227, 2012.

[39] G. Kail, P. Maechler, N. Preyss, and A. Burg, "Robust asynchronous indoor localization using LED lighting," in *2014 IEEE International Conference on Acoustics, Speech and Signal Processing (ICASSP)*, 2014, pp. 1866–1870.

[40] P. Lou, H. Zhang, X. Zhang, M. Yao, and Z. Xu, "Fundamental analysis for indoor visible light positioning system," in *2012 1st IEEE International Conference on Communications in China Workshops (ICCC)*, 2012.

[41] S. Y. Jung, S. Hann, and C. S. Park, "TDOA-based optical wireless indoor localization using LED ceiling lamps," IEEE Transactions *on Consumer Electronics,* vol. 57, pp. 1592–1597, 2011.

[42] D. Trong-Hop, H. Junho, and Y. Myungsik, "TDoA based indoor visible light positioning systems," in 2013 Fifth International Conference on Ubiquitous and Future Networks (ICUFN), 2013, pp. 456–458.

[43] W. Zhang and M. Kavehrad, "A 2-D indoor localization system based on visible light LED," in *2012 IEEE Photonics Society Summer Topical Meeting Series*, 2012, pp. 80–81.

[44] Y. S. Erogluy, I. Guvency, N. Palay, and M. Yukselz, "AOA-based localization and tracking in multi-element VLC systems," in Wireless and Microwave Technology Conference (WAMICON), 2015 IEEE 16th Annual, 2015, pp. 1–5.

[45] M. T. Taylor and S. Hranilovic, "Angular diversity approach to indoor positioning using visible light," in *Globecom Workshops (GC Wkshps), 2013 IEEE*, 2013, pp. 1093–1098.

[46] Kristina Gligorit'c, Manisha Ajmani, Dejan Vukobratovit'c and Sinan Sinanovit'c, "Visible Light Communications Based Indoor Positioning via Compressed Sensing" [Online] https://arxiv.org/abs/1805.01001v1.

References and Bibliography

[1] G. Arfken, and H. J. Weber, Mathematical methods for Physicists, 4th ed. Boston, MA: Academic Press, 1995.

[2] F. B. Hildebrand, Advanced Calculus for Applications, 2nd ed. Englewood Cliffs, NJ: Prentice Hall, 1976.

[3] G. B. Thomas, Jr., and Finney, R. L., Calculus and Analytic Geometry, 9th ed., Reading, MA: Addison-Wesley, 1996.

[4] M. N. Khan, S. K. Hasnain, and M. Jamil, Digital Signal Processing: a breadth-first Approach, International ed., Rivers Publishers, Denmark, 2016.

[5] G. Birkhoff and G. -C. Rota, Ordinary Differential Equations, 3rd ed., New York, NY: John Wiley, 1978.

[6] W. E. Boyce and R. C. DiPrima, Elementary Differential Equations, 3rd ed. New York, NY: John Wiley, 1977.

[7] G. F. Carrier, M. Krook, and Pearson, C. E., Functions of a Complex Variables: Theory and Technique, Ithaca, New York, NY: Hod Books, 1983.

[8] R. V. Churchill, J. W. Brown, and R. F. Verhey, Complex Variables and Applications, 5th ed. New York, NY: McGraw-Hill, 1990.

[9] R. M. Gray and J. W. Goodman, Fourier Transforms: An Introduction for Engineers, Boston, MA: Kluwer Academic Publishers, 1995.

[10] E. D. Rainville, The Laplace Trnsform: An Introduction, New York, NY: Macmillan, 1963.

[11] E. I. Jury, Theory and Application of the Z-Transform Method, Malabar, FL: R.E. Krieger, 1982.

[12] G. Strang, Introduction to Linear Algebra, Wellesley, MA: Wellesley-Cambridge Press, 1993.

[13] L. B. Jackson, Signals, Systems and Transforms, Reading, MA: Addison-Wesley, 1991.

[14] J. A. Cadzow and H.F. Van Landingham, Signals and Systems, Englewood Cliffs, NJ: Prentice Hall, 1985.

[15] E. Kamen, Introduction to Signals and Systems, New York, NY: Macmillan, 1987.

[16] R. J. Mayhan, Discrete-time and Continuous-time Linear Systems, Reading, MA: Addison-Wesley, 1984.

[17] H. P. Neff, Continuous and Discrete Linear Systems, New York, NY: Harper and Row, 1984.

[18] R. E. Ziemer, Tranter, W. H., and D. R. Fannin, Signals and Systems: Continuous and Discrete, 2nd ed. New York, NY: Macmillan, 1989.

[19] O. E. Brigham, The Fast Fourier Transform and its Applications, Englewood Cliffs, NJ: Prentice Hall, 1988.

[20] B. Gold and C. M. Rader, Digital Processing of Signals, McGraw-Hill, New York, 1969.

[21] A. V. Oppenheim and R. W. Schafer, Discrete-Time Signal Processing, Prentice Hall, Englewood Cliffs, NJ, 1989.

[22] A. V. Oppenheim and R. W. Schafer, Digital Signal Processing, Prentice Hall, Englewood Cliffs, NJ, 1975.

[23] L. R. Rabiner and B. Gold, Theory and Application of Digital Signal Processing, Prentice Hall, Englewood Cliffs, NJ, 1975.

[24] S. K. Mitra and J. F. Kaiser, eds., Handbook of Digital Signal Processing, Wiley, New York, 1993.

[25] S. K. Mitra, Digital Signal Processing A computer-based approach, McGraw Hill, 2015.

[26] T. W. Parks and C. S. Burrus, Digital Filter Design, Wiley, New York, 1987.

[27] A. Antoniou, Digital Filters: Analysis and Design, 2nd ed., McGraw-Hill, New York, 1993.

[28] D. F. Elliott, ed., Handbook of Digital Signal Processing, Academic Press, New York, 1987.

[29] L. R. Rabiner and C. M. Rader, eds., Digital Signal Processing, IEEE Press, New York, 1972.

[30] Selected Papers in Digital Signal Processing, II, edited by the Digital Signal Processing Committee and IEEE ASSP, IEEE Press, New York, 1976.

[31] R. A. Roberts and C. T. Mullis, Digital Signal Processing, Addison-Wesley, Reading, MA, 1987.

[32] S. A. Tretter, Introduction to Discrete-Time Signal Processing, New York, NY: John Wiley, 1976.

[33] B. D. O. Anderson and J. B. Moore, Optimal Control: Linear Quadratic Methods, Englewood Cliffs, NJ: Prentice Hall, 1990.

[34] R. C. Dorf and R. H. Bishop, Modern Control Systems, 7th ed., Reading, MA: Addison-Wesley 1995.

[35] R. J. Vaccaro, Digital Control: A State-Space Approach, New York, NY: McGraw-Hill 1995.

[36] R. N. Bracewell, Two-Dimensional Imaging, Englewood Cliffs, NJ: Prentice Hall, 1995.

[37] K. R. Castleman, Digital Image Processing, Englewood Cliffs, NJ: Prentice Hall, 1996.

[38] P. A. Lynn and W. Fuerst, Introductory Digital Signal Processing with Computer Applications, Wiley, New York, 1989.

[39] J. G. Proakis and D. G. Manolakis, Introduction to Digital Signal Processing, 2nd ed., Macmillan, New York, 1988.

[40] E. C. Ifeachor and B. W. Jervis, Digital Signal Processing: A Practical Approach, Addison-Wesley, Reading, MA, 1993.

[41] R. A. Haddad and T. W. Parsons, Digital Signal Processing: Theory, Applications, and Hardware, Computer Science Press, W. H. Freeman, New York, 1991.

[42] D. J. DeFatta, J. G. Lucas, and W. S. Hodgkiss, Digital Signal Processing: A System Design Approach, Wiley, New York, 1988.

[43] E. Robinson and S. Treitel, Geophysical Signal Analysis, Prentice Hall, Englewood Cliffs, NJ, 1980.

[44] S. M. Kay, Modern Spectral Estimation: Theory and Application, Prentice Hall, Englewood Cliffs, NJ, 1988.

[45] S. L. Marple, Digital Spectral Analysis with Applications, Prentice Hall, Englewood Cliffs, NJ, 1987.

[46] B. Widrow and S. D. Stearns, Adaptive Signal Processing, Prentice Hall, Englewood Cliffs, NJ, 1985.

[47] H. R. Chillingworth, Complex Variables, Pergamon, Oxford, 1973.

[48] P. H. Scholfield, The Theory of Proportion in Architecture, Cambridge University Press, London, 1958.

[49] J. Kappraff, Connections: The Geometric Bridge Between Art and Science, McGraw-Hill, New York, 1990.

[50] D. A. Linden, "A Discussion of Sampling Theorems," Proc. IRE, **47**, 1219 (1959).

[51] A. J. Jerri, "The Shannon Sampling Theorem—Its Various Extensions and Applications: A Tutorial Review," Proc. IEEE, **65**, 1565 (1977).

[52] P. L. Butzer and R. L. Strauss, "Sampling Theory for not Necessarily Band-Limited Functions: A Historical Overview," SIAM Review, **34**, 40 (1992).

[53] D. H. Sheingold, ed., Analog-Digital Conversion Handbook, 3d ed., Prentice Hall, Englewood Cliffs, NJ, 1986.

[54] A. VanDoren, Data Acquisition Systems, Reston Publishing, Reston, VA, 1982.

[55] W. R. Bennett, "Spectra of Quantized Signals," Bell Syst. Tech. J., **27**, 446 (1948).

[56] B. Widrow, "Statistical Analysis of Amplitude-Quantized Sampled-Data Systems," AIEE Trans. Appl. Ind., pt.2, **79**, 555 (1961).

[57] G. W. Wornell, Signal Processing with Fractals: A Wavelet-Based Approach, Upper Saddle River, NJ: Prentice Hall, 1996.

[58] P. F. Swaszek, ed., Quantization, Van Nostrand Reinhold, New York, 1985.

[59] A. B. Sripad and D. L. Snyder, "A Necessary and Sufficient Condition for Quantization Errors to Be Uniform and White," IEEE Trans. Acoust., Speech, Signal Process., ASSP-25, 442 (1977).

[60] C. W. Barnes, et al., "On the Statistics of Fixed-Point Roundoff Error," IEEE Trans. Acoust., Speech, Signal Process., **ASSP-33**, 595 (1985).

[61] R. M. Gray, "Quantization Noise Spectra," IEEE Trans. Inform. Theory, **IT-36**, 1220 (1990), and earlier references therein. Reprinted in Ref. [276], p. 81.

[62] L. Schuchman, "Dither Signals and Their Effect on Quantization Noise," IEEE Trans. Commun., COM-12, 162 (1964).

[63] N. S. Jayant and P. Noll, Digital Coding of Waveforms, Prentice Hall, Englewood Cliffs, NJ, 1984.

[64] J. F. Blinn, "Quantization Error and Dithering," IEEE Comput. Graphics & Appl. Mag., (July 1994), p. 78.

[65] A. W. Drake, Fundamental of Applied Probability Theory, New York, NY: McGraw Hill, 1967.

[66] S. P. Lipshitz, R. A. Wannamaker, and J. Vanderkooy, "Quantization and Dither: A Theoretical Survey," J. Audio Eng. Soc., **40**, 355 (1992).

[67] J. Vanderkooy and S. P. Lipshitz, "Resolution Below the Least Significant Bit in Digital

[68] Systems with Dither," J. Audio Eng. Soc., **32**, 106 (1984).

[69] J. Vanderkooy and S. P. Lipshitz, "Dither in Digital Audio," J. Audio Eng. Soc., **35**, 966 (1987).

[70] J. Vanderkooy and S. P. Lipshitz, "Digital Dither: Signal Processing with Resolution Far Below the Least Significant Bit," Proc. 7th Int. Conf.: Audio in Digital Times, Toronto, May 1989, p. 87.

[71] R. A. Wannamaker, "Psychoacoustically Optimal Noise Shaping," J. Audio Eng. Soc., 40, 611 (1992).

[72] M. A. Gerzon, P. G. Graven, J. R. Stuart, and R. J. Wilson, "Psychoacoustic Noise Shaped Improvements in CD and Other Linear Digital Media," presented at 94th Convention of the AES, Berlin, May 1993, AES Preprint no. 3501.

[73] R. van der Waal, A. Oomen, and F. Griffiths, "Performance Comparison of CD, Noise-Shaped CD and DCC," presented at 96th Convention of the AES, Amsterdam, February 1994, AES Preprint no. 3845.

[74] J. A. Moorer and J. C. Wen, "Whither Dither: Experience with High-Order Dithering Algorithms in the Studio," presented at 95th Convention of the AES, New York, October 1993, AES Preprint no. 3747.

[75] R. A. Wannamaker, "Subtractive and Nonsubtractive Dithering: A Comparative Analysis," presented at 97th Convention of the AES, San Francisco, November 1994, AES Preprint no. 3920.

[76] D. Ranada, "Super CD's: Do They Deliver The Goods?," Stereo Review, July 1994, p. 61.

[77] C. T. Mullis and R. A. Roberts, "An Interpretation of Error Spectrum Shaping in Digital Filters," IEEE Trans. Acoust., Speech, Signal Process., **ASSP-30**, 1013 (1982).

[78] J. Dattoro, "The Implementation of Digital Filters for High-Fidelity Audio," Proc. AES 7th Int. Conf., Audio in Digital Times, Toronto, 1989, p. 165.

[79] R. Wilson, et al., "Filter Topologies," J. Audio Eng. Soc., **41**, 455 (1993).

[80] U. Z··olzer, "Roundoff Error Analysis of Digital Filters," J. Audio Eng. Soc., **42**, 232 (1994).

[81] S. Ross, Introduction to Probability Models, 5th ed., Boston, MA: Academic Press, 1993.

[82] S. M. Kay, Fundamental of Statistical Signal Processing: Estimation Theory, Englewood Cliffs, NJ: Prentice Hall, 1993.

[83] W. Chen, "Performance of the Cascade and Parallel IIR Filters," presented at 97th Convention of the AES, San Francisco, November 1994, AES Preprint no. 3901.

[84] D. W. Horning and R. Chassaing, "IIR Filter Scaling for Real-Time Signal Processing," IEEE Trans. Educ., **34**, 108 (1991).

[85] K. Baudendistel, "Am Improved Method of Scaling for Real-Time Signal Processing Applications," IEEE Trans. Educ., **37**, 281 (1994).

[86] R. Chassaing, Digital Signal Processing with C and the TMS320C30, Wiley, New York, 1992.

[87] M. El-Sharkawy, Real Time Digital Signal Processing Applications with Motorola's DSP56000 Family, Prentice Hall, Englewood Cliffs, NJ, 1990.

[88] M. El-Sharkawy, Signal Processing, Image Processing and Graphics Applications with Motorola's DSP96002 Processor, vol. I, Prentice Hall, Englewood Cliffs, NJ, 1994.

[89] V. K. Ingle and J. G. Proakis, Digital Signal Processing Laboratory Using the ADSP-2101 Microcomputer, Prentice Hall, Englewood Cliffs, NJ, 1991.

[90] J. Tow, "Implementation of Digital Filters with the WE-R DSP32 Digital Signal Processor," AT&T Application Note, 1988.

[91] S. L. Freeny, J. F. Kaiser, and H. S. McDonald, "Some Applications of Digital Signal Processing in Telecommunications," in Ref. [12], p. 1.

[92] J. R. Boddie, N. Sachs, and J. Tow, "Receiver for TOUCH-TONE Service," Bell Syst. Tech. J., 60, 1573 (1981).

[93] A. Mar, ed., Digital Signal Processing Applications Using the ADSP-2100 Family, Prentice Hall, Englewood Cliffs, NJ, 1990.

[94] IEEE ASSP Mag., **2**, no. 4, October 1985, Special Issue on Digital Audio.

[95] Y. Ando, Concert Hall Acoustics, Springer-Verlag, New York, 1985.

[96] D. Begault, "The Evolution of 3-D Audio," MIX, October 1993, p. 42.

[97] B. Blesser and J. M. Kates, "Digital Processing of Audio Signals," in Ref. [12], p. 29.

[98] N. Brighton and M. Molenda, "Mixing with Delay," Electronic Musician, **9**, no. 7, 88 (1993).

[99] J. M. Eargle, Handbook of Recording Engineering, 2nd ed., Van Nostrand Reinhold, New York, 1992.

[100] P. Freudenberg, "All About Dynamics Processors, Parts 1 & 2," Home & Studio Recording, March 1994, p. 18, and April 1994, p. 44.

[101] D. Griesinger, "Practical Processors and Programs for Digital Reverberation," Proc. AES 7th Int. Conf., Audio in Digital Times, Toronto, 1989, p. 187.

[102] D. G. Childers, "Biomedical Signal Processing," in Selected Topics in Signal Processing, S. Haykin, ed., Prentice Hall, Englewood Cliffs, NJ, 1989.

[103] A. Cohen, Biomedical Signal Processing, vols. 1 and 2, CRC Press, Boca Raton, FL, 1986.

[104] H. G. Goovaerts and O. Rompelman, "Coherent Average Technique: A Tutorial Review," J. Biomed. Eng., 13, 275 (1991).

[105] P. Horowitz and W. Hill, The Art of Electronics, 2nd ed., Cambridge University Press, Cambridge, 1989.

[106] O. Rompelman and H. H. Ros, "Coherent Averaging Technique: A Tutorial Review, Part 1: Noise Reduction and the Equivalent Filter," J. Biomed. Eng., **8**, 24 (1986); and "Part 2: Trigger Jitter, Overlapping Responses, and Non-Periodic Stimulation," ibid., p. 30.

[107] V. Shvartsman, G. Barnes, L. Shvartsman, and N. Flowers, "Multichannel Signal Processing Based on Logic Averaging," IEEE Trans. Biomed. Eng., **BME-29**, 531 (1982).

[108] H. L. Van Trees, Detection, Estimation, and Modulation Theory: Part I, New York, NY: Jown Wiley and Sons, Inc, 1968.

[109] C. W. Therrien, Discrete Random Signals and Statistical Signal Processing, Englewood Cliffs, NJ: Prentice Hall, Inc, 1992.

[110] C. W. Thomas, M. S. Rzeszotarski, and B. S. Isenstein, "Signal Averaging by Parallel Digital Filters," IEEE Trans. Acoust., Speech, Signal Process., **ASSP-30**, 338 (1982).

[111] T. H. Wilmshurst, Signal Recovery from Noise in Electronic Instrumentation, 2nd ed., Adam Hilger and IOP Publishing, Bristol, England, 1990.

[112] M. Bromba and H. Ziegler, "Efficient Computation of Polynomial Smoothing Digital Filters," Anal. Chem., **51**, 1760 (1979).

[113] M. Bromba and H. Ziegler, "Explicit Formula for Filter Function of Maximally Flat Nonrecursive Digital Filters," Electron. Lett., **16**, 905 (1980).

[114] M. Bromba and H. Ziegler, "Application Hints for Savitzky-Golay Digital Smoothing Filters," Anal. Chem., **53**, 1583 (1981).

[115] T. H. Edwards and P. D. Wilson, "Digital Least Squares Smoothing of Spectra," Appl. Spectrosc., **28**, 541 (1974).

[116] T. H. Edwards and P. D. Wilson, "Sampling and Smoothing of Spectra," Appl. Spectrosc. Rev., **12**, 1 (1976).

[117] C. G. Enke and T. A. Nieman, "Signal-to-Noise Ratio Enhancement by Least-Squares Polynomial Smoothing," Anal. Chem., **48**, 705A (1976).

[118] R. R. Ernst, "Sensitivity Enhancement in Magnetic Resonance," in Advances in Magnetic Resonance, vol. 2, J. S. Waugh, ed., Academic Press, 1966.

[119] R. W. Hamming, Digital Filters, 2nd ed., Prentice Hall, Englewood Cliffs, NJ, 1983.

[120] M. Kendall, Time-Series, 2nd ed., Hafner Press, Macmillan, New York, 1976.

[121] M. Kendall and A. Stuart, Advanced Theory of Statistics, vol. 3, 2nd ed., Charles Griffin & Co., London, 1968.

[122] H. H. Madden, "Comments on the Savitzky-Golay Convolution Method for Least-Squares Fit Smoothing and Differentiation of Digital Data," Anal. Chem., **50**, 1383 (1978).

[123] A. Savitzky and M. Golay, "Smoothing and Differentiation of Data by Simplified Least Squares Procedures," Anal. Chem. **36**, 1627 (1964).

[124] C. S. Williams, Designing Digital Filters, Prentice Hall, Englewood Cliffs, NJ, 1986.

[125] H. Ziegler, "Properties of Digital Smoothing Polynomial (DISPO) Filters," Appl. Spectrosc., **35**, 88 (1981).

[126] J. F. Kaiser and W. A. Reed, "Data Smoothing Using Lowpass Digital Filters," Rev. Sci. Instrum., **48**, 1447 (1977).

[127] J. F. Kaiser and R. W. Hamming, "Sharpening the Response of a Symmetric Nonrecursive Filter by Multiple Use of the Same Filter," IEEE Trans. Acoust., Speech, Signal Process., **ASSP-25**, 415 (1975).

[128] F. J. Harris, "On the Use of Windows for Harmonic Analysis with the Discrete Fourier Transform," Proc. IEEE, **66**, 51 (1978).

[129] N. C. Ge, ckinli and D. Yavuz, "Some Novel Windows and a Concise Tutorial Comparison of Window Families," IEEE Trans. Acoust., Speech, Signal Process., **ASSP-26**, 501 (1978).

[130] J. F. Kaiser and R. W. Schafer, "On the Use of the I0-Sinh Window for Spectrum Analysis," IEEE Trans. Acoust., Speech, Signal Process., **ASSP-28**, 105 (1980).

[131] A. H. Nuttal, "Some Windows with Very Good Sidelobe Behavior," IEEE Trans. Acoust., Speech, Signal Process., **ASSP-29**, 84 (1981).

[132] E. O. Brigham, The Fast Fourier Transform, Prentice Hall, Englewood Cliffs, NJ, 1988.

[133] R. W. Ramirez, The FFT, Fundamentals and Concepts, Prentice Hall, Englewood Cliffs, NJ, 1985.

[134] J. A. Richards, Analysis of Periodically Time Varying Systems, New York, NY: Springler-Verlag, 1983.

[135] M. Vidyasager, Nonlinear Systems Analysis, 2nd ed. Englewood Cliffs, NJ: Prentice Hall, 1993.

[136] C. S. Burrus and T. W. Parks, DFT/FFT and Convolution Algorithms, Wiley, New York, 1985.

[137] C. Van Loan, Computational Frameworks for the Fast Fourier Transform, SIAM, Philadelphia, 1992.

[138] G. D. Bergland, "A Guided Tour of the Fast Fourier Transform," IEEE Spectrum, **6**, 41, July 1969.

[139] J. W. Cooley, P. A. W. Lewis, and P. D. Welch, "The Fast Fourier Transform Algorithm: Programming Considerations in the Calculation of Sine, Cosine, and Laplace Transforms," J. Sound Vib., **12**, 315 (1970). Reprinted in Ref. [9].

[140] F. J. Harris, "The Discrete Fourier Transform Applied to Time Domain Signal Processing," IEEE Commun. Mag., May 1982, p. 13.

[141] J. W. Cooley, P. A. W. Lewis, and P. D. Welch, "Historical Notes on the Fast Fourier Transform," IEEE Trans. Audio Electroacoust., **AU-15**, 76 (1967). Reprinted in Ref. [9].

[142] M. T. Heideman, D. H. Johnson, and C. S. Burrus, "Gauss and the History of the Fast Fourier Transform," IEEE ASSP Mag., **4**, no. 4, 14 (1984).

[143] J. W. Cooley, "How the FFT Gained Acceptance," IEEE Signal Proc. Mag., **9**, no. 1, 10 (1992).

[144] P. Kraniauskas, "A Plain Man's Guide to the FFT," IEEE Signal Proc. Mag., **11**, no. 2, 36 (1994).

[145] J. R. Deller, "Tom, Dick, and Mary Discover the DFT," IEEE Signal Proc. Mag., **11**, no. 2, 36 (1994).

[146] J. F. Kaiser, "Design Methods for Sampled Data Filters," Proc. 1st Allerton Conf. Circuit System Theory, p. 221, (1963), and reprinted in Ref. [9], p. 20.

[147] J. F. Kaiser, "Digital Filters," in F. F. Kuo and J. F. Kaiser, eds., System Analysis by Digital Computer, Wiley, New York, 1966, p. 228.

[148] J. F. Kaiser, "Nonrecursive Digital Filter Design Using the I0-Sinh Window Function," Proc. 1974 IEEE Int. Symp. on Circuits and Systems, p. 20, (1974), and reprinted in [10], p. 123.

[149] H. D. Helms, "Nonrecursive Digital Filters: Design Methods for Achieving Specifications on Frequency Response," IEEE Trans. Audio Electroacoust., **AU-16**, 336 (1968).

[150] H. D. Helms, "Digital Filters with Equiripple or Minimax Response," IEEE Trans. Audio Electroacoust., **AU-19**, 87 (1971), and reprinted in Ref. [9], p. 131.

[151] C. L. Dolph, "A Current Distribution Which Optimizes the Relationship Between Beamwidth and Side-lobe Level," Proc. I.R.E, **34**, 335 (1946).

[152] D. Barbiere, "A Method for Calculating the Current Distribution of Tschebyscheff Arrays," Proc. I.R.E, **40**, 78 (1952).

[153] R. J. Stegen, "Excitation Coefficients and Beamwidths of Tschebyscheff Arrays," Proc. I.R.E, **41**, 1671 (1953).

[154] R. C. Hansen, "Linear Arrays," in A. W. Rudge, et al., eds., Handbook of Antenna Design, vol. 2, 2nd ed., P. Peregrinus and IEE, London, 1986.

[155] T. Saram··aki, "Finite Impulse Response Filter Design," in Ref. [5], p. 155.

[156] K. B. Benson and J. Whitaker, Television and Audio Handbook, McGraw-Hill, New York, 1990.

[157] P. L. Schuck, "Digital FIR Filters for Loudspeaker Crossover Networks II: Implementation Example," Proc. AES 7th Int. Conf., Audio in Digital Times, Toronto, 1989, p. 181.

[158] G. E. P. Box and G. M. Genkins, Time Series Analysis: Forecasting and Control, Rev ed. San Francisco, CA: Holden-Day, 1976.

[159] J.D. Hamilton, Time Series Analysis, Princeton, NJ: Princeton University Press, 1994.

[160] R. Wilson, et al., "Application of Digital Filters to Loudspeaker Crossover Networks," J. Audio Eng. Soc., **37**, 455 (1989).

[161] K. Steiglitz, T. W. Parks, and J. F. Kaiser, "METEOR: A Constraint-Based FIR Filter Design Program," IEEE Trans. Acoust., Speech, Signal Process., **ASSP-40**, 1901 (1992). This program is available via anonymous ftp from princeton.edu.

[162] K. Steiglitz and T. W. Parks, "What is the Filter Design Problem?," Proc. 1986 Princeton Conf. Inform. Sci. Syst., p. 604 (1986).

[163] K. Hirano, S. Nishimura, and S. Mitra, "Design of Digital Notch Filters," IEEE Trans. Commun., **COM-22**, 964 (1974).

[164] M. N. S. Swami and K. S. Thyagarajan, "Digital Bandpass and Bandstop Filters with Variable Center Frequency and Bandwidth," Proc. IEEE, **64**, 1632 (1976).

[165] J. A. Moorer, "The Manifold Joys of Conformal Mapping: Applications to Digital Filtering in the Studio," J. Audio Eng. Soc., **31**, 826 (1983).

[166] S. A. White, "Design of a Digital Biquadratic Peaking or Notch Filter for Digital Audio Equalization," J. Audio Eng. Soc., **34**, 479 (1986).

[167] P. A. Regalia and S. K. Mitra, "Tunable Digital Frequency Response Equalization Filters," IEEE Trans. Acoust., Speech, Signal Process., **ASSP-35**, 118 (1987).

[168] D. J. Shpak, "Analytical Design of Biquadratic Filter Sections for Parametric Filters," J. Audio Eng. Soc., **40**, 876 (1992).

[169] D. C. Massie, "An Engineering Study of the Four-Multiply Normalized Ladder Filter," J. Audio Eng. Soc., **41**, 564 (1993).

[170] F. Harris and E. Brooking, "A Versatile Parametric Filter Using Imbedded All-Pass Sub-Filter to Independently Adjust Bandwidth, Center Frequency, and Boost or Cut," presented at the 95th Convention of the AES, New York, October 1993, AES Preprint 3757.

[171] R. Bristow-Johnson, "The Equivalence of Various Methods of Computing Biquad Coefficients for Audio Parametric Equalizers," presented at the 97th Convention of the AES, San Francisco, November 1994, AES Preprint 3906.

[172] P. P. Vaidyanathan, Multirate Systems and Filter Banks, Prentice Hall, Englewood-Cliffs, NJ, 1993.

[173] J. C. Candy and G. C. Temes, eds., Oversampling Delta-Sigma Data Converters, IEEE Press, Piscataway, NJ, 1992.

[174] J. C. Candy and G. C. Temes, "Oversampling Methods for A/D and D/A Conversion," in Ref. [113].

[175] R. M. Gray, "Oversampled Sigma-Delta Modulation," IEEE Trans. Commun., **COM-35**, 481 (1987). Reprinted in Ref. [276], p. 73.

[176] D. Goedhart, et al., "Digital-to-Analog Conversion in Playing a Compact Disc," Philips Tech. Rev., **40**, 174–179, (1982).

[177] M. W. Hauser, "Principles of Oversampling A/D Conversion," J. Audio Eng. Soc., **39**, 3-26, (1991).

[178] D. E. Knuth, The Art of Computer Programming, vol. 2, 2nd ed., Addison-Wesley, Reading, MA, 1981.

[179] P. Bratley, B. L. Fox, and L. Schrage, A Guide to Simulation, Springer-Verlag, New York, 1983.

[180] W. H. Press, B. P. Flannery, S. A. Teukolsky, and W. T. Vetterling, Numerical Recipes in C, 2nd ed., Cambridge Univ. Press, New York, 1992.

[181] S. K. Park and K. W. Miller, "Random Number Generators: Good Ones Are Hard to Find," Comm. ACM, **31**, 1192 (1988).

[182] B. D. Ripley, "Computer Generation of Random Variables: A Tutorial," Int. Stat. Rev., **51**, 301 (1983).

[183] A. C. Kak, and M. Slaney, Principles of Computerized Tomography, Englewood Cliffs, NJ: Prentice Hall, 1989.

[184] S. M. Kay, Modern Spectral Estimation: Theory and Application, Englewood Cliffs, NJ: Prentice Hall, 1988.

[185] D. J. Best, "Some Easily Programmed Pseudo-Random Normal Generators," Austr. Comput. J., **11**, 60 (1979).

[186] G. Marsaglia, "Toward a Universal Random Number Generator," Statist. & Prob. Lett., **8**, 35 (1990).

[187] G. Marsaglia and A. Zaman, "A New Class of Random Number Generators," Annals Appl. Prob., **1**, 462 (1991).

[188] G. Marsaglia and A. Zaman, "Some Portable Very-Long Random Number Generators," Comput. Phys., **8**, 117 (1994).

[189] A. van der Ziel, Noise in Solid State Devices and Circuits, Wiley, New York, 1986.

[190] M. Reichbach, "Modeling 1/f Noise by DSP," Graduate Special Problems Report, ECE Department, Rutgers University, Fall 1993.

[191] R. Lyons, Understanding Digital Signal Processing. Addison-Wesley, 1997.

[192] S. W. Smith, Digital Signal Processing A Practical Guide for Engineers and Scientists. Newnes, 2003.

[193] R. E. Walpole and R. H. Myers, Probability and Statistics for Engineers and Scientists, Third Edition. New York: MacMillan Publishing Company, 1985.

[194] E. C. Ifeachor and B. W. Jervis, Digital Signal Processing A Practical Approach, Second Edition. Harlow, England: Prentice-Hall, 2002.

[195] C. B. Boyer, The History of the Calculus and Its Conceptual Development. New York: Dover Publications, Inc., 1959.

[196] C. J. Richard, Twelve Greeks and Romans Who Changed The World. Rowman & Little_eld, 2003.

[197] L. Gonick, The Cartoon History of the World, Volumes 1–7. New York: Doubleday, 1990.

[198] M. N. Geselowitz, Hall of Fame: Heinrich Hertz," IEEE-USA News & Views, November 2002.

[199] D. D. Andrew Bruce and H.-Y. Gao, Wavelet Analysis," IEEE Spectrum, pp. 26–35, October 1996.

[200] E. A. Lee and P. Varaiya, Structure and Interpretation of Signals and Systems. Addison-Wesley, 2003.

[201] R. W. Hamming, Digital Filters, Third Edition. Mineola, New York: Dover Publications, 1998.

[202] D. Sheffield, Equalizers," Stereo Review, pp. 72–77, April 1980.

[203] G. Kaiser, A Friendly Guide to Wavelets. Boston: Birkhauser, 1994.

[204] S. P. Thompson and M. Gardner, Calculus Made Easy. New York: St. Martin's Press, 1998.

[205] E. W. Swokowski, Elements of Calculus with Analytic Geometry. Boston, Massachusetts: Prindle, Weber and Schmidt, 1980.

[206] W. K. Chen, ed., The Electrical Engineering Handbook. Burlington, MA: Elsevier Academic Press, 2005.

[207] J. Ozer, New Compression Codec Promises Rates Close to MPEG," CD-ROM Professional, September 1995.

[208] A. Hickman, J. Morris, C. L. S. Rupley, and D. Willmott, Web Acceleration," PC Magazine, June 10, 1997.

[209] W. W. Boles and Q. M. Tieng, Recognition of 2D Objects from the Wavelet Transform Zero-crossing Representations," in Proceedings SPIE, Mathematical Imaging, (San Diego, California), pp. 104–114, July 11–16, 1993. Volume 2034.

[210] M. Vishwanath and C. Chakrabarti, A VLSI Architecture for Real-time Hierarchical Encoding/decoding of Video Using the Wavelet Transform," in IEEE International Conference on Acoustics, Speech and Signal Processing (ICASSP '94), (Adelaide, Australia), pp. 401–404, April 19–22, 1994. Volume 2.

[211] B. Widrow, and S.D. Stearns, Adaptive Signal Processing, Englewood Cliffs, NJ: Prentice Hall, 1985.

[212] E. A. Robinson, et al., Geophysical Signal Processing, Englewood Cliffs, NJ: Prentice Hall, 1986.

[213] M. Weeks and M. Bayoumi, Discrete Wavelet Transform: Architectures, Design and Performance Issues," Journal of VLSI Signal Processing, vol. 35, pp. 155–178, September 2003.

[214] A. Haar, Zur theorie der orthogonalen funktionensysteme," Mathematische Annalen, vol. 69, pp. 331–371, 1910.

[215] M. J. T. Smith and T. P. Barnwell III, Exact Reconstruction Techniques for Tree-structured Subband Coders," IEEE Transactions on Acoustics, Speech, and Signal Processing, pp. 434–441, June 1986.

[216] S. Jaffard, Y. Meyer, and R. D. Ryan, Wavelets Tools for Science & Technology. Philadelphia: Society for Industrial and Applied Mathematics (SIAM), 2001.

[217] H. Anton, Elementary Linear Algebra, 6th Edition. New York: John Wiley & Sons, Inc., 1991.

[218] C. Chakrabarti and M. Vishwanath, Efficient Realizations of the Discrete and Continuous Wavelet Transforms: From Single Chip Implementations to Mappings on SIMD Array Computers," IEEE Transactions on Signal Processing, vol. 43, pp. 759–771, March 1995.

[219] D. Goldberg, What Every Computer Scientist Should Know About Floating Point Arithmetic," Computing Surveys, pp. 171–264, March 1991.

[220] S. -W. Wu, Additive Vector Decoding of Transform Coded Images," IEEE Transactions on Image Processing, vol. 7, pp. 794–803, June 1998.

[221] O. K. Al-Shaykh and R. M. Mersereau, Lossy Compression of Noisy Images," IEEE Transactions on Image Processing, vol. 7, pp. 1641–1652, December 1998.

[222] E.-B. Fgee, W. J. Phillips, and W. Robertson, Comparing Audio Compression Using Wavelets with Other Audio Compression Schemes," IEEE Canadian Conference on Electrical and Computer Engineering, vol. 2, pp. 698–701, 1999.

[223] T. Rabie, Robust Estimation Approach for Blind Denoising," IEEE Transactions on Image Processing, vol. 14, pp. 1755–1765, November 2005.

Index

About the Authors

Muhammad Nasir Khan received his B.E. degree in electronic engineering from Dawood College of Engineering and Technology, Pakistan in 2003 and the M.Sc. degree in electrical engineering from the Delft University of Technology, Delft, The Netherlands in 2009 and the Ph.D. degree in electrical engineering at the Institute for Telecommunications Research (ITR), University of South Australia, Australia in 2013. He worked with Pakistan Telecommunication Industries (PCI) Pty. Ltd. from 2003 to 2005 as an assistant manager, specializing in telecommunication. He worked with the SUPARCO, SRDC Lahore from 2005 to 2007 as an RF communication engineer, specializing in modem and codec design for satellite communications. Currently, he is a Professor in Electrical Engineering Department, The University of Lahore, Pakistan. His research interests include communication theory, signal processing and channel decoding and free space adaptive communications systems.

Syed K. Hasnain worked in different national level engineering colleges with an excellent teaching record. He worked with the Pakistan Naval Engineering College as a professor, specializing in digital signal processing for communications. He served as an adjunct professor in Electrical Engineering Department, Aalborg University, Denmark. His research interests include detection schemes and channel coding.

Mohsin Jamil (Senior Member, IEEE) received the B.Eng. degree in industrial electronics from NED University, Pakistan, in 2004, two master's degrees from the National University of Singapore and Dalarna University Sweden, in 2008 and 2006, respectively, and the Ph.D. degree from the University of Southampton, U.K., in 2012. He was an associate professor with the Department of Electrical Engineering, Islamic University Madinah, Saudi Arabia and an assistant professor with the Department of Robotics and AI, National University of Sciences and Technology (NUST), Islamabad, Pakistan. He is currently an assistant professor with the Department of Electrical and Computer Engineering, Memorial University of Newfoundland, St John's, NL, Canada. His research interests include control system design for power electronic converter and mechatronic systems. His additional interests include myoelectric control, system identification and renewable energy systems for smart grid technologies. He is the author and co-author of several IEEE publications in different journals and peer-reviewed conferences. He is a recipient of different awards and funding grants. He is an associate editor of the IEEE ACCESS.

Sameeh Ullah has been with the technology industry since 1994 and has a Doctorate (Ph.D.) in electrical and computer engineering with a specialization in artificial intelligence from the University of Waterloo, Ontario, Canada. Dr. Ullah is a technology consultant and worked with several different verticals, including healthcare, financial institutions, education, auto, retail, and the insurance sector. Dr. Ullah has developed several architecture solutions and strategies based on artificial intelligence, machine learning, data analytics, micro-services, Blockchain, and the Internet of Things (IoT). Dr. Ullah has a wide variety of academic experience, including teaching in the School of Information Technology at Illinois State University from 2017 to present and actively engaged in publishing his research work in the IEEE journals and conferences. Dr. Ullah's research interests are in the areas of operational artificial intelligence, natural language processing, autonomous and intelligent systems, Blockchain, and the Internet of Things. Industrial and academic applications of Professor Ullah's research include cognitive and self-aware machines and self-healing infrastructure systems, machine learning, predictive analytics in the big data and infrastructure management systems. Currently, Dr. Sameeh Ullah is working at the School of Information Technology Illinois State University, USA.